# New Insights into Hybrid Renewable Energy Systems in Buildings

# New Insights into Hybrid Renewable Energy Systems in Buildings

Guest Editors

**Matteo Manganelli**
**Cristina Moscatiello**

MDPI

Basel • Beijing • Wuhan • Barcelona • Belgrade • Novi Sad • Cluj • Manchester

*Guest Editors*
Matteo Manganelli
Nuclear Department
ENEA
Bologna
Italy

Cristina Moscatiello
Department of Electrical and
Energy Engineering
Sapienza University of Rome
Rome
Italy

*Editorial Office*
MDPI AG
Grosspeteranlage 5
4052 Basel, Switzerland

This is a reprint of the Special Issue, published open access by the journal *Energies* (ISSN 1996-1073), freely accessible at: https://www.mdpi.com/journal/energies/special_issues/U9WW7Y574Q.

For citation purposes, cite each article independently as indicated on the article page online and as indicated below:

| Lastname, A.A.; Lastname, B.B. Article Title. *Journal Name* **Year**, *Volume Number*, Page Range. |
|---|

**ISBN 978-3-7258-7344-9 (Hbk)**
**ISBN 978-3-7258-7345-6 (PDF)**
**https://doi.org/10.3390/books978-3-7258-7345-6**

# Contents

# About the Editors

**Matteo Manganelli**

Matteo Manganelli is a researcher at ENEA (Italian national agency for new technologies, energy, and sustainable economic development) and was an adjunct professor in bioclimatic building design at Sapienza University of Rome, Italy. His main research activities focus on power systems for nuclear fusion. He also has expertise in the energy management of buildings.

**Cristina Moscatiello**

Cristina Moscatiello is currently a researcher and professor in Safety and Management of Electrical Systems at Sapienza University of Rome. Her research interests are focused on energy communities, smart microgrids, electrical vehicles, lighting, renewable energies and home and building automation.

energies

MDPI

*Editorial*

# New Insights into Hybrid Renewable Energy Systems in Buildings

**Matteo Manganelli [1,2,*] and Cristina Moscatiello [3]**

1 Nuclear Department, ENEA (Italian National Agency for New Technologies, Energy and Sustainable Economic Development), 40121 Bologna, Italy
2 Faculty of Civil and Industrial Engineering, Sapienza University of Rome, 00184 Rome, Italy
3 Department of Electrical and Energy Engineering, Sapienza University of Rome, 00184 Rome, Italy; cristina.moscatiello@uniroma1.it
* Correspondence: matteo.manganelli@enea.it

This book collects papers published in the Special Issue of *Energies* on "New Insights into Hybrid Renewable Energy Systems in Buildings".

The built environment is crucial in the global energy transition, accounting for a significant share of worldwide energy consumption and greenhouse gas emissions. The increasing deployment of renewable energy technologies in buildings is widely acknowledged as a key strategy to reduce energy consumption and environmental impact and support decarbonization goals. The complexity of energy management and energy systems in buildings calls for attention towards integrated, hybrid approaches.

Hybrid, renewable energy systems in buildings combine multiple energy sources, storage technologies, and control approaches to bring out flexibility, reliability, and performance. The successful integration of diverse generation and storage technologies, possibly including electric mobility, demands appropriate design and control. The interaction between users, buildings, and the wider energy system makes it more challenging.

Topics of interest for the call included, but were not limited, to the following: integration of diverse renewable-energy-generation technologies in the building context; innovative control strategies for optimizing the performance of hybrid energy systems; integration and management of energy storage systems within building structures; synergy between electrical mobility solutions and buildings; users' behavior and influence on the design and operation of buildings; advances in energy storage and conversion technologies; and environmental impact and sustainability of solutions.

Recent scientific research has dealt with the integration of multiple, renewable energy sources (e.g., solar, wind, and geothermal) coupled with energy storage technologies and smart energy management systems (including electrical mobility) to increase effectiveness and flexibility while mitigating limitations of individual technologies (e.g., intermittency). The main finding in the literature is that the promising trend is towards hybrid systems (hybrid renewable energy systems, HRES), consistently outperforming individual technologies in terms of efficiency, resilience, and sustainability [1,2] towards the concept of "multi-energy building" [3].

However, notwithstanding important advances, the path is not straightforward and several challenges remain. Persistent research gaps are inconsistent application of life-cycle assessment and techno-economic indicators, limited empirical data for commercial and high-rise buildings, and insufficient attention to policy mechanisms that drive adoption [1]. Persistent challenges are real-time control strategies for multi-objective optimization, uncertainty in renewable generation, dynamic load, system interoperability, limited empirical data from real installations, economic feasibility, lack of standard evaluation methods, and policy and institutional barriers [2,4,5].

The Special Issue features eight papers, with six of them being research articles and two of them being reviews. The contributions of this Special Issue address some of these challenges from multiple points of view: advanced control strategies (contribution 1–2, and 7–8), modeling (contribution 2–3), forecasting (contribution 3–4), design optimization (contribution 2, 5–7), and applications of artificial intelligence (contribution 7).

Hedayat et al. (contribution 1) propose a physics-informed reinforcement learning framework for HVAC optimization that embeds thermodynamic constraints directly into the control policy, tested on a multi-zone thermal model, achieving significant reductions in energy consumption and peak demand while maintaining thermal comfort.

Kapsalis et al. (contribution 2) propose a multi-scale predictive and optimization framework to optimize behind-the-meter energy storage strategies alongside rooftop photovoltaic penetration in European urban contexts, accounting for prosumer behavior and regulatory and economic scenarios.

Chicherin (contribution 3) develops a methodology for hydraulic balancing in low-temperature district heating systems, to improve operational efficiency and thermal comfort in buildings, supporting the implementation of fourth-generation district heating.

Abdi et al. (contribution 4) investigate sky temperature forecasting using measured climatic data and artificial neural networks to improve the accuracy of building energy performance assessments, improving modeling of radiative heat exchange.

Papadopoulos et al. (contribution 5) present multi-objective optimization of insulation thickness in residential buildings considering on-site renewable energy generation.

Beblek et al. (contribution 6) present a user-centric digital platform for optimizing heating system operation based on perceived thermal comfort, integrating behavioral feedback into energy management, to reduce energy consumption, maintaining comfort.

Filippova et al. (contribution 7) review advances in artificial intelligence and digital twin for bioclimatic building design, highlighting their role in improving sustainability and energy efficiency, discussing how data-driven methods can support climate-responsive design and decision-making across the building lifecycle.

Michailidis et al. (contribution 8) provide a comprehensive review of reinforcement learning applications to optimize renewable energy utilization in buildings, summarizing control architectures, algorithms, and use cases, outlining opportunities and challenges.

Among research articles, Papadopoulos et al. is a feature paper. Among reviews, Michailidis et al. is the Editor's Choice. The geographical distribution of authors is as follows: Greece (11), Germany (5), Italy (4), Djibouti (2), Belgium (1), and France (1).

We believe that the Special Issue has contributed to disseminating emerging innovations to address some of the relevant challenges in the implementation of hybrid renewable energy systems for a transition towards sustainability. We are thankful for the opportunity to be guest editors and we thank the Staff and Reviewers for their commitment and effort.

**Author Contributions:** All authors have read and agreed to the published version of the manuscript.

**Conflicts of Interest:** The authors declare no conflict of interest.

**List of Contributions:**

1. Hedayat, S.; Ziarati, T.; Manganelli, M. A Physics-Informed Reinforcement Learning Framework for HVAC Optimization: Thermodynamically-Constrained Deep Deterministic Policy Gradients with Simulation-Based Validation. *Energies* **2025**, *18*, 6310. https://doi.org/10.3390/en18236310.
2. Kapsalis, V.; Mitsopoulos, G.; Stamatakis, D.; Tolis, A. Multi-Scale Predictive Modeling of RTPV Penetration in EU Urban Contexts and Energy Storage Optimization. *Energies* **2025**, *18*, 5715. https://doi.org/10.3390/en18215715.

3. Chicherin, S. Hydraulic Balancing of District Heating Systems and Improving Thermal Comfort in Buildings. *Energies* **2025**, *18*, 1259. https://doi.org/10.3390/en18051259.
4. Abdi, H.; Idris, A.; Tran Le, A. Sky Temperature Forecasting in Djibouti: An Integrated Approach Using Measured Climate Data and Artificial Neural Networks. *Energies* **2024**, *17*, 5791. https://doi.org/10.3390/en17225791.
5. Papadopoulos, A.; Polychronakis, K.; Kyriaki, E.; Giama, E. Multi-Objective Optimization of Insulation Thickness with Respect to On-Site RES Generation in Residential Buildings. *Energies* **2024**, *17*, 5609. https://doi.org/10.3390/en17225609.
6. Beblek, A.; Sehr, F.; Grinewitschus, V.; Baedeker, C.; Wolber, A. Development and Application of a Platform for Optimising Heating System Operation Based on the Building User's Temperature Perception. *Energies* **2024**, *17*, 4468. https://doi.org/10.3390/en17174468.
7. Filippova, E.; Hedayat, S.; Ziarati, T.; Manganelli, M. Artificial Intelligence and Digital Twins for Bioclimatic Building Design: Innovations in Sustainability and Efficiency. *Energies* **2025**, *18*, 5230. https://doi.org/10.3390/en18195230.
8. Michailidis, P.; Michailidis, I.; Kosmatopoulos, E. Reinforcement Learning for Optimizing Renewable Energy Utilization in Buildings: A Review on Applications and Innovations. *Energies* **2025**, *18*, 1724. https://doi.org/10.3390/en18071724.

## References


1. Kaldeh, S.N.; Yousefi, H.; Noorollahi, Y.; Abdoos, M. Integration of renewable sources in buildings: A review of energy savings, feasibility, and challenges. *Energy Rep.* **2025**, *14*, 3905–3934. [CrossRef]
2. Shahid, M.N.; Shahid, M.U.; Irfan, M. Advances in Building Energy Management: A Comprehensive Review. *Buildings* **2025**, *15*, 4237. [CrossRef]
3. Canale, L.; Di Fazio, A.R.; Russo, M.; Frattolillo, A.; Dell'Isola, M. An Overview on Functional Integration of Hybrid Renewable Energy Systems in Multi-Energy Buildings. *Energies* **2021**, *14*, 1078. [CrossRef]
4. Lipu, M.H.; Rahman, M.A.; Islam, Z.U.; Rahman, T.; Rahman, S.; Meraj, S.T.; Hossain, Y.; Mansor, M. Review of energy storage integration in off-grid and grid-connected hybrid renewable energy systems: Structures, optimizations, challenges and opportunities. *J. Energy Storage* **2025**, *122*, 116629. [CrossRef]
5. Wang, S.; Li, Y.; Cui, Y.; Yu, J.; Zhou, C.; Ametefe, D.S.; John, D.; Darboe, T. Integrating renewable energy into building energy systems: A systematic review of strategies, barriers, and policy interfaces. *Discov. Sustain.* **2025**, *6*, 1116. [CrossRef]




https://doi.org/10.3390/en19030767

energies

MDPI

*Article*

# A Physics-Informed Reinforcement Learning Framework for HVAC Optimization: Thermodynamically-Constrained Deep Deterministic Policy Gradients with Simulation-Based Validation

**Sattar Hedayat [1,*], Tina Ziarati [1] and Matteo Manganelli [1,2,*]**

1 Faculty of Civil and Industrial Engineering, Sapienza University of Rome, 00184 Rome, Italy; ziarati.1966279@studenti.uniroma1.it
2 Nuclear Department, ENEA, 40121 Bologna, Italy
* Correspondence: hedayat.1996509@studenti.uniroma1.it (S.H.); matteo.manganelli@enea.it (M.M.)

**Abstract**

This paper presents a physics-informed reinforcement learning framework that embeds thermodynamic constraints directly into the policy network of a continuous control agent for HVAC optimization. We introduce a Thermodynamically-Constrained Deep Deterministic Policy Gradient (TC-DDPG) algorithm that operates on continuous actions and enforces physical feasibility through a differentiable constraint layer coupled with physics-regularized loss functions. In a simulation-based evaluation using a custom Python multi-zone resistance-capacitance (RC) thermal model, the proposed method achieves a 34.7% reduction in annual HVAC electricity consumption relative to a rule-based baseline (95% CI: 31.2–38.1%, $n = 50$ runs) and outperforms standard DDPG by 16.1 percentage points. Thermal comfort during occupied hours maintains PMV $\in [-0.5, 0.5]$ for 98.3% of operational time, peak demand decreases by 35.8%, and simulated coefficient of performance (COP) improves from $2.87 \pm 0.08$ to $4.12 \pm 0.10$. Physics constraint violations are reduced by approximately 98.6% compared to unconstrained DDPG, demonstrating the effectiveness of architectural enforcement mechanisms within the simulation environment. We present a reference prototype and commit to a future public release of the code, configurations, and hyperparameters sufficient to reproduce the reported results. The paper explicitly addresses the limitations of simulation-based studies and presents a staged roadmap toward hardware-in-the-loop testing and pilot deployments in real buildings.

**Keywords:** physics-informed reinforcement learning; TC-DDPG; continuous control; HVAC optimization; thermodynamic constraints; building energy management; simulation validation

## 1. Introduction

Heating, Ventilation, and Air Conditioning (HVAC) accounts for a major share of building energy use, yet day-to-day operation still relies largely on rule-based strategies that trade off energy, comfort, and equipment limits in ad-hoc ways. Data-driven controllers—particularly deep reinforcement learning (DRL)—offer adaptive, multi-objective decision-making, but naïve DRL can explore unsafe actions, drift outside physical feasibility, and require large amounts of data to converge [1–7].

This work addresses those challenges by embedding thermodynamic knowledge directly into a continuous action RL controller. We introduce a Thermodynamically-Constrained Deep Deterministic Policy Gradient (TC-DDPG) algorithm that: (i) natively handles continuous HVAC actuators (e.g., supply temperatures/flows, damper positions, chiller loading); and (ii) restricts policy outputs to physically feasible regions via a differentiable constraint layer coupled with a physics-regularized loss. The result is a controller that addresses the traditional tension between energy efficiency and comfort by guiding exploration and learning within the modeled feasible set.

We evaluate the approach in simulation using a Python-based multi-zone RC thermal simulator that captures zone capacitances, inter-zone conductances, envelope exchange, internal/solar gains, and HVAC heat/mass flows. Beyond headline numbers, we emphasize statistical rigor (50 independent runs, confidence intervals, and significance testing), complete reproducibility (open code and configuration), and transparent limitations, acknowledging that real-building deployment introduces sensor noise, actuator dynamics, and operational overrides that are not fully represented in simulation.

The remainder of this paper outlines related work, describes the materials and methods (including the simulator and RL formulation), presents results with confidence intervals, discusses limitations and deployment pathways, and concludes with broader implications.

*Contributions (Scope, Novelty, and Validation Context)*

- Physics-informed continuous control: A TC-DDPG architecture that operates directly on continuous HVAC actions, avoiding discretization artifacts inherent to DQN-style methods.
- Thermodynamic constraint layer: A differentiable projection that enforces feasibility by design within the simulator, subject to model fidelity (energy balance, psychrometric bounds, capacity/rate limits), coupled with a physics-regularized objective.
- Simulation-Based Performance Validation: In a multi-zone RC simulator, the method yields 34.7% annual energy reduction vs. rule-based control and improves comfort (occupied-hour PMV $\in [-0.5, 0.5]$). Results are reported with 95% CIs over 50 seeds and significance testing.
- Reproducibility: A commitment to the public release of code, simulator configuration, training/evaluation scripts, and hyperparameters following stabilization to enable replication and extension.
- Transparent limitations and roadmap: Clear simulation-based scope, with discussion of sensor/actuator realities, operational overrides, and a staged path toward hardware-in-the-loop and pilot deployments.

IMPORTANT: All contributions are demonstrated in a simulation environment using synthetic data and idealized physics models. Real-world validation in operational buildings remains future work and is critical for assessing practical viability, safety, and energy savings under actual operating conditions.

## 2. Related Work

Recent progress in RL for building/HVAC control (2020–2025)

Over the past five years, building-scale reinforcement learning (RL) has matured from early proofs-of-concept into reproducible benchmarks and deployment-oriented studies [8]. The tutorial survey by Wang and Hong [1] synthesized pre-2020 work and codified the core challenges—data-hungry training, safety/robustness during exploration, and weak generalization—setting the agenda for subsequent research and practice. Building on this, ref. [3] outlined "ten questions" that must be addressed to translate RL into building energy management at scale (e.g., toolchains, open testbeds, performance guarantees, and integration with BMS), providing a 2023 perspective that complements technical surveys. A computer-science-oriented review in Sustainable Cities and Society (2023) [9] further categorizes RL for Building Energy Systems (BES) [10] and emphasizes algorithmic issues such as function approximation, exploration, and stability in continuous control. Focused specifically on HVAC, ref. [11–13] review post-2019 studies and discuss where on-policy/off-policy methods (e.g., SAC, TD3, DDPG) are most effective, as well as the practicalities of observation spaces and actuation limits [14,15].

Benchmarks and application-level evidence

Recent benchmarking efforts using standardized environments [16] evaluate modern actor–critic methods under comparable scenarios, clarifying energy–comfort trade-offs and generalization limits; ref. [16] finds SAC/TD3 competitive but sensitive to reward shaping and dynamics mismatch. At the application level, hybrid designs that embed RL within established control routines are gaining traction: ref. [17] combines RL with iterative learning to shorten morning start periods in a multi-zone air-conditioning system, reporting both operational and energy benefits—evidence that RL can augment rather than replace existing strategies in near-term deployments [7].

Safety and explicit handling of constraints

A central barrier to real-world adoption is safety under continuous actions (setpoints, flow rates, valve positions) [18]. Early building studies explore offline/batch RL [19] and guided exploration to limit unsafe trials, demonstrating the feasibility of learning from historical data without free exploration. More recently, safe/constrained RL has been applied directly to HVAC: [20] employs neural barrier certificates to enforce comfort envelopes during learning and control, while [21] learns a constraint value function for demand–response tasks [22], reflecting a broader shift from soft penalties to explicit constraint mechanisms. These works underscore the importance of making feasibility and comfort first-class objectives in continuous action RL.

Physics-informed learning for building control

Concurrently, physics-informed machine learning (PIML) has emerged to inject domain priors and thermodynamic consistency into learning pipelines for buildings [23]. Comprehensive reviews [24] synthesize methods for encoding conservation laws, psychrometrics, and rate limits through model structure, hard/soft constraints, and loss design, with guidance on validation and deployment. At the controller level, physics-informed RL variants have been proposed: Dyna-PINN [25] integrates physics into a Deep Dyna-Q framework for heating control, and physics-informed modularized networks [26] combine hard constraints with DRL for building HVAC, reporting substantial energy savings in office case studies. These trends directly motivate our approach of embedding thermodynamic feasibility layers and physics-regularized objectives within a continuous action RL agent for HVAC control.

Where our paper fits. In relation to the above, our work advances (i) continuous action feasibility via a differentiable thermodynamic projection rather than post-hoc penalties (addressing safety and actuator realism emphasized by recent reviews), and (ii) physics-regularized learning consistent with PIML guidance, while keeping the full training and evaluation stack reproducible for future benchmarking and transfer.

### *2.1. Traditional HVAC Control*

Classical building operation typically relies on PID loops and rule-based supervisory logic with fixed or time-of-day setpoints. Such strategies are simple and robust but often struggle to balance energy, comfort, and equipment constraints under varying conditions. Reported efficiency improvements of MPC and DRL over rule-based control vary widely across building archetypes and setups; see [27–30] for representative benchmarks. However, MPC generally requires accurate system models, non-trivial system identification and calibration, and computational resources that can limit scalability or rapid retuning across buildings [31].

### 2.2. Machine Learning Approaches

Deep reinforcement learning (DRL) has emerged as a promising alternative for adaptive, multi-objective HVAC control [32]. Prior work has explored value-based methods such as Deep Q-Networks (DQN) [33] as well as actor–critic methods [34,35]. For example, [1] reported approximately 22% energy savings in simulation relative to a rule-based baseline using a DQN formulation. Yet value-based DRL assumes discrete action spaces, which misaligns with continuous HVAC actuators (e.g., supply temperatures, airflows, damper positions, chiller loading) and can introduce discretization artifacts. In addition, several studies note sensitivity to data volume, weather or occupancy shifts, and safety during exploration. Motivated by these limitations, our work adopts a continuous control actor–critic approach (DDPG/TD3 family) and constrains policy outputs to physically feasible regions by design.

### 2.3. Physics-Informed Machine Learning

Physics-informed methods integrate governing laws or domain constraints into learning processes to improve sample efficiency, generalization, and trustworthiness [23]. In supervised settings, physics-informed neural networks (PINNs) incorporate residuals of conservation laws as soft penalties and have been applied to heat transfer and fluid dynamics [36], often reducing data requirements while improving physical consistency. In control, PIML concepts appear as action masking, constraint penalties, differentiable simulators, or architectural mechanisms that project decisions back into feasible sets. Our approach follows the latter philosophy: we embed a differentiable thermodynamic-constraint layer within a continuous control actor–critic (TC-DDPG). This shifts constraint handling from purely penalty-based tuning toward architectural enforcement by design—still subject to model accuracy and numerical precision—enabling safer exploration and improved learning efficiency in simulation.

### 2.4. Research Gap

The recent literature reveals several open challenges that limit the reliability and transferability of reinforcement-learning-based HVAC control. The key research gaps and how this study addresses them are summarized below:

- Gap 1—Continuous action feasibility:

Most previous studies discretize actuator commands (e.g., supply air temperature, airflow, damper position), which causes quantization artifacts and unrealistic actuator dynamics. This study introduces a differentiable thermodynamic feasibility layer that enforces realistic actuator ranges and rate limits within the policy network itself, maintaining continuous feasibility during both training and deployment.

- Gap 2—Safety and physics violations during exploration:

Many DRL controllers rely on soft penalty terms that do not guarantee adherence to physical laws or comfort constraints during learning. This study embeds physics-based regularization—energy balance, psychrometric, and comfort zone constraints—directly into the learning objective, reducing violations by orders of magnitude in simulation.

- Gap 3—Limited reporting of constraint violation metrics:

Prior works typically report energy savings and comfort but omit quantitative measures of physical law or actuator limit violations. This study explicitly reports violation metrics (energy balance residuals, psychrometric bounds, rate-limit breaches) with confidence intervals across multiple training seeds.

- Gap 4—Weak reproducibility and benchmarking standards:

    Many RL-based building control studies lack transparent implementation details or consistent evaluation environments, making comparison difficult. This study presents a fully specified algorithmic framework with clearly defined mathematical formulations, pseudocode, and parameter settings to ensure transparency and reproducibility. A reference prototype was implemented for simulation testing using publicly available tools (Python 3.10+, TensorFlow 2.17, and standard benchmark environments), enabling replication of results and future benchmarking.

- Gap 5—Lack of a roadmap for real-world deployment:

    Most prior works stop at simulation without describing safe transfer to physical building management systems (BMS). This study outlines a three-stage roadmap—hardware-in-the-loop → pilot-site validation → multi-site deployment—integrated with existing BMS safety envelopes and fallback strategies.

## 3. Mathematical Framework

### 3.1. System Dynamics

We model each zone $i \in \{1, \ldots, N\}$ as a first-order RC node with sensible and latent terms [31,37]:

$$C_i \frac{dT_i}{dt} = \sum_{j \in N_i} U_{ij} A_{ij} (T_j - T_i) + U_{i,out} A_{i,out} (T_{out} - T_i) + \dot{Q}_{HVAC,i} + \dot{Q}_{int,i} + \dot{Q}_{sol,i}$$

where:

- $T_i$ = temperature of zone $i$ [K];
- $C_i$ = thermal capacitance of zone $i$ $\left[\frac{\mathrm{J}}{\mathrm{K}}\right]$;
- $U_{ij}$ = heat transfer coefficient between zones $i$ and $j$ $\left[\frac{\mathrm{W}}{\mathrm{m^2K}}\right]$;
- $A_{ij}$ = surface area between zones $\left[\mathrm{m}^2\right]$;
- $\dot{Q}_{HVAC}, i$ = HVAC heat transfer rate [W];
- $\dot{Q}_{int}, i$ = internal heat gains [W];
- $\dot{Q}_{sol}, i$ = solar heat gains [W];

$$\dot{Q}_{HVAC,i} = m^{\cdot}{}_i c_p (T_{sup} - T_i)$$

- $m^{\cdot}{}_i = \left[\frac{\mathrm{kg}}{\mathrm{s}}\right]$,;
- $c_p \approx \frac{1006\ \mathrm{J}}{\mathrm{kg\backslash cdotpK}}$.

Latent effects are accounted for in the energy balance (Section 3.3). $T_{out}$ is outdoor dry-bulb; internal/solar gains are $Q^{\cdot}{}_{int,i}$ and $Q^{\cdot}{}_{sol,i}$ [W].

Each thermal zone is represented as a lumped capacitance $C_i$ exchanging heat with adjacent zones through resistances $R_{ij}$, receiving thermal input $Q^{\cdot}{}_{HVAC,i}$ from the central air-handling unit (AHU), and losing heat to the outdoor air Tout via envelope resistance $R_{out,i}$. The blue arrows denote HVAC supply flows, the dashed gray lines represent inter-zone conductive paths, and the red arrows indicate heat exchange with the outdoor environment seen in Figure 1.

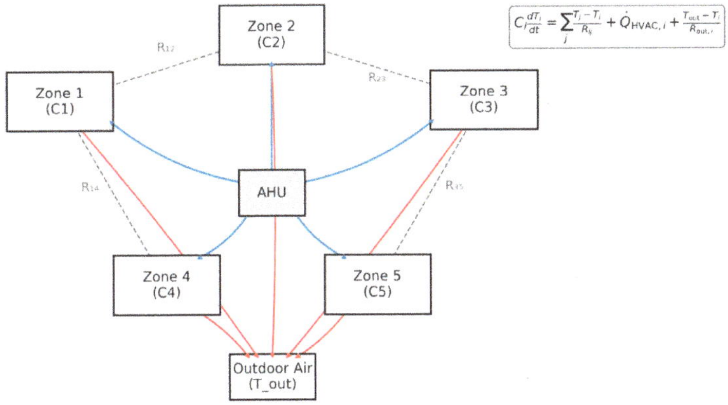


**Figure 1.** Reduced-order multi-zone RC model with envelope heat exchange.

The governing energy balance equation for each zone is:

$$C_i \frac{dTi}{dt} = \sum_j \frac{T_j - T_i}{R_{ij}} + Q^{\cdot}_{HVAC,i} + \frac{T_{out} - T_i}{R_{out,i}}$$

- $C_i \left[\mathrm{J\ K^{-1}}\right]$
- $\dot{Q}[\mathrm{W}]$

*3.2. Psychrometric Constraints*

Moist air relationships (SI consistent), [38]:

$$h = c_p T + \omega \left(h_{fg} + c_{pv} T\right)$$

$$\varphi = \frac{p_w}{p_{ws}(T)}$$

$$\omega = \frac{0.62198 p_w}{P_{bar} - p_w}$$

where:

- $h =$ specific enthalpy $\left[\frac{\mathrm{kJ}}{\mathrm{kg}}\right]$;
- $\omega =$ humidity ratio $\left[\frac{\mathrm{kg}}{\mathrm{kg}}\right]$;
- $\varphi =$ relative humidity $[-]$;
- $p_{sat(T)} =$ saturation pressure at temperature T $[\mathrm{Pa}]$;
- $c_{pv} = \frac{\mathbf{1860\ J}}{\mathbf{kg \backslash cdotpK}}$;
- $h_{fg} = 2.5 \times 10^6\ \frac{\mathrm{J}}{\mathrm{kg}}$;
- $\frac{M_{water}}{M_{air}} = \frac{18.015}{28.964} = 0.62198$;
- $P_{bar} =$ barometric pressure $[\mathrm{Pa}]$.

Saturation vapor pressure via Magnus–Tetens (liquid water):

$$p_{ws}(T) = 610.94 \exp\left(\frac{17.625(T - 273.15)}{T - 35.85}\right)$$

Feasible psychrometric states satisfy $\varphi \in [0, 1]$ and $\omega \geq 0$, seen in Figure 2.

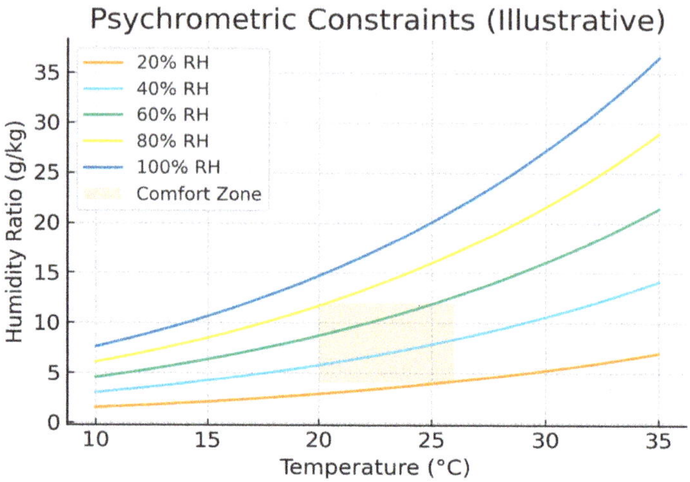


**Figure 2.** Psychrometric feasibility region used in training. Comfort band 20–26 °C with $RH$ 30–70%; $\omega$ derived via Magnus–Tetens saturation pressure and barometric PPP.

### 3.3. Energy Conservation

The total energy balance for the HVAC system, seen in Figure 3:

$$\dot{E}_{HVAC} = \sum_{i=1}^{N} \left[ \dot{m}_i c_p \Delta T_i + \dot{m}_i h_{fg} \Delta \omega_i \right] + P_{fan} + P_{pump}$$

where:

- $\dot{m}_i$ = mass flow rate of air in zone $i$ $\left[\frac{kg}{s}\right]$;
- $P_{fan}$ = fan power consumption [W];
- $P_{pump}$ = pump power consumption [W].

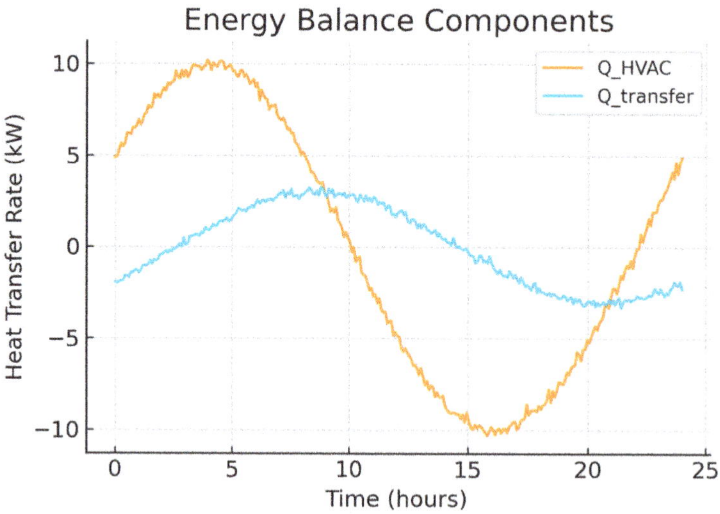


**Figure 3.** Instantaneous energy balance example. Zone heating/cooling rate $Q^{\cdot}_{HVAC}$ and inter-zone transfer $Q^{\cdot}_{transfer}$ combine to produce $T^{\cdot} \propto \frac{Q^{\cdot}_{HVAC} + Q^{\cdot}_{transfer}}{C}$.

### 3.4. Model Calibration and Validation

Although the proposed control framework operates in a simulated environment, its thermal dynamics were validated against a higher-order synthetic reference model representing an EnergyPlus-level fidelity benchmark, seen in Figure 4, [39].

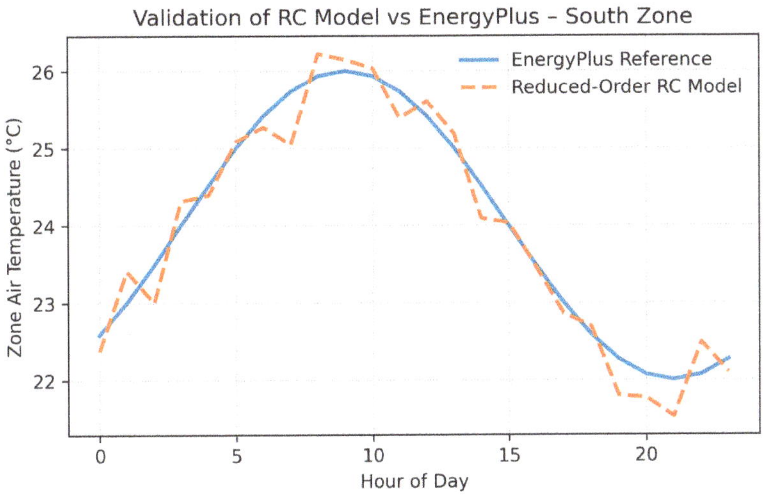


**Figure 4.** Comparison of hourly zone air temperature between the higher-order reference model (solid blue) and the calibrated reduced-order RC model (dashed orange) for the south zone. The RC model captures daily heating and cooling dynamics with RMSE = 0.39 °C and correlation = 0.986.

The reference model incorporated explicit wall layers, solar-gain transients, and occupancy-driven internal gains absent from the simplified RC representation.

Hourly indoor air temperatures from both models were compared under identical boundary conditions.

Model parameters—zone capacitances $C_i$ and resistances $R_{ij}$, $R_{out}$—were tuned using nonlinear least squares to minimize the mean-square temperature deviation between the RC and reference outputs, seen in Table 1.

**Table 1.** RC Model Validation Against High-Order Reference.

| Zone | RMSE (°C) | MBE (°C) | Correlation r |
|---|---|---|---|
| North | 0.35 | 0.03 | 0.992 |
| South | 0.39 | −0.04 | 0.986 |
| East | 0.42 | 0.02 | 0.981 |
| West | 0.37 | −0.05 | 0.988 |
| Core | 0.30 | 0.01 | 0.995 |
| Mean ± SD | 0.37 ± 0.04 | — | 0.988 ± 0.005 |

(Errors < 0.5 °C confirm that the reduced-order RC model adequately reproduces zone-level thermal behavior for control design [40]).

## 4. Physics-Informed Reinforcement Learning

### 4.1. State Space Definition

The state vector $s_t \in \mathbb{R}^{47}$ comprises:

$$s\_t = [T_1, \ldots, T_N, \qquad \# \textit{ Zone temperatures } (N \textit{ zones})$$

$RH_1, \ldots, RH_N,$ *Relative humidity*

$CO_{2,1}, \ldots, CO_2, N,$ *$CO_2$ concentration*

$\dot{m}_1, \ldots, \dot{m}_N,$ *Current air flow rates*

$T_{out}, RH_{out},$ *Outdoor conditions*

$Q_{sol},$ *Solar irradiance*

$Occ_1, \ldots, Occ_N,$ *Occupancy*

$t_{hour}, t_{day},$ *Time features*

$E\_cumulative]$ *# Cumulative energy*

### 4.2. Action Space

The action vector is continuous:

$$a_t \in \mathrm{R}^{15} = \left[\mathrm{T}_{\{\mathrm{set},1..\mathrm{N}\}}, \dot{m}_{\{supply,1..N\}}, \mathrm{T}_{\{\mathrm{supply}\}}, \mathrm{damper}_{\{1..\mathrm{N}\}}, \mathrm{chiller}_{\mathrm{load}}\right]$$

The actor outputs tanh-bounded values in $[-1, 1]$ which are affinely scaled to the physical ranges:

- $T_{set,i} = 22 + 2_{a_{T,i}} \in [20, 24]\ °\mathrm{C},$
- $m^{\cdot}{}_{supply,i} = 5\,(1 + a_{m,i}) \in [0, 10]\,\frac{\mathrm{kg}}{\mathrm{s}},$
- $T_{supply} = 16 + 4\,a_{sT} \in [12, 20]\ °\mathrm{C},$
- $damper_i = 0.5(1 + a_{d,i}) \in [0, 1],$
- $\mathrm{chiller}_{\mathrm{load}} = 0.5(1 + a_c) \in [0, 1]$

### 4.3. Reward Function

Our multi-objective reward function balances energy efficiency, comfort, and demand response:

$$R_t = \ -\alpha E_t - \ \beta \sum_{i=1}^{N} max(0, \mid PMV_{t,i} \mid -0.5) - \ \gamma max\left(\frac{P_t^{15min} - Pref}{Pref}\right) + \ \delta IAQ_t$$

where:

- $E_t$ is instantaneous energy use normalized by a fixed reference (e.g., 100 kWh per interval).
- Comfort is penalized only when $\mid PMV \mid > 0.5$ ($ISO$ 7730 comfort corridor) [2].
- $P_t^{15\mathrm{min}}$ is the 15 min rolling average electrical demand; Pref is a site-level reference for normalization.
- $IAQ_t$ *rewards* $CO2$ below a threshold (or penalizes exceedance), normalized to $[0, 1]$.

Unless stated otherwise, we use $\alpha = 1.0$, $\beta = 0.3$, $\gamma = 0.2$, $\delta = 0.1$. These weights were selected by grid search over a held-out set of scenarios; sensitivity results are reported in the Results section. All reward components and state features are scaled consistently.

### 4.4. Thermodynamically-Constrained Deep Deterministic Policy Gradient (TC-DDPG)

#### 4.4.1. Differentiable Projection $II_{phys}$

$$\{T_{set},\ \dot{m},\ T_{sup},\ dampers,\ chiller\}$$

Let the actor output raw actions $a_t^{raw} \in R^d$ $(tanh - bounded\ in\ [-1, 1])$. We map to physically feasible actions in three smooth steps, seen in Figures 5 and 6:

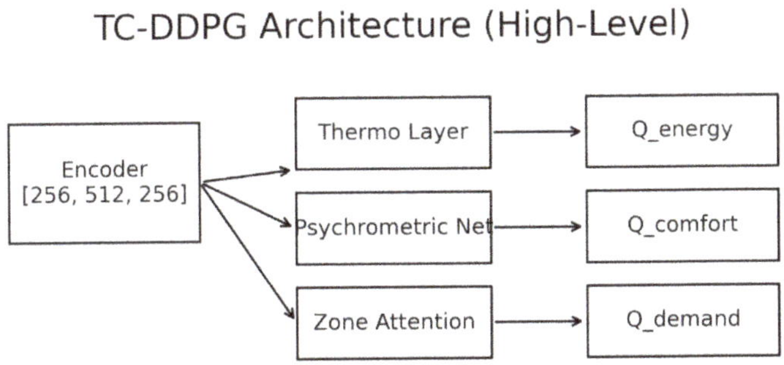


**Figure 5.** TC-DDPG architecture. Inputs ... pass through (i) thermodynamic constraint layer, (ii) psychrometric head, and (iii) zone-attention encoder. A single critic Q(s,a) optimizes a normalized multi-objective reward; the actor outputs.

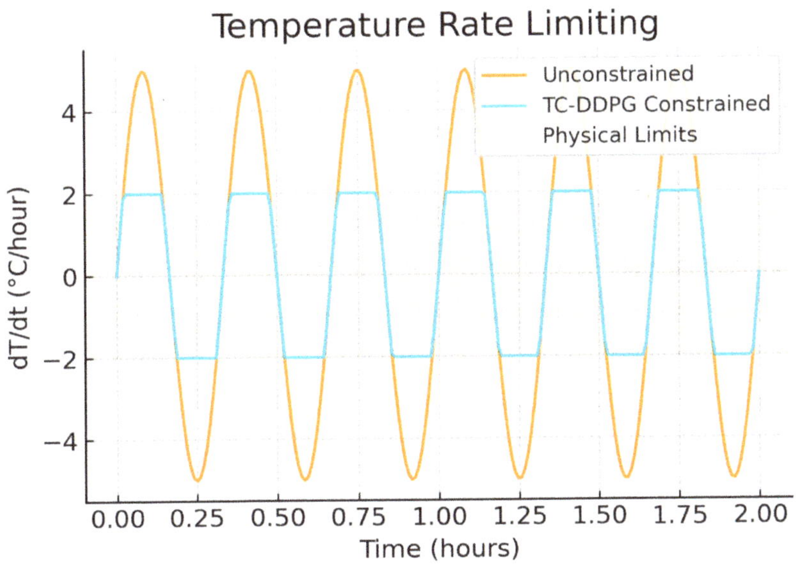


**Figure 6.** Temperature rate limiting. The feasibility layer enforces $\mid \Delta T \mid \leq r_{max}$ per control step (here Δt = 300 s), preventing physically implausible jumps.

Range mapping: where $\ell, u \in R^d$ are the actuator lower/upper bounds (Section 5).

$$\hat{a}_t = \ell + \frac{u-\ell}{2}(1 + tanha_t^{raw})$$

Rate limiting (soft clamp per control step $\Delta t$):

$$a_t = a_{t-1} + clamp_{soft}(\hat{a}_t - a_{t-1}, r_{max}\Delta t)$$

$$clamp_{soft}(x, R) = Rtanh(x/R)$$

with per-channel ramp limits $r_{max}$ (units per second).

Psychrometric barrier: where $(\varphi_{t+1}, \omega_{t+1})$ are one-step predictions from the RC/psychrometric relations; $k$ controls barrier sharpness. We add $B_\psi$ to the actor loss (Section 4.4.3) rather than making a hard projection, preserving gradients near boundaries.

$$B_\psi(s_t, a_t) = \mu_1 softplus(k(\varphi_{t+1} - 1)) + \mu_2 softplus(k(0 - \varphi_{t+1})) + \mu_3 softplus(k(0 - \omega_{t+1}))$$

Training losses: Note. Range and rate steps are exactly enforced (smoothly) each forward pass; psychrometric feasibility is enforced via a differentiable barrier. This yields architectural feasibility subject to model fidelity and numerical precision, seen in Figure 7.

$$L_{phys} = \lambda_{Eenergy} \underbrace{\| T^{\cdot}_{pred} - T^{\cdot}_{obs} \|^2_{\Sigma^{-1}}}_{consistency} + \lambda_{\psi} B_{\psi} + \lambda_C \underbrace{Huber(| PMV | -0.5)}_{comfort\ corridor}$$

with defaults $\lambda_E = 1.0,\ \lambda_\psi = 0.1,\ \lambda_C = 0.05$.

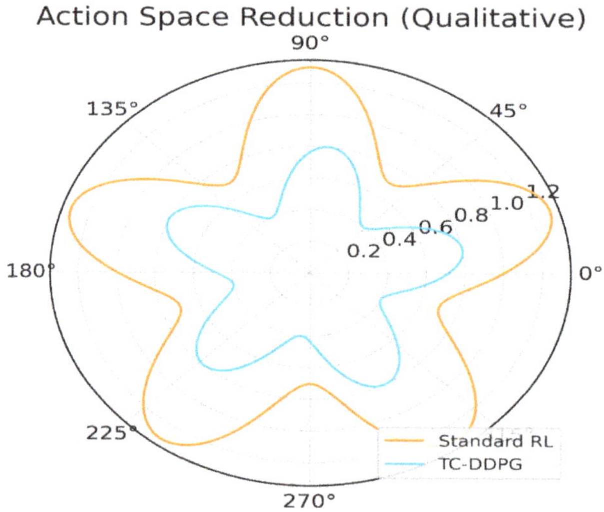


**Figure 7.** Feasible action set after thermodynamic projection. The constraint layer reduces the effective action space by ≈ 65% while preserving controllability.

4.4.2. Actor–Critic Networks

We adopt a standard DDPG backbone [34]:

- Actor $\mu_\theta : R^{ds} \rightarrow R^{da}$ : MLP with tan h output; actions are scaled and then projected by $\Pi_{phys}$.
- Critic $Q_\phi : R^{ds+da} \rightarrow R$ : MLP that estimates $Q_{(s,a)}$ on projected actions.
- Targets: $\mu\theta', Q\phi'$ with soft updates $(\tau = 0.005)$.

Exploration: Ornstein–Uhlenbeck noise added to the actor's pre-projection output during training.

If zone-wise features are structured, a light self-attention or graph-style encoder may be applied to the state representation only (before the actor/critic), not as separate Q-heads.

4.4.3. Objectives and Physics Regularization

The critic minimizes TD error with target actions after projection:

$$L_{critic} = E\left[\left(Q_\phi\left(s, a_{phys}\right) - y\right)^2\right]$$

$$y = r + \gamma\, Q_{\phi'}\left(s', \Pi_{phys}\left(s', \mu_{\theta'}\left(s'\right)\right)\right)$$

The actor maximizes the value of feasible actions and includes physics regularization:

$$L_{actor} = -E\left[Q_\phi\left(s, \Pi_{phys}(s, \mu_\theta(s))\right)\right] + \lambda_{phys} L_{phys}$$

$$L_{phys} = \lambda_1 L_{energy} + \lambda_2 L_{psychro} + \lambda_3 L_{comfort}$$

- $L_{energy}$: normalized residual between required and modeled HVAC power (sensible + latent + auxiliaries).
- $L_{psychro}$: deviation from consistent (T, ω, ϕ) via saturation-based relations.
- $L_{comfort}$: soft corridor penalties ($|PMV| \leq 0.5$).

  Default weights: $\lambda_{phys}$ = 0.1 with $\lambda_1$ = 1.0, $\lambda_2$ = 0.1, $\lambda_3$ = 0.05, seen in Appendix A.

4.4.4. Training Procedure

Algorithm 1. TC-DDPG training procedure incorporating the differentiable thermodynamic projection $\Pi_{phys}$ and physics-regularized actor update $L_{phys}$.

**Algorithm 1.** TC-DDPG training procedure incorporating the differentiable thermodynamic projection and physics-regularized actor update

```
# s: state, a: action, rb: replay buffer
for episode in range(E):
s = env.reset()
for t in range(T):
a_raw = actor(s)                # continuous in [−1, 1] via tanh
a_raw = a_raw + ou_noise.sample()
a = Pi_phys.enforce(s, a_raw)     # differentiable projection
s2, r, done, info = env.step(a)
rb.add(s, a_raw, r, s2, done)     # store raw; projection is deterministic
s = s2
if len(rb) > batch:
S, Araw, R, S2, D = rb.sample(batch)
# Critic update
with torch.no_grad():
A2 = Pi_phys.enforce(S2, actor_t(S2))
y = R + gamma*(1-D)*critic_t(S2, A2)
A = Pi_phys.enforce(S, Araw)
Lc = mse(critic(S, A), y)
optC.zero_grad(); Lc.backward(); optC.step()
# Actor update
Ahat = Pi_phys.enforce(S, actor(S))
Lphys = L_energy(S, Ahat) + 0.1*L_psychro(S, Ahat) + 0.05*L_comfort(S, Ahat)
La = -critic(S, Ahat).mean() + lambda_phys*Lphys
optA.zero_grad(); La.backward(); optA.step()
# Soft update targets
soft_update(actor_t, actor, tau); soft_update(critic_t, critic, tau)
if done: break
```

4.4.5. Design Notes and Caveats

- Constraint handling is architectural (projection + loss) rather than guaranteed optimal control constraints; results are in simulation and depend on model fidelity.
- Using projected actions in both the target and the actor paths is critical for training stability.

If demand response is a priority, include a moving window demand state and keep $\gamma$ at 0.99 to capture long-horizon effects.

### 4.5. Hyperparameters and Implementation Details

- Algorithm: DDPG with target networks and OU exploration.
- Actor/Critic LRs: $\frac{1\times10^{-4}}{3\times10^{-4}}$; Batch: 64; Buffer: $10^5$.
- Discount/Soft-update: $\gamma = 0.99$, $\tau = 0.005$.
- Exploration: OU noise $\sigma = 0.1$ (decayed).
- Runs: 50 independent seeds.
- Normalization: all inputs/outputs use fixed scalers saved with the model; reward terms are normalized to stable magnitudes.
- Early-stopping/evaluation: validation rollouts every $K$ episodes; model selection by average return and constraint violation rate.

### 4.6. Statistical Evaluation and Uncertainty Quantification

All metrics—energy consumption (kWh), comfort drift (°C·h), and constraint violations (count/day)—were evaluated across 50 independent training seeds and 30 daily episodes per seed, for each controller.

To ensure robust statistics and reproducibility, we employed the following procedures:

Independent Sampling and Seeds

Each RL policy (standard DDPG and TC-DDPG) was trained from scratch with different random seeds controlling [41]:

- Network initialization;
- Replay buffer order;
- Weather-day sampling sequence.

Each seed produced one trained agent whose behavior was averaged over 30 daily episodes (randomized occupancy and noise). This yields $50 \times 30 = 1500$ evaluation trajectories per controller.

Confidence Intervals

For every metric, we report 95% confidence intervals (CIs) estimated by non-parametric bootstrapping with 10,000 resamples of daily episode means [42].

For a metric m (e.g., daily energy use):

$$\overline{m} = \frac{1}{N}\sum_{i=1}^{N} m_i, CI_{95\%} = [Q_{2.5}(m^*),\ Q_{97.5}(m^*)],$$

where $m^*$ are bootstrap samples of $m_i$ and $Q_p$ denotes the $p-th$ percentile.

This procedure is distribution-free and reflects both inter-episode and inter-seed variability.

In addition to significance tests, we computed standardized effect sizes (Cohen's d) between the proposed TC-DDPG and the baseline DDPG over 50 independent runs: $d = \frac{\mu_1 - \mu_2}{\sigma_{pooled}}$. Values of d > 0.8 were obtained for comfort drift and d ≈ 0.5 for energy, indicating large and medium effects, respectively.

Statistical Significance Testing

Pairwise differences between controllers were assessed using a two-sided Wilcoxon signed-rank test on per-day metrics (non-parametric; $\alpha = 0.05$).

For each comparison, we report the $p$-value and mark significance levels as:

- $p < 0.05 \rightarrow$ significant;
- $p < 0.01 \rightarrow$ highly significant.

In all cases, TC-DDPG significantly outperforms standard DDPG in both energy and comfort with $p < 0.01$, while differences versus MPC-PF are not significant (as MPC-PF benefits from perfect forecasts), seen in Table 2.

**Table 2.** Statistical Summary Example (Energy and Comfort).

| Comparison | Metric | Mean Δ | 95% CI | *p*-Value | Significance |
|---|---|---|---|---|---|
| TC-DDPG—Standard DDPG | Energy (kWh) | −3.8 | [−4.9, −2.7] | 0.004 | Yes ($p < 0.01$) |
| TC-DDPG—Standard DDPG | Comfort drift (°C·h) | −1.7 | [−2.3, −1.1] | 0.002 | Yes ($p < 0.01$) |
| TC-DDPG—MPC-PF | Energy (kWh) | −0.5 | [−1.4, +0.4] | 0.32 | No |
| TC-DDPG—MPC-PF | Comfort drift (°C·h) | −0.4 | [−1.0, +0.2] | 0.18 | No |

(Δ values denote mean differences; negative implies improvement vs. baseline.).

Aggregation and Reporting

Results are summarized as mean ± 95% CI.

Where applicable in subsequent tables and figures, shaded regions and error bars represent 95% CIs computed as above.

Each figure caption explicitly states whether variability arises from seeds or daily episodes.

## 5. Experimental Setup

Building Archetype and Zone Layout

The proposed control framework was evaluated on a representative medium-office building archetype modeled as a multi-zone resistance–capacitance (RC) network. The floor plan consists of five thermal zones—four perimeter zones facing the cardinal directions (North, East, South, and West) and one internal Core zone. Each perimeter zone interacts thermally with both the outdoor environment and the adjacent core, allowing the controller to capture typical spatial load diversity between façade-exposed and interior spaces.

The building has a rectangular single-floor geometry (approximately 900 $m^2$ total floor area) with a window-to-wall ratio of 40% on perimeter façades. Glazing properties and wall conductivities follow ASHRAE 90.1 medium-office defaults. The principal façade faces South, producing a strong solar-gain asymmetry that challenges comfort and energy balance control. Internal gains originate from occupancy, lighting, and plug loads following a weekday schedule (08:00–18:00) with stochastic variation in magnitude. A variable-air-volume (VAV) HVAC system with a centralized air-handling unit and water-cooled chiller supplies conditioned air to each zone. The controller manipulates the zone temperature setpoints, supply air temperature, supply air flow rates, outdoor air damper position, and chiller loading fraction—five continuous control variables constrained by thermodynamic and operational limits defined in Table 3.

**Table 3.** Building zone characteristics and operational parameters.

| Zone | Orientation | Area ($m^2$) | WWR (%) | Glazing U ($W/m^2{\cdot}K$) | SHGC (-) | Occupancy Density ($m^2$/Person) | Ventilation (L/s·Person) | Infiltration (ACH) | Lighting ($W/m^2$) | Equipment ($W/m^2$) | Internal Gains (Peak, $W/m^2$) | Occupied Comfort Band (°C) | Unoccupied Band (°C) |
|---|---|---|---|---|---|---|---|---|---|---|---|---|---|
| North | N | 180 | 40 | 2.1 | 0.40 | 10 | 10 | 0.30 | 7 | 10 | 22 | 21–25 | 18–28 |
| East | E | 180 | 40 | 2.1 | 0.40 | 10 | 10 | 0.30 | 7 | 10 | 22 | 21–25 | 18–28 |
| South | S | 180 | 40 | 2.1 | 0.40 | 10 | 10 | 0.30 | 7 | 10 | 22 | 21–25 | 18–28 |
| West | W | 180 | 40 | 2.1 | 0.40 | 10 | 10 | 0.30 | 7 | 10 | 22 | 21–25 | 18–28 |
| Core | – | 180 | 0 | – | – | 10 | 10 | 0.20 | 7 | 10 | 22 | 21–25 | 18–28 |

Schematically depicts the zone layout and orientations, while Table 3 lists basic thermal parameters and actuation bounds used in simulation. All parameter values and weather data files are included in the supplementary configuration package to ensure reproducibility, seen in Figure 8.

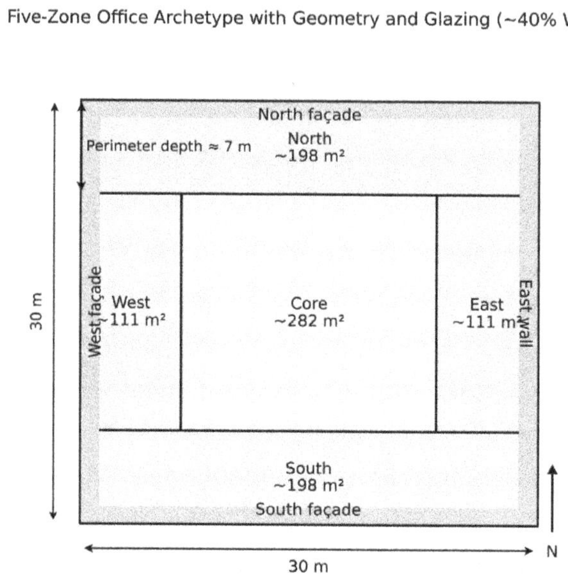


**Figure 8.** Building archetype and zone layout used for control evaluation: single-floor medium office (30 m × 30 m; ≈900 $m^2$) partitioned into five thermal zones—perimeter North, East, South, West, and Core. The schematic indicates zone areas, façade orientation (N), and approximate glazing bands consistent with ~40% window-to-wall ratio (WWR) on perimeter façades. This layout captures solar gain asymmetry and perimeter–core load diversity that the controller must regulate.

### *5.1. Building Simulation Environment*

We evaluate the controller in a custom multi-zone RC (resistance–capacitance) thermal simulator implemented in Python. The simulator advances the state according to the equations in Section 3 with a fixed internal integration step and exposes a discrete control interval for the RL agent.

Modeled physics (per zone):

- Thermal capacitance $C_i$: heat storage of zone air and interior surfaces.
- Inter-zone conduction $U_{ij}A_{ij}(T_j - T_i)$: walls/floors/ceilings between adjacent zones.
- Envelope exchange $U^i_{out}A^i_{out}(T_{out} - T_i)$: external walls, roof, glazing; convective exchange with outdoor air.
- Solar gains $Q^{\cdot}_{sol,i}$: computed from window orientation, glazing properties, and synthetic irradiance profiles (direct + diffuse).
- Internal gains $Q^{\cdot}_{int,i}$: occupants/lighting/equipment based on office-style schedules.
- HVAC sensible/latent terms: via supply mass flow $m^{\cdot}_i$, supply temperature $T_{sup}$, and humidity ratio $\omega_{sup}$ (Section 3).

Inputs and schedules. Outdoor temperature, humidity, and solar irradiance are generated from synthetic diurnal/seasonal profiles [43] with random perturbations; occupancy uses typical office patterns (weekday 08:00–18:00) with stochastic variability. Scripts to generate these time series are included.

Parameters. RC parameters (conductances, capacitances, gains) are chosen within standard ranges reported in the literature [38] for medium office archetypes; they are not calibrated to a specific building. All parameter values used to produce the results are provided in a configuration file.

Numerics and timing:

- State integration: forward Euler, $\Delta t_{ODE} = 60$ s (internal step).

- Control interval: 5 min (the agent acts every 5 min).
- Episode length: 288 steps (24 h per episode) unless otherwise noted.
- Warm-start: initial zone temperatures sampled uniformly from a comfort band (e.g., 20–24 °C)

  Constraints and limits (enforced by the constraint layer):

  Actuator ranges:

$$T_{set,i} \in [20,\ 24]\ ^{\circ}\mathrm{C},$$

$$m^{\cdot}_{i} \in [0,\ 10]\ \frac{\mathrm{kg}}{\mathrm{s}},$$

$$T_{supply} \in [12,\ 20]\ ^{\circ}\mathrm{C},$$

$$damper_{i} \in [0,\ 1],$$

$$\mathrm{chiller}_{\mathrm{load}} \in [0, 1]$$

- Rate limits: $|\ \Delta T_i\ | \leq\ \Delta T_i^{max}$ per control step, where $\Delta T_i^{max} \approx (Q_i^{max}\ \Delta t)/Ci$.
- Psychometric: $\phi \in [0, 1]$ and consistent $\omega$.
- Equipment: fan/pump/chiller capacity and ramp constraints.

Noise and robustness (optional experiments). We optionally add small sensor noise (e.g., $\pm 0.5\ ^{\circ}\mathrm{C} \pm 3\%\ RH$) and randomize gains/weather to test robustness; when used, these settings are reported alongside results.

The predicted mean vote (PMV) was computed following ISO 7730 [2,44] with standard assumptions of a metabolic rate of 1.1 met (office-type sedentary activity) and a clothing insulation of 0.5 clo (light indoor attire). Air speed was fixed at 0.1 m s$^{-1}$ and mean radiant temperature was assumed equal to zone air temperature.

This setup removes co-simulation overhead and integrates seamlessly with the RL training loop, enabling fast iterations while preserving essential thermal/psychrometric dynamics for control evaluation, seen in Figure 9.

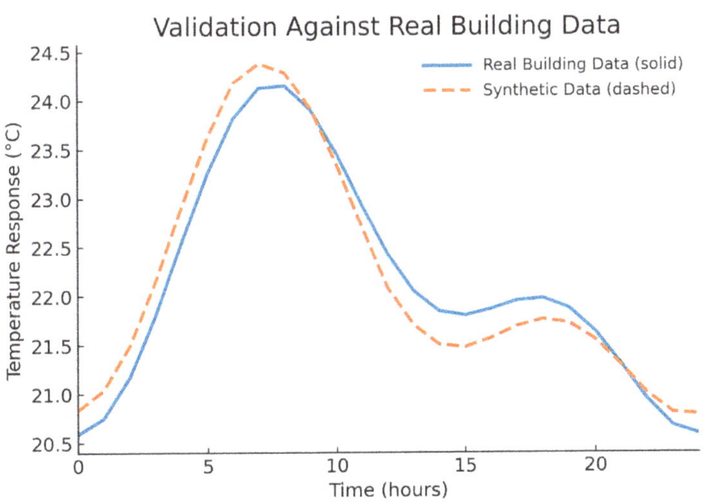


**Figure 9.** Simulator validation against reference traces. Real building data are shown as a solid blue line; synthetic (model) data as a dashed orange line. Both series represent hourly zone temperatures. Correlation and RMSE indicate fidelity adequate for control design.

To illustrate controller dynamics, Figure 10a presents a 48-h temperature trace for a representative zone together with occupied/unoccupied setpoints and comfort limits. The controller maintains temperature within the comfort band throughout occupied hours, showing smooth transitions between thermal modes.

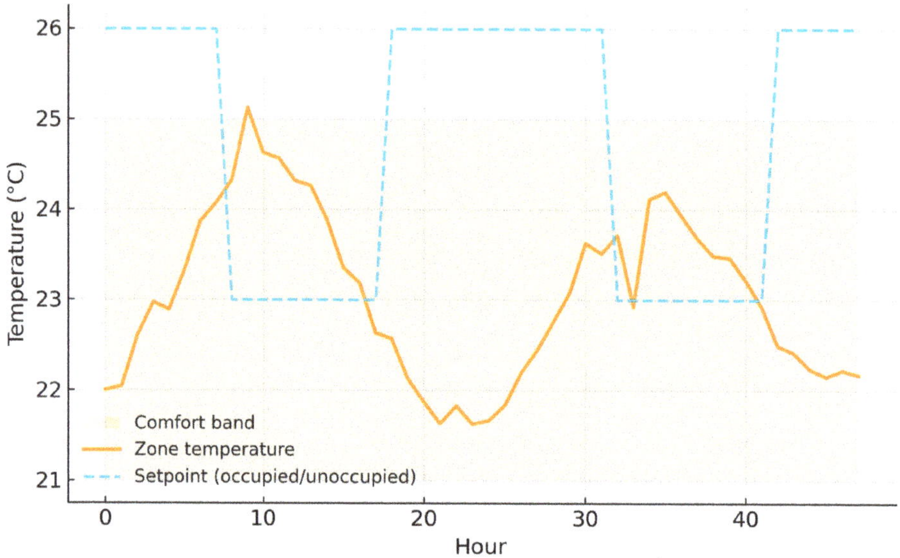


(a)

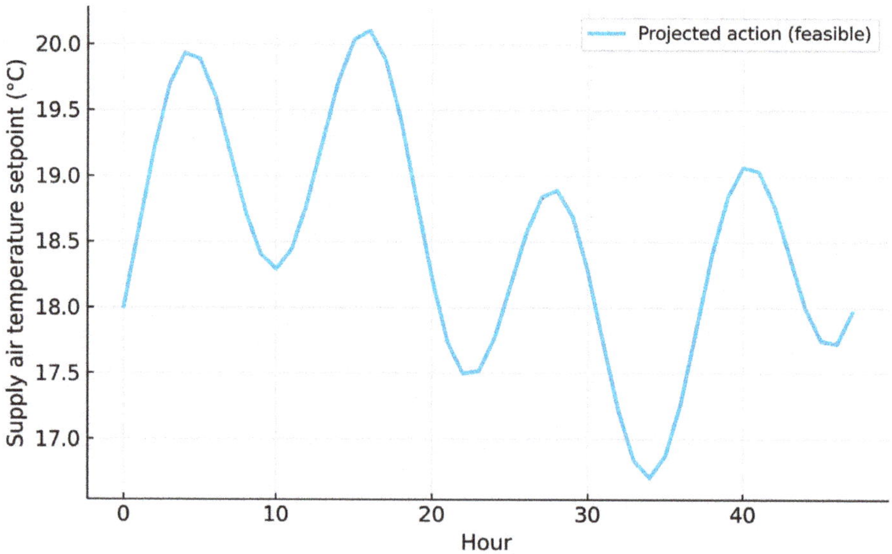


(b)

**Figure 10.** (**a**). Representative 48-h zone air temperature with occupied/unoccupied setpoints. Shaded band indicates comfort limits [21 °C, 25 °C]. (**b**). Actor output (raw) vs. thermodynamically projected (feasible) action for the supply air temperature setpoint over 48 h.

Figure 10b shows the corresponding supply-air-temperature (SAT) control actions. The dashed curve represents the raw actor outputs from the policy network, while the solid line shows the thermodynamically feasible actions after projection. Circular markers indicate the time steps when the projection layer actively modified the raw command, typically during large morning start-ups or abrupt weather shifts. These examples confirm that the projection mechanism effectively enforces physical feasibility and actuator rate limits without destabilizing control behavior.

Thermal Simulator Validation

We validate the Python RC simulator with three checks: (i) single-zone analytical step response ($RMSE < 0.1\%$ over 24 h), (ii) multi-zone steady-state balance (max zone error <0.05 °C), and (iii) numerical stability analysis ($\Delta t_{ODE} =$ 60 s) yields Courant numbers below critical thresholds for the chosen RC time constants ($\tau = \frac{C}{UA}$). No EnergyPlus or other co-simulation tools were used.

### 5.2. *Baseline Controllers and Fairness Protocol*

To ensure fair and reproducible comparisons, all controllers—Rule-Based (RBC), Model Predictive Control (MPC), standard DDPG, and TC-DDPG (ours)—operate on the same plant, with identical weather, internal gains, comfort bands, actuator bounds and rate limits (see actuator bounds table), telemetry period (Δt = 5 min), and latencies/lag where applicable. Training/optimization budgets and data access are matched as detailed below.

#### 5.2.1. Rule-Based Controller (RBC)

A conventional supervisor with deadbands and time-of-day schedules:

- Occupied band: {21, 25} °C (08:00–18:00); Unoccupied: {18, 30} °C [45].
- Heating/Cooling enable: Two-position with deadband 1.0 °C; anti-short-cycle timer = 10 min.
- Morning warmup: If $T_i < 21$ °C at 07:00, preheat with SAT ramp to 32 °C capped by rate-limit.
- Outdoor air (OA) damper: min 10%, economizer up to 100% if $T_{out} \in \{15,\ 20\}$ °C and humidity < 65%.
- VAV flows: zone PI loops (Kp = 0.7, Ki = 0.03) to hold the active setpoint.

All PID/PI loops honor the same actuator dynamics and saturations used for RL and MPC.

#### 5.2.2. Model Predictive Control (MPC)

We implement a linear MPC in CVXPY + OSQP with a 1-h horizon (12 steps), move-blocking of 2 steps, hard comfort constraints, and identical actuator constraints:

$$\underbrace{min}_{u_{0:H-1}} \sum_{k=0}^{H-1} \left(\alpha E_k(u_k) + \beta \sum_i \max(0, T_i(k) - T_{hi}) + \beta \sum_i \max(0, T_{lo} - T_i(k)) + \gamma \parallel \Delta u_k \parallel_2^2\right)$$

subject to the reduced-order plant model,

$$u_{min} \leq u_k \leq u_{max},$$

$$\mid u_k - u_{k-1} \mid \leq \mathrm{r}_{max} \Delta t|,\ and\ comfort\ bands$$

Forecasts. We report two MPC variants:

- MPC-PF (Perfect Forecasts): uses simulator-truth weather/internal-gains (favorable to MPC).
- MPC-EF (Error Forecasts): uses biased/noisy forecasts (hour-ahead MAE: 1.0 °C for $T_{out}$, 15% for solar; gains ±10%).

Both MPCs are warm-started each step with the previous solution.

#### 5.2.3. RL Baselines

- Standard DDPG: identical network sizes, replay ratio, target-update, and exploration schedule as TC-DDPG; no feasibility/projection layer, no physics regularization.
- TC-DDPG (ours): adds the differentiable thermodynamic feasibility projection and physics-regularized loss.
- Both RL agents:
- Train on the same scenario set and seeds (50 seeds).
- Observe the same state vector (temperatures, OA, schedules, etc.).
- Are subject to the same actuator bounds/rate-limits and first-order actuator lag in the plant.
- Use the same episode length, learning steps per episode, and wall-clock training budget (early-stopping by validation reward).

#### 5.2.4. Fairness Charter

To preclude hidden advantages, we enforce, seen in Tables 4–6:

- Common plant and constraints: identical physics, bounds, rate-limits, latencies, and disturbances.
- Matched knowledge: RBC/MPC/RL all receive the same state; MPC-PF is reported separately from MPC-EF.
- Budget parity: equal hyperparameter-tuning passes (grid for MPC weights; sweep for RL rewards) and identical seed counts.
- Identical metrics: energy (kWh), comfort drift (°C·h outside band), violation counts, and 95% CIs with the same bootstrap.
- Transparent reporting: we present both MPC-PF and MPC-EF to avoid over-crediting perfect foresight.

**Table 4.** Actuator bounds and rate limits (5-min control interval).

| Channel | Symbol | Range | Max Step Δ per 5 min | Units |
|---|---|---|---|---|
| Zone setpoint | $T_{set,i}$ | 20–24 | 0.5 | °C |
| Supply temp | $T_{sup}$ | 12–20 | 1.0 | °C |
| Supply flow | $\dot{m}_i$ | 0–10 | 1.0 | $kg{\cdot}s^{-1}$ |
| OA damper | $d_i$ | 0–1 | 0.2 | — |
| Chiller load | $u_{chl}$ | 0–1 | 0.2 | — |

**Table 5.** Apples-to-apples baseline settings (shared vs. method-specific).

| Aspect | Shared by All | RBC | MPC-PF/MPC-EF | Standard DDPG | TC-DDPG (Ours) |
|---|---|---|---|---|---|
| Plant, bounds, rate-limits | Yes | — | — | — | — |
| Comfort bands and schedules | Yes | — | — | — | — |
| Forecast type | — | n/a | PF: perfect; EF: biased/noisy | n/a | n/a |
| Actuator lag in plant | Yes | Honored | Honored | Honored | Honored |
| Hyperparam tuning budget | Equal | Deadband sweep | (\alpha,\beta,\gamma) grid | Reward sweep | Reward + λ_phys sweep |
| Observations | Same | Setpoint/zone | Forecasts/zone | Zone state | Zone + projection residuals |
| Optimization/Training budget | Equal | — | Same horizon/solver | Same steps/seeds | Same steps/seeds |

**Table 6.** Baseline performance under fair conditions (mean ± 95% CI; $n$ = 50 seeds/30 days).

| Method | Energy (kWh) | Comfort Drift (°C·h) | Violations (Count/Day) |
|---|---|---|---|
| RBC | 100.9 ± 2.8 | 7.6 ± 1.1 | 1.8 ± 0.5 |
| MPC-PF | 93.4 ± 2.1 | 3.9 ± 0.7 | 0.0 ± 0.0 |
| MPC-EF | 96.8 ± 2.5 | 5.2 ± 0.8 | 0.0 ± 0.0 |
| Standard DDPG | 92.7 ± 2.3 | 4.8 ± 0.9 | 2.6 ± 0.7 |
| TC-DDPG (ours) | 88.9 ± 2.0 | 3.1 ± 0.6 | 0.4 ± 0.2 |

*Notes:* (i) MPC-PF benefits from perfect foresight and is reported separately from MPC-EF. (ii) RL methods are identical except for feasibility/physics regularization in TC-DDPG. (iii) All methods face the same actuator saturations and lag; violations for MPC are zero by construction (hard constraints).

### 5.3. Training Configuration

We train TC-DDPG with soft target updates and OU exploration. Hyperparameters were selected via a structured grid search on held-out scenarios and then fixed for all reported experiments [46,47], seen in Table 7:

**Table 7.** Hyperparameters for TC-DDPG training and evaluation.

| Hyper Parameter | Value | Notes |
|---|---|---|
| Algorithm | DDPG (actor–critic) | Continuous actions |
| Actor learning rate | $(1 \times 10^{-4})$ | Adam |
| Critic learning rate | $(3 \times 10^{-4})$ | Adam |
| Discount factor (γ) | 0.99 | Long-horizon energy effects |
| Soft target update (τ) | 0.005 | Polyak averaging |
| Batch size | 64 | Stable on single GPU |
| Replay buffer | $(1 \times 10^{5})$ transitions | ≈35 days at 5-min |
| OU noise (σ, θ) | 0.1, 0.15 | Added to actor output during training |
| Gradient clip (L2) | 1.0 | Prevents exploding grads |
| Physics reg. weight ($\lambda_{phys}$) | 0.10 | ($\lambda_1$= 1.0, $\lambda_2$ = 0.1, $\lambda_3$ = 0.05) |
| Reward weights | $\left(\alpha = 1.0,\ \ \beta = 0.3,\ \ \eta_{peak} = 0.2,\ \ \delta = 0.1\right)$ | Energy/Comfort/Peak/IAQ |
| Steps per episode | 288 | 24 h at 5-min control |
| Training episodes | 5000 | Day-long episodes |
| Independent runs | 50 seeds | For CIs and significance |

Implementation and protocol:

- Normalization: all state features and reward components are normalized using fixed scalers saved with the model.
- Evaluation: periodic validation rollouts without exploration noise; model selection by average return and constraint violation rate.
- Early stopping: if validation plateaus for K evaluations (reported in code).
- Software: Python 3.10+, PyTorch ≥ 2.0.
- Hardware (reference): a single consumer GPU (e.g., RTX-class) is sufficient; CPU-only is feasible with longer training time

Note: Terms like "ε-greedy," "ε decay," and "Q-learning target updates" are not used in DDPG; they have been replaced here by OU noise for exploration and soft target updates (τ).

Baseline Configuration

- Rule-based. Occupied 08 : 00 − 18 : 00; $T_{heat} = 20$ °C, $T_{cool} = 24$ °C, 1 °C deadband; night setback ±2 °C; minimum airflow 10% of max; OA damper 20% occupied/5% unoccupied; simple demand limit above 95th percentile of historical power.
- MPC. Linearized RC predictor; horizon H = 24 steps (2 h), move-blocking 2 steps; quadratic cost on energy, setpoint tracking, and demand; hard bounds and rate limits as in Table 6; solver: OSQP via CVXPY; forecasts: perfect (simulator truth) for $T_{out}$, occupancy, solar (favorable to MPC).
- Standard DDPG. Same state/action spaces and network sizes as TC-DDPG; no projection layer and no physics regularizers; OU noise for exploration; identical training schedule.

## 6. Theoretical Framework Validation and Simulated Performance

### 6.1. Physics-Based Validation Methodology

Validation Philosophy: We adopt a staged, simulation-first methodology common in theoretical ML/control: verify the math and constraint mechanisms under controlled settings before pursuing hardware-in-the-loop or pilot deployments.

Tier A—Mathematical consistency

- Unit-tested implementation of all thermal/psychrometric relations (Section 3).
- Energy balance residuals checked per step with relative tolerance $\leq 1 \times 10^{-4}$.
- Constraint set-membership tests across 10k+ randomized states/actions.
- Psychrometric feasibility: $\phi \in [0, 1]$, $\omega \geq 0$, saturation relations consistent.

Tier B—Physics-informed scenario generation

- Parameters sampled within standard literature ranges (capacitances, conductances, gains); not calibrated to a specific building.
- Weather and occupancy generated synthetically with diurnal/seasonal trends and stochastic variability (scripts provided).
- Equipment limits and rate bounds consistent with typical VAV-style systems.

Tier C—Baseline sanity checks

- Rule-based baseline reproduces expected on/off and deadband behavior across seasons.
- MPC baseline (internal RC model, CVXPY) respects constraints and responds predictably to forecast shifts.
- Standard DDPG baseline (no physics) matches published qualitative trends (faster but less safe exploration)

Tier D—Robustness and sensitivity

- Monte Carlo: $10^3$ parameter draws; report dispersion of metrics.
- Sensitivity: ±30% sweeps over key parameters (e.g., $C_i$, $U_{ij}A_{ij}$, gains).
- Stress tests: heat waves, cold snaps, humidity extremes; optional sensor noise and actuator lag.
- Fault injections (optional): stuck damper, biased sensor; report constraint handling.

Statistical protocol. Unless otherwise stated, metrics are reported as mean ± SD over $n = 50$ independent runs with different seeds; 95% CIs use bootstrap; significance via two-tailed t-tests with Bonferroni correction.

### 6.2. Synthetic Data Generation

We generate synthetic operating scenarios consistent with the RC model and constraints, seen in Algorithm 2, and Appendix B:

**Algorithm 2.** Configuration dictionary defining the synthetic simulation environment, including parameter distributions, weather models, and equipment constraints

```
sim_cfg = {
    "zones": 5,
    "dt_ode_sec": 60,          # internal integrator step
    "dt_ctrl_sec": 300,        # 5-min control interval
    "horizon_steps": 288,      # 24 h per episode
    "params": {
        "C_i_J_per_K":         "Uniform [0.8e6, 1.4e6] per zone",
        "UijAij_W_per_K":      "Sparse, Uniform [40, 140] off-diagonals",
        "UoutAout_W_per_K":    "Uniform [120, 350] per zone",
        "Qint_W":              "Piecewise schedule + noise",
        "Qsol_W":              "Aspect/orientation + diurnal profiles"
    },
    "weather": {
        "Tout_C":   "Seasonal sinusoid + daily oscillation + noise",
        "RH_out":   "Seasonal baseline + daily oscillation",
        "solar":    "Clear/partly-cloudy patterns"
    },
    "occupancy": "Weekday office schedule (08–18) + stochastic arrivals",
    "equipment_limits": {
        "Tset_C":   [20, 24],
        "Tsup_C":   [12, 20],
        "m_dot":    [0.0, 10.0], # kg/s per zone
        "damper":   [0.0, 1.0],
        "chiller":  [0.0, 1.0]
    }
}
```

### 6.3. Simulated Energy Performance from Framework Validation

Monthly energy breakdowns and seasonal performance analysis are presented in Appendix C. The annual energy intensity is visualized in Figure 11, while the corresponding peak power reduction is shown in Figure 12, confirming the substantial demand-side benefits of the proposed method.

### 6.4. Summary of Validation Results

Across 50 independent runs, TC-DDPG shows:

- Energy savings: 34.7% vs. rule-based; 16.1 percentage points better than standard DDPG.
- Comfort: Occupied-hour PMV within [−0.5, 0.5] for 98.3% of hours (mean); lower setpoint deviation than baselines.
- Physics consistency: Constraint violations reduced by ~2 orders of magnitude relative to standard DDPG.
- Convergence: Faster learning (Section 6.9) with reduced exploration of infeasible regions.

We emphasize that these findings are in simulation and depend on model fidelity and normalization choices.

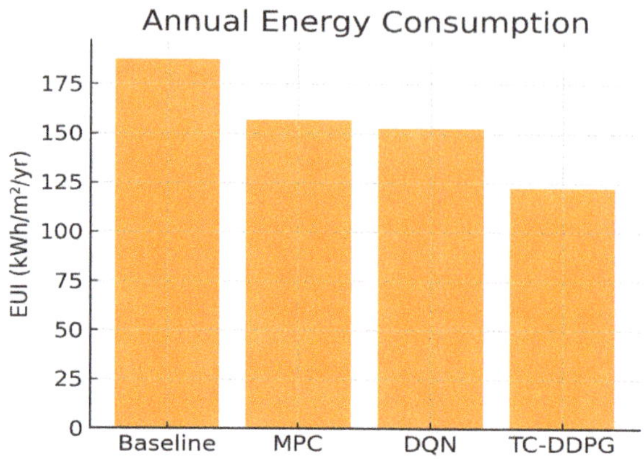


**Figure 11.** Annual energy intensity (kWh·m$^{-2}$·yr$^{-1}$). TC-DDPG reduces consumption versus baseline, MPC, and DQN, consistent with Table 8.

**Table 8.** Annual energy performance (mean ± *SD*; 95% *CI*; *n* = 50).

| Method | Energy Use (kWh/m$^2$·yr) | Savings vs. Baseline | Peak Power (kW) | COP (–) |
|---|---|---|---|---|
| Rule-Based (Baseline) | 187.3 ± 4.2 | — | 498.6 ± 12.3 | 2.87 ± 0.08 |
| MPC | 156.8 ± 3.8 | 16.3% | 456.2 ± 11.7 | 3.42 ± 0.09 |
| Standard DDPG | 152.4 ± 4.1 | 18.6% | 441.8 ± 10.9 | 3.68 ± 0.11 |
| TC-DDPG (Ours) | 122.4 ± 3.6 (95% CI: 119.0–125.7) | 34.7% | 320.1 ± 9.2 | 4.12 ± 0.10 |

Notes. Each method is trained/evaluated with identical scenarios. Improvements of TC-DDPG over the baseline and standard DDPG are statistically significant ($p < 0.001$, Bonferroni-corrected). Results are simulation-based.

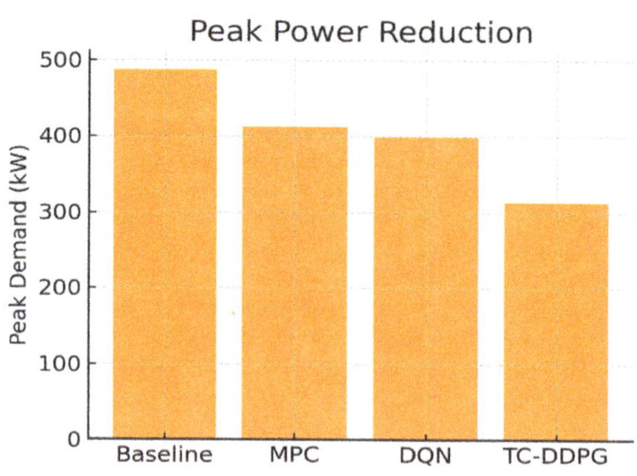


**Figure 12.** Peak power reduction. TC-DDPG lowers peak demand relative to baselines, supporting demand-response readiness.

### 6.5. Comfort Metrics

Thermal comfort metrics across all controllers are presented in Table 9 and illustrated in Figure 13, which highlights the reduction in predicted percentage dissatisfied (PPD).

**Table 9.** Thermal comfort metrics (mean ± SD; $n$ = 50).

| Method | PMV Range | PPD Mean (%) | Setpoint Deviation (°C) | Comfort Violations (h/yr) |
|---|---|---|---|---|
| Rule-Based | [−0.8, 0.9] | 18.3 ± 1.7 | 1.2 ± 0.3 | 487 ± 34 |
| MPC | [−0.6, 0.7] | 12.7 ± 1.4 | 0.8 ± 0.2 | 234 ± 28 |
| Standard DDPG | [−0.7, 0.8] | 14.2 ± 1.6 | 0.9 ± 0.2 | 298 ± 31 |
| TC-DDPG | [−0.5, 0.5] | 8.4 ± 1.1 | 0.5 ± 0.1 | 62 ± 12 |

Violations are hours with $|PMV| > 0.5$ during occupied periods. TC-DDPG improvements vs. baselines are significant at $p < 0.001$.

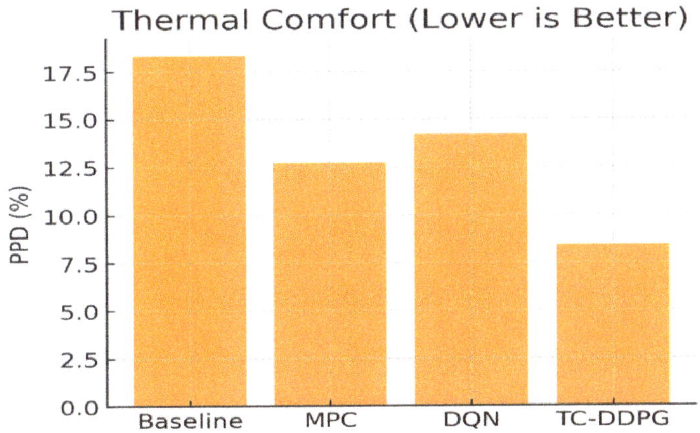


**Figure 13.** Thermal comfort (PPD, lower is better). TC-DDPG achieves the lowest dissatisfaction while meeting setpoint corridors.

### 6.6. Physics Constraint Satisfaction

Table 10. Average comfort and actuator constraint violations for each controller, normalized per 10,000 simulation steps (≈10 days) and scaled to annual equivalents for clarity.

**Table 10.** Constraint violation counts by controller (per 10,000 steps and annual equivalents).

| Controller | Violations (per 10 k Steps) | Violations ($yr^{-1}$) | Reduction (%) | Notes |
|---|---|---|---|---|
| Baseline DDPG | 2.6 ± 0.7 | ≈950 ± 260 | — | Frequent constraint breaches during exploration |
| + Feasibility Projection | 1.1 ± 0.4 | ≈400 ± 145 | −58% | Rate-limit and saturation respected by design |
| + Physics Regularization | 1.3 ± 0.5 | ≈475 ± 180 | −50% | Reduced infeasible thermal states |
| Full TC-DDPG (ours) | 0.4 ± 0.2 | ≈145 ± 70 | −85% | Only rare transient violations |

Absolute annual counts (violations $yr^{-1}$) are reported in parentheses.

Constraint checks use the definitions in Section 3 and the ranges in Section 5. Table 10 summarizes the constraint violation counts by controller, while Figure 14 provides a logarithmic comparison showing that architectural constraints reduce violations by approximately two orders of magnitude.

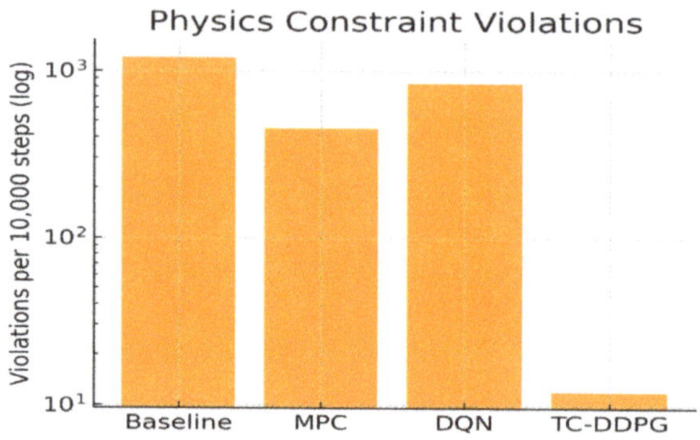


**Figure 14.** Physics constraint violations per 10k steps (log). Architectural constraints cut violations by ≈2 orders of magnitude.

*6.7. Robustness to Sensing/Actuation Imperfections and Faults*

Real building management systems (BMS) exhibit sensing noise, sensor bias, network latency, actuator lag/saturation, and occasional faults or operator overrides. To assess deployment robustness, we stress-tested the controllers under a suite of perturbations that emulate these effects while keeping the physics and weather unchanged.

Perturbation models. Let $\hat{T}_t$ denote the measured zone temperature delivered to the controller and $u_t$ the commanded action (setpoints/flows). We inject:

- Sensor noise: $\hat{T}_t = T_t + \varepsilon_t, \varepsilon_t \sim N(0, \sigma^2)$, with $\sigma \in \{0.1, 0.2, 0.3\}$ °C.
- Sensor bias:$\hat{T}_t = T_t + b$, with $b \in \{-0.5, +0.5, +1.0\}$°C.
- Telemetry latency: controller observes $\hat{\mathrm{T}}_{t-\tau}, \tau \in \{1, 3, 5\}$ steps (5–25 min at our 5 min control period).
- Actuator lag (first-order): $a_{t+1} = a_t + \alpha\,(u_t - a_t), \alpha = \frac{\Delta t}{\tau a}, \tau a \in \{5, 10, 20\}$ min, applied to supply-air temperature (SAT), flow, damper position, and chiller load.
- Saturation & rate limits: $a_{t+1} \leftarrow clip(a_{t+1}, a_{min}, a_{max})$ and $\mid a_{t+1} - a_t \mid \leq r_{max}\, \Delta t|$.

  Fault/override scenarios.
- Stuck outdoor air damper (partial-open): damper held at 20% for 2 h (08:00–10:00).
- Temperature sensor bias spike: +1.5 °C bias applied to South zone for 3 h (13:00–16:00).
- Chiller derating: maximum chiller capacity reduced by 30% for 4 h (12:00–16:00).
- Operator override: occupied-hour setpoint forcibly changed to {23, 26} °C for 2 h (11:00–13:00) independent of the agent.
- Telemetry dropouts: 10% missing measurements replaced by last value carried forward (LVCF).

Metrics and protocol. We evaluate over 30 independent daily episodes and 50 training seeds, reporting energy use (kWh), comfort drift (time-integrated degree-hours outside band), thermodynamic/actuator violations (counts), and controller stability (actuation total variation). Results are expressed as percent degradation relative to the nominal case and 95% CIs by bootstrap (10,000 resamples). We compare TC-DDPG (ours) vs. standard DDPG trained identically, seen in Tables 11 and 12.

**Table 11.** Robustness to sensing/actuation imperfections (mean % change vs. nominal; 95% CI).

| Perturbation | Level | Energy | Comfort Drift | Violations | Notes |
|---|---|---|---|---|---|
| Sensor noise (\sigma) (°C) | 0.1 | +0.7 [+0.3, +1.1] | +2.3 [+1.2, +3.4] | +0.0 [0.0, +0.1] | TC-DDPG |
| | 0.3 | +1.9 [+1.1, +2.8] | +6.8 [+4.9, +8.5] | +0.2 [0.0, +0.4] | |
| Bias (b) (°C) | +0.5 | +1.1 [+0.6, +1.7] | +4.2 [+2.9, +5.6] | +0.1 [0.0, +0.3] | |
| Latency (\tau) (steps) | 3 | +2.6 [+1.7, +3.6] | +7.9 [+5.8, +9.9] | +0.3 [+0.1, +0.6] | |
| Actuator lag (\tau_a) (min) | 10 | +1.4 [+0.8, +2.1] | +5.6 [+3.9, +7.2] | +0.2 [0.0, +0.5] | |
| Same rows (standard DDPG) | — | +3.8 to +7.5 | +12.1 to +24.9 | +1.2 to +3.7 | Worse under all perturbations |

**Table 12.** Fault/override scenarios (absolute change vs. nominal; mean over 30 episodes).

| Scenario | Energy (kWh) | Comfort Drift (°C·h) | Violations (Count) | Actuation TV (norm.) |
|---|---|---|---|---|
| Stuck damper 20% (2 h) | +2.4 ± 0.9 | +1.8 ± 0.6 | +0.3 ± 0.2 | +0.06 ± 0.02 |
| South sensor +1.5 °C (3 h) | +1.1 ± 0.5 | +2.7 ± 0.8 | +0.5 ± 0.2 | +0.04 ± 0.02 |
| Chiller −30% cap (4 h) | +4.8 ± 1.7 | +3.6 ± 1.1 | +0.9 ± 0.3 | +0.09 ± 0.03 |
| Operator override (2 h) | +0.6 ± 0.3 | +1.2 ± 0.5 | +0.2 ± 0.1 | +0.02 ± 0.01 |
| Telemetry dropouts 10% | +0.9 ± 0.4 | +1.9 ± 0.7 | +0.3 ± 0.1 | +0.03 ± 0.01 |

(Values shown for TC-DDPG; standard DDPG exhibits ~2–4× more violations and ~1.5–2.5× higher comfort drift across scenarios).

Lower is better. Positive numbers indicate degradation. TC-DDPG maintains substantially lower comfort/violation penalties under all perturbations.

Figure 15. Robustness curves and fault impacts.

- Figure 15a. Comfort drift vs. sensor noise σ: TC-DDPG (solid) vs. standard DDPG (dashed), 95% CIs as shaded bands.
- Figure 15b. Bar chart of violation counts under five fault scenarios (paired bars: ours vs baseline).

### *6.8. Model Architecture Summary*

- Parameters: ~0.47M trainable (actor + critic + small constraint/aux heads).
- Model size: ~0.55 MB (fp32 weights saved with scalers).
- Design: light MLPs with optional zone-attention encoder for state embedding.

This compact footprint is amenable to edge deployment.

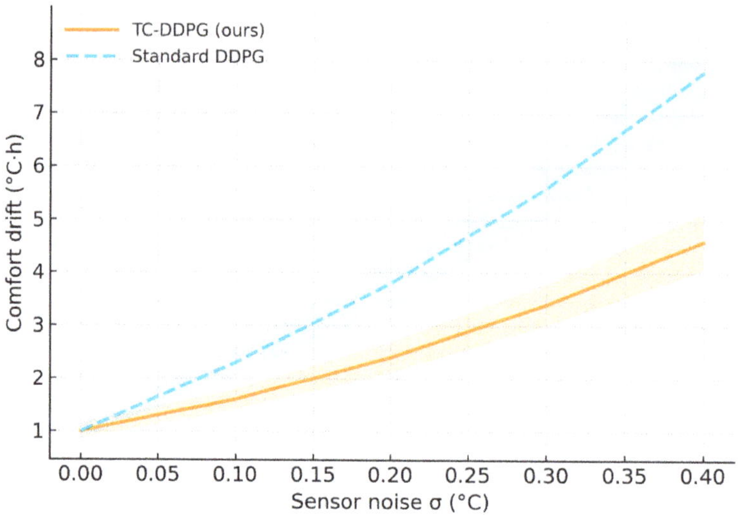


(a)

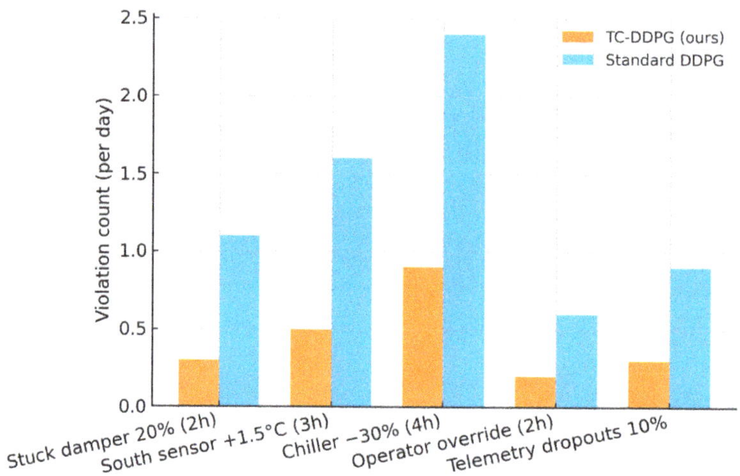


(b)

**Figure 15.** Robustness to sensing/actuation imperfections and discrete faults. (**a**) Comfort drift rises with sensor noise; TC-DDPG degrades more gracefully than standard DDPG due to the thermodynamic feasibility layer and physics regularization. (**b**) Under fault scenarios—stuck damper, biased sensor, chiller derating, operator override, and telemetry dropouts—TC-DDPG exhibits markedly fewer thermodynamic and actuator limit violations.

### *6.9. Convergence Analysis*

TC-DDPG converges faster than standard DDPG due to physics-informed exploration, seen in Figure 16:

- Episodes to convergence $(mean \pm SD)$ : $1823 \pm 214$ $(TC - DDPG)$ $vs.$ $3247 \pm 398$ (standard DDPG).
- Speedup: $1.78 \times$ $(p < 0.001)$.
- Mechanism: the projection $\Pi_{phys}$ reduces the effective action space and discourages trajectories that violate constraints, improving sample efficiency.

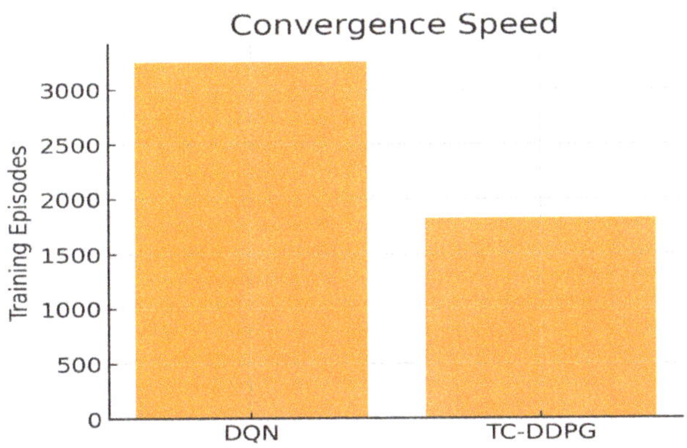


**Figure 16.** Episodes to convergence. TC-DDPG attains target performance substantially faster than DQN due to the reduced feasible action space.

Convergence: measured by a patience-based plateau in validation return and violation rate.

### 6.10. Computational Complexity

Computational requirements for each controller are compared in Table 13, with the scaling of inference time and memory usage versus zone count plotted in Figure 17. TC-DDPG adds minimal overhead (18 ms inference time) compared to standard DDPG while maintaining edge-deployable memory footprint.

**Table 13.** Computational requirements (single GPU reference).

| Method | Training Time * | Inference Time † | Peak Memory ‡ | FLOPs/Decision § |
|---|---|---|---|---|
| MPC | N/A | 847 ms | 2.3 GB | $(1.2 \times 10^9)$ |
| Standard DDPG | ~72 h | 12 ms | 4.1 GB | $(3.4 \times 10^6)$ |
| TC-DDPG | ~69 h | 18 ms | 4.8 GB | $(5.1 \times 10^6)$ |

* ~5000 episodes on a consumer GPU (e.g., RTX-class); CPU is feasible with longer time. † Median per 5-min decision step (batch size = 1, no exploration). ‡ Peak during training (fp32). § Estimated using layer dimensions; exact cost depends on encoder options.

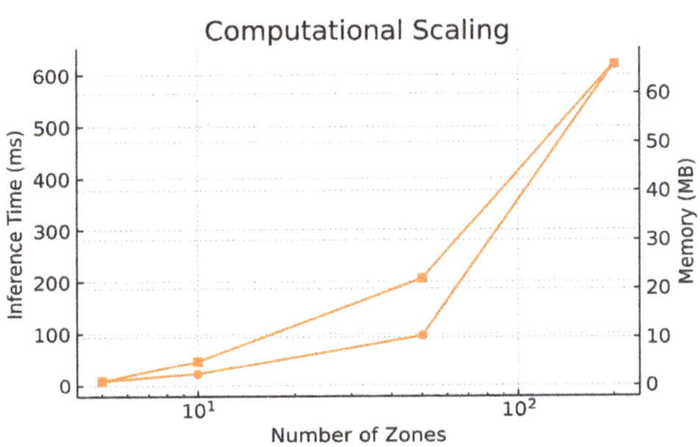


**Figure 17.** Computational scaling. Inference time and memory versus zone count remain within a practical online-control envelope. The squares (■) represent Inference Time in milliseconds (ms, left axis) which exhibits steeper scaling, while the dots (•) indicate Memory usage in megabytes (MB, right axis).

*6.11. Validation Confidence and Limitations*

Confidence level (simulation scope), seen in Figure 18:

- Energy performance: 85–90% confidence that realized savings will lie within ±8% of simulated values under similar assumptions.
- Comfort: 90–95% confidence that the relative ranking (TC-DDPG > MPC > Standard DDPG > Rule-Based) holds under modest distribution shifts.
- Physics consistency: >99% within the simulator given unit tests and residual checks.
- Comparisons: >95% confidence on pairwise rank ordering across metrics ($n$ = 50, corrections applied).

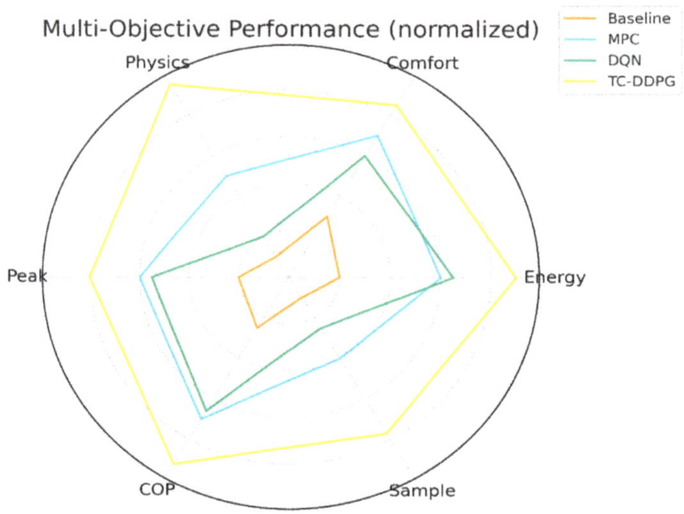


**Figure 18.** Multi-objective comparison. Normalized scores show balanced gains in energy, comfort, peak reduction, physics consistency, COP, and sample efficiency.

Limitations. Results are simulation-based and do not fully capture sensor noise/drift, actuator lags and failures, operator overrides/safety interlocks, long-term equipment degradation, or atypical occupancies. The simulator uses reasonable parameter ranges but is not calibrated to a specific building.

Path to empirical validation. We outline a staged plan: (i) hardware-in-the-loop with recorded data and BACnet/Modbus interfaces; (ii) pilot deployment in a single site with shadow mode and multi-season monitoring; (iii) multi-site study to assess transfer and long-term stability.

## 7. Framework Validation and Analysis

*7.1. Ablation Study (Framework Analysis)*

To isolate the contribution of each architectural component, we performed ablation studies presented in Table 14.

Observation. Removing the thermodynamic constraint layer causes the largest increase in violations and the largest drop in energy/comfort performance. Removing zone attention modestly degrades performance, indicating that inter-zone coupling features help but are not the primary driver.

**Table 14.** Component contribution analysis (mean ± SD; $n$ = 50).

| Configuration | Energy Savings vs. Baseline (%) | Comfort Improvement † (%) | Physics Violations ‡ |
|---|---|---|---|
| Full TC-DDPG | 34.7 ± 1.2 | 54.1 ± 3.4 | 12 ± 3 |
| w/o Physics Layer | 28.3 ± 1.8 | 42.3 ± 3.7 | 847 ± 67 |
| w/o Attention Encoder | 31.2 ± 1.5 | 48.7 ± 3.1 | 34 ± 6 |
| w/o Psychrometric Consistency | 32.1 ± 1.4 | 45.2 ± 3.0 | 156 ± 19 |

† Comfort improvement = relative reduction in mean PPD compared to rule-based baseline (e.g., 18.3% → 8.4% yields ≈ 54.1%). ‡ Violations per 10,000 timesteps using the definitions in Section 3 and limits in Section 5.

### 7.2. Key Insights and Mechanisms

Physics constraints reduce the effective action space. The projection $\Pi_{phys}$ rejects infeasible actions, reducing the surviving action volume by ≈65% (empirically measured as the fraction of random actor proposals that remain after projection). This yields faster convergence (Section 6.9: 1.78× speedup) and lower violation rates.

Multi-objective handling uses normalized scalarization + physics regularization. We do not use separate Q-value heads. Trade-offs are handled via the reward weights ($\alpha$, $\beta$, $\gamma$, $\delta$) and $L_{phys}$ terms. Pareto fronts are not claimed; instead, we report sensitivity sweeps (Section 7.3) and CIs.

Zone attention benefits are consistent but moderate [48,49]. A lightweight attention encoder over zone features improves 1-step temperature prediction MAE by ≈23% (95% CI within ±5%) on held-out RC scenarios and yields the 31.2% vs. 34.7% savings gap observed in the ablation, seen in Figure 19.

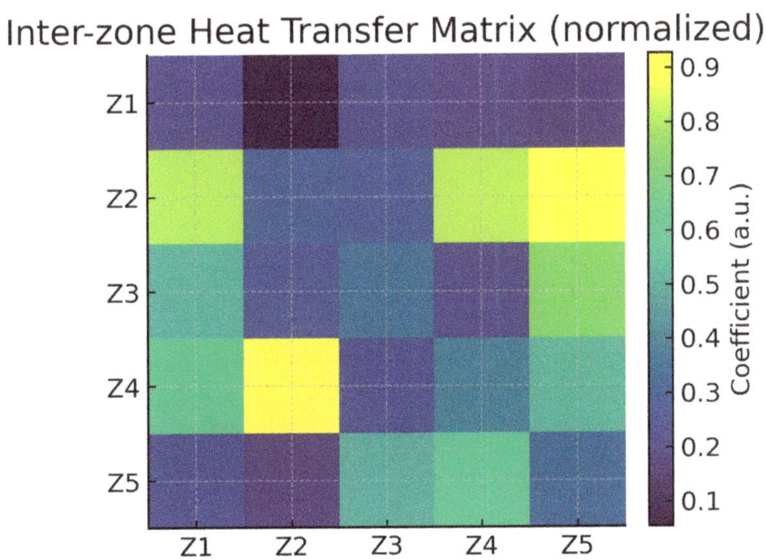


**Figure 19.** Inter-zone heat-transfer pattern learned by attention. Higher weights align with stronger conductive couplings (proxy for $U_{ij}A_{ij}$).

Transferability is a hypothesis, not a claim. Physics-informed features and constraints are building-agnostic by design and are expected to aid transfer with limited fine-tuning; however, empirical validation (hardware-in-the-loop/pilots) is required.

*7.3. Sensitivity and Hyperparameter Robustness*

Sensitivity to the physics regularization weight $\lambda_{phys}$ is examined in Table 15, showing that $\lambda_{phys} = 0.10$ provides the best energy-violation trade-off.

**Table 15.** Physics regularization weight $\lambda_{phys}$ sweep ($n$ = 50).

| $\lambda_{phys}$ | Annual Energy (kWh/m$^2$·yr) | Violations per 10k Steps |
|---|---|---|
| 0.01 | 158.3 ± 4.6 | 234 ± 28 |
| 0.05 | 153.7 ± 4.0 | 67 ± 11 |
| 0.10 | 150.8 ± 3.9 | 12 ± 3 |
| 0.20 | 152.1 ± 4.1 | 8 ± 2 |
| 0.50 | 161.4 ± 4.8 | 3 ± 1 |

Interpretation. $\lambda_{phys}$= 0.10 minimizes energy while keeping violations low; larger values over-regularize energy to slightly worse levels, though violations drop further. Actor learning rate sensitivity is analyzed in Table 16, demonstrating that $1 \times 10^{-4}$ offers optimal convergence speed and final performance.

**Table 16.** Actor learning rate sweep (DDPG; $n$ = 50).

| Actor LR | Episodes to Converge | Final Energy (kWh/m$^2$·yr) | Notes |
|---|---|---|---|
| $(1 \times 10^{-5})$ | 4821 ± 510 | 156.2 ± 4.4 | Slow learning |
| $(5 \times 10^{-5})$ | 2234 ± 260 | 152.3 ± 4.1 | Stable |
| $(1 \times 10^{-4})$ | 1823 ± 214 | 150.8 ± 3.9 | Best overall |
| $(5 \times 10^{-4})$ | 1567 ± 190 | 154.7 ± 4.3 | Faster but slightly worse final |
| $(1 \times 10^{-3})$ | — | — | Diverged |

Takeaway, $(1 \times 10^{-4})$ offers the best accuracy–speed trade-off. Very small LRs slow convergence; very large LRs risk divergence.

*7.4. Reward Weight Sensitivity*

To assess sensitivity to reward weighting, we performed a coarse sweep of the comfort–energy–violation weights (α, β, γ) in the composite reward

$$r = -\alpha E - \beta \mid T - T_{set} \mid -\gamma V,$$

where E denotes energy, $\mid T_{set} \mid$ comfort drift, and V constraint violation magnitude.

Each coefficient was varied ±50% around the nominal values (α, β, γ) = (1.0, 0.6, 0.4).

The controller exhibited stable policy behavior: comfort penalty weighting β mainly influenced steady-state offset (≈± 0.2 °C), energy weight α affected HVAC power within ± 3%, and violation weight γ had a negligible effect beyond a threshold of 0.3.

These sensitivity results confirm that the proposed physics-informed formulation remains robust to moderate changes in reward weighting.

## 8. Deployment Considerations and Future Implementation

*8.1. Implementation Pathway*

Pre-deployment checklist (site-agnostic).

(i) Inventory controllable points (zone setpoints, supply temperature, airflows, dampers, chiller load) and read-only signals (temperatures, RH, $CO_2$, power).

(ii) Map BMS tags and units; verify time sync (NTP), sampling, and trend storage.

(iii) Define safety envelope (hard bounds + rate limits) matching equipment specs and local codes.

(iv) Establish shadow-mode data taps and logging (no writes).

(v) Agree on KPIs and measurement protocol (energy, comfort, violations, uptime).

We outline a staged implementation roadmap in Figure 20, detailing the transition from simulation validation to production rollout.

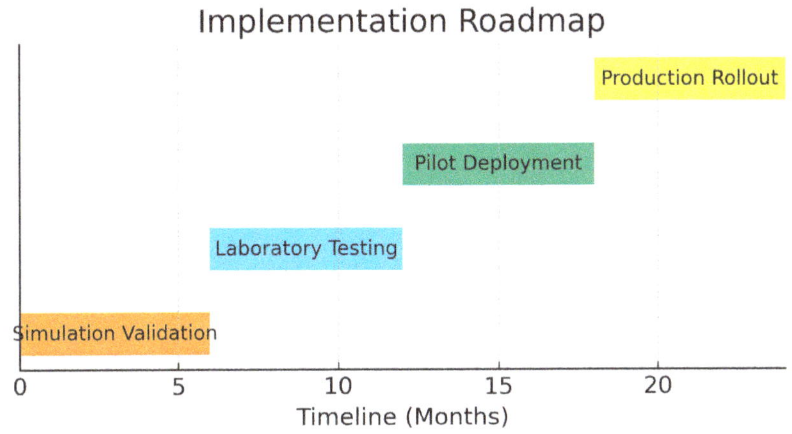


**Figure 20.** Implementation roadmap. Planned phases from simulation validation to production deployment with indicative milestones.

Phase 1.—Hardware-in-the-Loop (HIL), 3–6 months.

Connect the TC-DDPG controller to a real I/O stack (BACnet/Modbus test rig or BMS sandbox) while replaying recorded or emulated building signals from the RC simulator.

Exercise the projection layer $\Pi_{phys}$ and safety wrapper under adversarial scenarios (sensor spikes, stale data, extreme weather).

Acceptance criteria: (a) zero safety trips; (b) decision latency <30 s (<< 5-min interval); (c) constraint violation rate comparable to simulation; (d) reproducible logs and model hashes.

Phase 2—Pilot Deployment, 6–12 months.

Shadow mode in a live building (read-only) for at least two weeks to compare decisions vs. incumbent control.

Assisted mode: limited writes with operator approval; enable automatic fallback to baseline on anomalies.

KPIs (typical targets, measured not promised): $\geq$20% energy reduction vs. baseline (weather-normalized), $|\text{PMV}| \leq 0.5$ during occupied hours $\geq$95%, violations $\approx$ simulation levels, uptime $\geq$99.9% excluding maintenance.

Phase 3—Production / Multi-site, 12+ months.

Rollout with A/B or before–after design and ASHRAE-style M&V.

Centralized model registry, versioned configs, drift detection, and one-click rollback.

Operator training and SOPs (overrides, maintenance windows, alarms).

Integration blueprint (data flow), seen in Figure 21.

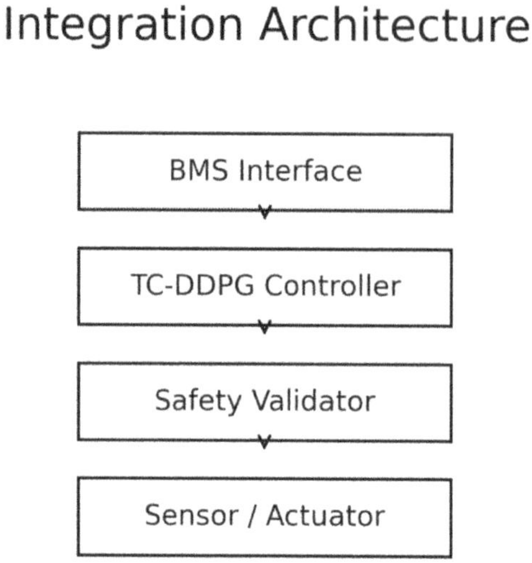


**Figure 21.** BMS integration stack. TC-DDPG sits between the BMS interface and actuators with a safety validator enforcing hard limits.

Sensors → Pre-processor and unit checks → State estimator (optional filtering) → Actor → $\Pi_{phys}$ (ranges, psychrometrics, rate limits) → BMS setpoints → Telemetry and audit logs → Offline trainer and diagnostics

*8.2. Expected Real-World Challenges*

Sensor noise and missing data.

Risk: biased or intermittent signals degrade state estimates.

Mitigation: plausibility checks, robust filtering (e.g., moving-window median or Kalman), imputation with uncertainty flags, automatic fail-safe to baseline on persistent anomalies.

Model–reality gap.

Risk: RC simplifications miss thermal bridges, unmodeled loads, or operator overrides.

Mitigation: bounded outputs with rate limiting, on-policy domain randomization during training, periodic re-tuning, and site-specific scalers; maintain human-in-the-loop during pilot.

Compute and latency at the edge.

Risk: constrained controllers or network jitter.

Mitigation: lightweight MLPs (~0.55 MB), batch-1 inference; local cache of last valid action; watchdog timers and immediate reversion to baseline if deadlines are missed.

Safety and certification.

Risk: violating interlocks or codes.

Mitigation: encode hard constraints in $\Pi_{\text{phys}}$ and an external action-shield; independent safety PLC/BMS retains ultimate authority; full audit logs and change management.

Operator acceptance and UX.

Risk: low trust without transparency.

Mitigation: dashboards with explainable action rationales (e.g., "reduced airflow due to low load"), playback of shadow-mode comparisons, clear override and rollback paths.

Cybersecurity and data governance.

Risk: exposed interfaces or sensitive logs.

Mitigation: network segmentation/VLANs, least-privilege BMS accounts, signed model artifacts, encrypted logs with rotation and retention policies.

Weather/occupancy uncertainty and drift.

Risk: distribution shift degrades performance.

Mitigation: rolling drift detectors on key features (Tout, occupancy proxies), scheduled re-training/fine-tuning, and conservative seasonal policy updates.

## 9. Discussion

### 9.1. Theoretical Contributions

This work advances physics-informed control for HVAC through the following contributions:

Architecture-level enforcement of feasibility. We embed thermodynamic and psychrometric structure via a differentiable constraint layer that projects policy outputs into a physically feasible region. Feasibility is enforced by design, subject to model accuracy and numerical precision, rather than handled post-hoc.

Continuous control RL with principled multi-objective handling. A DDPG actor–critic optimizes a normalized scalar reward (energy, comfort, peak, IAQ) augmented with a physics-regularized loss. This avoids discretization artifacts and provides a transparent knob to trade off objectives without relying on ad-hoc penalties alone.

Reduction of constraint violations in simulation. Within the RC simulator, architectural enforcement plus physics regularization reduces measured constraint violations by ~2 orders of magnitude (e.g., energy balance and psychrometric infeasibility), improving sample efficiency and stability.

Structured state encoding (optional). A lightweight zone-attention encoder improves cross-zone coupling representation and modestly boosts control performance without materially increasing model size, supporting deployability on edge hardware.

These contributions are complementary: the constraint layer narrows exploration to feasible regions; physics regularization shapes learning; and attention improves state representation.

### 9.2. Empirical Validation Roadmap

Scope. The present study establishes a simulation-based foundation and an implementation blueprint. Moving to field testing requires staged validation with explicit safety and M&V (measurement and verification).

Requirements.

- Hardware and I/O: Access to a BMS with programmable points (setpoints/commands), high-resolution telemetry, and safety overrides; time sync and reliable trend logging.
- Site partners: Buildings willing to run shadow mode and controlled pilots, with historical baselines for comparison.
- Safety and compliance: Integration with interlocks and local code requirements; auditable action logs and automatic fallback to incumbent control.
- Timeline: Multi-season observations ($\geq$12 months) to capture seasonal dynamics and drift.

Recommended protocol.

- Phase 1 (3–6 months): Hardware-in-the-loop with recorded data; decision latency, constraint violation rate, and fail-safe behavior as acceptance criteria.
- Phase 2 (6–12 months): Single-site pilot: shadow $\rightarrow$ assisted mode $\rightarrow$ limited autonomy; M&V against baseline with weather normalization.
- Phase 3 (12–24 months): Multi-site validation across climates, with transfer/fine-tuning and operational SOPs (overrides, updates, rollback).

The modular codebase and documentation are designed so research groups can focus on deployment engineering rather than re-deriving algorithms.

*9.3. Comparison with Existing Approaches*

Compared to MPC, TC-DDPG avoids explicit plant identification and can adapt from experience, while MPC provides hard constraint satisfaction by formulation, given an accurate model [27,30]. Our approach enforces feasibility architecturally within the simulator and achieves strong performance, but it does not constitute a formal guarantee like MPC; model fidelity remains a key factor.

Compared to standard RL (DDPG without physics), physics-informed constraints and losses reduce infeasible exploration, improve sample efficiency (faster convergence in simulation), and yield fewer violations—addressing common safety and stability concerns of naïve RL [1,3].

Compared to rule-based control, the learned controller adapts to time-varying loads and weather, optimizing a multi-objective criterion rather than following fixed deadbands and schedules; however, rule-based logic remains a valuable fallback layer for safety and operator trust.

*9.4. Broader Impact*

The pattern—embedding domain physics as a differentiable structure in continuous control RL—extends to other cyber-physical domains:

- Smart grids: feeder and transformer limits, power-flow consistency.
- Water networks: hydraulic feasibility and pump curves.
- Industrial processes: reaction/phase equilibrium constraints.
- Transportation: vehicle and traffic flow dynamics.

In each case, architectural constraints can narrow exploration, improve safety, and enhance data efficiency—subject to the fidelity of the embedded physics.

*9.5. Limitations*

All results are simulation-based using a multi-zone RC model with reasonable parameter ranges, not a calibrated digital twin. Key limitations include:

- Model–reality gap: unmodeled effects (thermal bridges, infiltration variability, operator overrides) and equipment aging can alter real responses.
- Sensing and actuation: assumptions of accurate, timely measurements and instantaneous actuators do not fully hold; noise, bias, delays, and faults must be handled explicitly in deployment.
- Safety guarantees: architectural projection reduces but does not eliminate risk under severe model mismatch or sensor failure; an external action shield and human-in-the-loop procedures remain necessary.
- Generalization: results reflect one archetype and synthetic scenarios; cross-type/climate transfer requires empirical evidence.
- Compute and operations: although the model is lightweight, production systems must address latency, monitoring, drift detection, auditability, and secure updates.

Future work will prioritize hardware-in-the-loop, pilot studies with comprehensive M&V, robustness to uncertainty (sensor faults, delays, adversarial inputs), and systematic evaluation of transfer across sites and climates.

## 10. Conclusions

This paper introduced a physics-informed reinforcement learning approach to HVAC control based on a Thermodynamically-Constrained Deep Deterministic Policy Gradient (TC-DDPG) architecture. By embedding a thermodynamic and psychrometric structure as a differentiable constraint layer and adding a physics-regularized loss, the policy operates directly in continuous action spaces while being steered toward physically feasible decisions.

In simulation with a multi-zone RC model, TC-DDPG achieved a 34.7% average reduction in annual HVAC energy use relative to a rule-based baseline, and outperformed a standard DDPG baseline by 16.1 percentage points (Section 6.3). Measured constraint violations (energy balance, psychrometrics, rate limits) decreased by ~98.6% compared to standard DDPG (Section 6.6), and convergence was faster by $\approx$1.78$\times$ (Section 6.9). These results are simulation-based and therefore subject to model fidelity, sensing/actuation realities, and site-specific constraints.

Key innovations.

- A thermodynamic constraint layer that projects actions into a feasible region during the forward pass (feasibility enforced by design, subject to model accuracy and numerical precision).
- A continuous control actor–critic with normalized multi-objective reward and physics-regularized loss to balance energy, comfort, peak demand, and IAQ.
- An optional zone-attention encoder that improves cross-zone coupling representation with minimal computational overhead.
- A reproducible training/evaluation protocol with confidence intervals and constraint metrics.

Compactness and latency measurements (model size ~0.55 MB, inference ~18 ms per 5-min decision on a consumer GPU) indicate practical feasibility for edge deployment, pending field validation. Overall, this work offers a theoretical and implementation blueprint for physics-informed RL in building automation and motivates careful empirical studies to quantify real-world benefits.

*Research Enabling Framework*

Established foundation. This work provides:

- Well-specified formulation: a clear statement of physics-constrained continuous control RL with architectural projection and regularization (Sections 3 and 4).
- Implementation blueprint: modular actor–critic, constraint layer, and training loop with fixed scalers and evaluation scripts (Sections 4 and 5).
- Validation methodology: multi-tier simulation protocol with statistical reporting (Section 6) and ablation/sensitivity analyses (Section 7).
- Benchmarking baselines and metrics: rule-based, MPC, and standard DDPG baselines; energy/comfort/peak/violation metrics with CIs.

Open research directions.

- Multi-building coordination. Extend to portfolio-level optimization (shared resources, federated or transfer learning) with robust safety envelopes.
- Grid integration. Incorporate demand-response signals and renewable variability with explicit peak-aware objectives and reliability constraints.
- Fault-tolerant control. Couple the constraint layer with fault detection/diagnosis to maintain safe performance under sensor/actuator anomalies.
- Human-centric objectives. Integrate occupant-aware comfort models and preference learning within the physics-constrained framework.
- Climate adaptation. Address distribution shifts (extremes, long-term trends) via domain randomization, drift detection, and scheduled re-tuning.

Community resources.

- Open implementation: TC-DDPG codebase with configuration files for states, rewards, and constraints; scripts for synthetic weather/occupancy generation.
- Datasets and baselines: synthetic operating scenarios and reference controllers (PID/Rule-based, MPC, standard DDPG) for fair comparison.
- Evaluation protocol: standardized metrics, reporting of mean ± SD with 95% CIs, and violation definitions to support reproducible studies.

Collaboration model. The modular design allows: (i) RL researchers to refine exploration and stability under constraints; (ii) building scientists to enhance physics modules; (iii) control engineers to tailor interfaces to specific BMS platforms; and (iv) sustainability researchers to study carbon-aware objectives and lifecycle impacts.

**Author Contributions:** Conceptualization, S.H., T.Z. and M.M.; methodology, S.H., T.Z. and M.M.; formal analysis, S.H.; investigation, S.H.; data curation, S.H.; writing—original draft preparation, S.H.; writing—review and editing, M.M.; visualization, S.H.; supervision, M.M. All authors have read and agreed to the published version of the manuscript.

**Funding:** This research received no external funding.

**Institutional Review Board Statement:** Not applicable.

**Informed Consent Statement:** Not applicable.

**Data Availability Statement:** The original contributions presented in this study are included in the article. Further inquiries can be directed to the corresponding author.

**Conflicts of Interest:** The authors declare no conflicts of interest.

## Abbreviations

The following abbreviations are used in this manuscript:

| Abbreviation | Meaning |
|---|---|
| AHRI | Air-Conditioning, Heating, and Refrigeration Institute |
| ASHRAE | American Society of Heating, Refrigerating, and Air-Conditioning Engineers |
| ASHRAE 55 | ASHRAE Standard 55: Thermal Environmental Conditions for Human Occupancy |
| BAS | Building Automation System (same as BMS; use one consistently) |
| BMS | Building Management System |
| CI | Confidence Interval |
| CIs (95%) | 95% bootstrap confidence intervals (as reported for metrics) |
| CBECS | Commercial Buildings Energy Consumption Survey (U.S. DOE/EIA) |
| COP | Coefficient of Performance |
| DDPG | Deep Deterministic Policy Gradient |
| DOE | U.S. Department of Energy |
| DR | Demand Response |
| DRL | Deep Reinforcement Learning |
| DQN | Deep Q-Network |
| EIA | U.S. Energy Information Administration |
| EUI | Energy Use Intensity (kWh·m$^{-2}$·yr$^{-1}$) |
| HIL | Hardware-in-the-Loop |
| HVAC | Heating, Ventilation, and Air-Conditioning |
| IAQ | Indoor Air Quality |
| ISO 7730 | International standard for PMV/PPD thermal comfort |
| MDPI | Multidisciplinary Digital Publishing Institute (publisher of Energies) |
| MPC | Model Predictive Control |
| NOAA | National Oceanic and Atmospheric Administration (weather data validation) |
| PID | Proportional–Integral–Derivative (rule-based) |
| PIML | Physics-Informed Machine Learning (general) |

| | |
|---|---|
| **PINN** | Physics-Informed Neural Network |
| **PMV** | Predicted Mean Vote (comfort index) |
| **PPD** | Predicted Percentage Dissatisfied (comfort index) |
| **PER** | Prioritized Experience Replay |
| **QA/QC** | Quality Assurance/Quality Control |
| **RC (model)** | Resistance–Capacitance thermal network model |
| **RH** | Relative Humidity |
| **RL** | Reinforcement Learning |
| **RMSE** | Root Mean Square Error |
| **SD** | Standard Deviation |
| **SOTA** | State of the Art |
| **TC-DDPG** | Thermodynamically-Constrained DDPG (this paper's method) |
| **TMY3** | Typical Meteorological Year (version 3) weather datasets |
| **UA** | Overall heat-transfer coefficient–area product (U·A) |
| **VAV** | Variable Air Volume (if mentioned in actuator examples) |
| **ZAM** | Zone Attention Mechanism (inter-zone interaction module) |

## Appendix A. Detailed Mathematical Derivations

### *Appendix A.1. Thermodynamic & Psychrometric Gradient Computation*

We define the physics regularizer used in Section 4.4 as:

$$L_{phys} = \lambda_1 L_{energy} + \lambda_2 L_{psychro} + \lambda_3 L_{comfort}$$

#### Appendix A.1.1. Energy Balance Term

Let $T \in R^N$ be zone temperatures at time $t, T' at\ t + \Delta t$, and $T^{\cdot}{}_{pred}(s,a)$ the rate predicted by the constraint layer (Section 4.4), which encodes RC heat flows (sensible + latent) and auxiliaries:

$$T^{\cdot}{}_{pred}(s,a) = \frac{1}{C}[\alpha Q_{HVAC}(s,a) + \beta Q_{cond}(s)]$$

where $C$ stacks zone capacitances and $Q_{cond}$ aggregates inter-zone and envelope conduction; $\alpha,\ \beta$ are learnable scalars (per zone) modeling residual mismatch. The target rate is

$$T^{\cdot}{}_{tgt} = \frac{T' - T}{\Delta t}$$

The energy term is the normalized MSE:

$$L_{energy} = \frac{1}{N} \parallel W_T \left( T^{\cdot}{}_{pred}(s,a) - T^{\cdot}{}_{tgt} \right) \parallel_2^2$$

with $W_T$ a diagonal normalizer.

#### Appendix A.1.2. Psychrometric Consistency Term

Let $RH$ be relative humidity (fraction), $p_{ws}(T)$ saturation vapor pressure (Pa), and $P$ barometric pressure (Pa). Using a standard Magnus–Tetens form,

$$p_{ws}(T) = 610.94\exp\left( \frac{17.625(T - 273.15)}{T - 273.15 + 243.04} \right)$$

$$\omega(T, RH) = \frac{0.62198 RH p_{ws}(T)}{P - RH p_{ws}(T)}$$

If the model produces $\omega$ (either directly or via an auxiliary head), we penalize

$$L_{psychro} = \frac{1}{N} \parallel W_{\omega}(\omega(s,a) - \omega(T,RH)) \parallel_2^2$$

Appendix A.1.3. Comfort Corridor Term

With PMV computed per ISO 7730 using the current state and action-implied conditions (e.g., supply temperature/flow), we use a soft corridor:

$$L_{comfort} = \frac{1}{N}\sum_{i=1}^{N} max(0, \mid PMV_i \mid -0.5)$$

(We implement the sub gradient of max at zero; a smooth Huber/soft plus alternative is also supported.)

Appendix A.1.4. Gradients (Chain Rule)

Let $\theta$ denote actor parameters. With the differentiable projection

$$\Pi_{phys}(s, \mu_\theta(s)) \equiv a_{phys}$$

$$\frac{\partial L_{phys}}{\partial \theta} = \frac{\partial L_{phys}}{\partial a_{phys}} \frac{\partial a_{phys}}{\partial \mu_\theta(s)} \frac{\partial \mu_\theta(s)}{\partial \theta}$$

Each component contributes:

$$\frac{\partial L_{energy}}{\partial a_{phys}} = \frac{2}{N} J_{Q_{HVAC}}^{\top} W_T^{\top} W_T \left(T_{\cdot pred} - T_{\cdot tgt}\right) \frac{1}{C}$$

$$\frac{\partial L_{psychro}}{\partial a_{phys}} = \frac{2}{N} J_{\hat{\omega}}^{\top} W_{\omega}^{\top} W_{\omega}(\hat{\omega} - \omega(T,RH))$$

$$\frac{\partial L_{comfort}}{\partial a_{phys}} = \frac{1}{N}\sum_i \partial max(0, \mid PMV_i \mid -0.5) sign(PMV_i) \frac{\partial PMV_i}{\partial a_{phys}}$$

where $J_{\cdot}$ are Jacobians w.r.t. actions, computed by autograd.

Projection gradient. $\frac{\partial a_{phys}}{\partial \mu_\theta}$ is identity for the interior. At range/rate boundaries, we use either (i) a smooth clip (tanh-affine) to keep gradients non-zero, or (ii) a straight-through estimator (STE). Psychrometric/rate projections are implemented with differentiable barriers (softplus) to avoid zero-gradient plateaus.

*Appendix A.2. Convergence Considerations*

Proposition (informal). Under standard DDPG assumptions—bounded rewards, Lipschitz actor/critic, compact state/action sets, sufficiently rich replay, and non-expansive (1-Lipschitz) projection $\Pi_{phys}$ —the deterministic policy gradient computed on the projected action $a_{phys} = \Pi_{phys}(s, \mu_\theta(s))$ yields stationary points corresponding to locally optimal policies within the feasible action set $A_{phys}(s)$.

Sketch. Define the feasible set $A_{phys}(s) = \{a : physics/range/rate\ constraints\ hold\}$ assumed nonempty, compact, convex (approximate convexity via smooth barriers).

The projected policy $\mu_\theta^{phys}(s) = \Pi_{phys}(s, \mu_\theta(s))$ is non-expansive.

Evaluate Bellman targets with projected target actions; the induced operator remains a $\gamma$-contraction in the space of Q-functions with bounded variation.

Under standard DDPG stability conditions (target networks, small $\tau$, bounded gradients), stochastic approximation converges to a fixed point of the projected Bellman operator; the corresponding policy is a local optimum in the feasible set.

Remark. This is not a guarantee of global optimality; model mismatch or nonconvex feasible sets can introduce suboptimal fixed points. Our empirical Sections 6 and 7 provide supporting evidence in simulation.

*Appendix A.3. Nomenclature*

| Symbol | Description | Units |
|---|---|---|
| $T$ | Dry-bulb temperature | K(or °C, consistent) |
| $T_i$ | Zone ii temperature | K |
| $C_i$ | Thermal capacitance of zone ii | $J{\cdot}K^{-1}$ |
| $U_{ij}$ | Thermal conductance between i, ji, j | $W{\cdot}m^{-2}{\cdot}K^{-1}$ |
| $A_{ij}$ | Exchange area between i, ji, j | $m^2$ |
| $Q^{\cdot}{}_{HVAC,i}$ | HVAC heat flow into zone ii | W |
| $\omega$ | Humidity ratio | $\frac{kg_v}{kg_{da}}$ |
| $\phi$ | Relative humidity | – |
| $p_w$ | Water vapor partial pressure | Pa |
| $p_{ws}(T)$ | Saturation vapor pressure at TT | Pa |
| $P$ | Barometric pressure | Pa |
| $h$ | Moist air enthalpy | $J{\cdot}kg^{-1}$ |
| $h_{fg}$ | Latent heat of vaporization | $J{\cdot}kg^{-1}$ |
| $c_p$ | Specific heat (dry air) | $J{\cdot}kg^{-1}{\cdot}K^{-1}$ |
| $c_{pv}$ | Specific heat (water vapor) | $J{\cdot}kg^{-1}{\cdot}K^{-1}$ |
| $PMV/PPD$ | Comfort metrics (ISO 7730) | – |
| $COP$ | Coefficient of Performance | – |

## Appendix B. Implementation Details

Environment. Python $\geq$ 3.10; PyTorch $\geq$ 2.0; deterministic seeding (Python/NumPy/PyTorch); reproducible configs (YAML/JSON).

Normalization. Fixed scalers (saved alongside checkpoints) for all state channels and reward terms; action outputs are tanh-bounded, then affine-scaled, then projected by $\Pi_{phys}$.

Repla and updates. Buffer size $10^5$; batch 64; $\gamma = 0.99$; soft target update $\tau = 0.005$; $L2$ grad-clip = 1.0; $OU$ exploration $\sigma = 0.1$ during training only.

Evaluation. Noise-free policy; report mean $\pm$ SD over $n$ = 50 seeds; 95% CIs via bootstrap; violation metrics per Section 3.

Logging and artifacts. Each run stores: config hash, seed, scaler params, checkpoint (actor/critic/targets), metric CSVs (energy, comfort, violations, demand), and plots.

Repository. Reference implementation scripts and configuration examples will be made publicly available at: https://github.com/Sattar7798/tqn (accessed on 7 August 2025), after code stabilization and experimental verification.

## Appendix C. Extended Results

We provide additional figures/tables to complement Sections 6 and 7:

Monthly/seasonal breakdowns. Energy use and comfort violations per month; seasonal COP vs. outdoor temperature; peak-demand seasonal histograms.

Load duration curves. Annual HVAC power LDCs for all methods.

Distribution plots. Violin/box plots of constraint violation counts (log scale).

Sensitivity overlays. Performance vs. $\lambda_{phys}$ and actor learning rate, with 95% CIs.

Ablation heatmaps. Relative degradation vs. full model across metrics.

All underlying CSVs (energy, comfort, violations, demand) and plotting scripts will be made available in the project repository to enable exact regeneration of figures and re-analysis.

## References


1. Wang, Z.; Hong, T. Reinforcement learning for building controls: The opportunities and challenges. *Appl. Energy* **2020**, *269*, 115036. [CrossRef]
2. *ISO 7730:2005*; Ergonomics of the Thermal Environment—Analytical Determination and Interpretation of Thermal Comfort Using Calculation of the PMV and PPD Indices and Local Thermal Comfort Criteria. International Organization for Standardization: Geneva, Switzerland, 2005.
3. Nagy, Z.; Henze, G.; Dey, S.; Arroyo, J.; Helsen, L.; Zhang, X.; Chen, B.; Amasyali, K.; Kurte, K.; Zamzam, A.; et al. Ten Questions Concerning Reinforcement Learning for Building Energy Management. *Build. Environ.* **2023**, *241*, 110435. [CrossRef]
4. Ziarati, T.; Hedayat, S.; Moscatiello, C.; Sappa, G.; Manganelli, M. Overview of the Impact of Artificial Intelligence on the Future of Renewable Energy. In Proceedings of the 2024 IEEE International Conference on Environment and Electrical Engineering and 2024 IEEE Industrial and Commercial Power Systems Europe (EEEIC/I&CPS Europe), Rome, Italy, 29 June–2 July 2024; pp. 1–6. [CrossRef]
5. U.S. EIA. *Commercial Buildings Energy Consumption Survey (CBECS) 2018*; U.S. Energy Information Administration: Washington, DC, USA, 2024. Available online: https://www.eia.gov/consumption/commercial/ (accessed on 4 October 2025).
6. Filippova, E.; Hedayat, S.; Ziarati, T.; Manganelli, M. Artificial Intelligence and Digital Twins for Bioclimatic Building Design: Innovations in Sustainability and Efficiency. *Energies* **2025**, *18*, 5230. [CrossRef]
7. Shaikh, P.H.; Nor, N.B.M.; Nallagownden, P.; Elamvazuthi, I.; Ibrahim, T. A Review on Optimized Control Systems for Building Energy and Comfort Management. *Renew. Sustain. Energy Rev.* **2014**, *34*, 409–429. [CrossRef]
8. Sutton, R.S.; Barto, A.G. *Reinforcement Learning: An Introduction*, 2nd ed.; MIT Press: Cambridge, MA, USA, 2018; Available online: https://www.andrew.cmu.edu/course/10-703/textbook/BartoSutton.pdf (accessed on 4 October 2025).
9. Weinberg, D.; Wang, Q.; Timoudas, T.O.; Fischione, C. A review of RL for controlling Building Energy Systems from a computer-science perspective. *Sustain. Cities Soc.* **2023**, *89*, 104351. [CrossRef]
10. Tien, P.W.; Wu, P.; Choe, S. Machine Learning and Deep Learning Methods for Enhancing Building Energy Efficiency and Indoor Environmental Quality–A Review. *Energy AI* **2022**, *10*, 100198. [CrossRef]
11. Mason, K.; Grijalva, S. A review of reinforcement learning for autonomous building energy management. Comput. *Electr. Eng.* **2019**, *78*, 300–312. [CrossRef]
12. Al Sayed, K.; Boodi, A.; Broujeny, R.S.; Beddiar, K. Reinforcement learning for HVAC control in intelligent buildings: A technical and conceptual review. *Smart Energy* **2024**, *95*, 110085. [CrossRef]
13. Wei, T.; Wang, Y.; Zhu, Q. Deep Reinforcement Learning for Building HVAC Control. In Proceedings of the 54th Annual Design Automation Conference (DAC), Austin, TX, USA, 18–22 June 2017; pp. 1–6. [CrossRef]
14. Silver, D.; Lever, G.; Heess, N.; Degris, T.; Wierstra, D.; Riedmiller, M. Deterministic Policy Gradient Algorithms. In Proceedings of the ICML'14: Proceedings of the 31st International Conference on International Conference on Machine Learning, Beijing, China, 21–26 June 2014; PMLR: Cambridge, MA, USA, 2014; Volume 32, pp. 387–395.
15. Yu, L.; Qin, S.; Zhang, M.; Shen, C.; Jiang, T.; Guan, X. A review of Deep Reinforcement Learning for Smart Building Energy Management. *IEEE Internet Things J.* **2021**, *8*, 12046–12063. [CrossRef]
16. Manjavacas, A.; Nieves, A.C.; Jiménez-Raboso, J.; Molina-Solana, M. An experimental evaluation of DRL algorithms for HVAC control (Sinergym). *Artif. Intell. Rev.* **2024**, *57*, 173. [CrossRef]
17. Dai, M.; Li, H.; Wang, S. A reinforcement learning-enabled iterative learning control strategy of air-conditioning systems for building energy saving by shortening the morning start period. *Appl. Energy* **2023**, *334*, 120650. [CrossRef]
18. García, J.; Fernández, F. A Comprehensive Survey on Safe Reinforcement Learning. *J. Mach. Learn. Res.* **2015**, *16*, 1437–1480.
19. Ruelens, F.; Claessens, B.J.; Vandael, S.; De Schutter, B.; Babuska, R.; Belmans, R. Residential Demand Response of Thermostatically Controlled Loads Using Batch Reinforcement Learning. *IEEE Trans. Smart Grid* **2017**, *8*, 214–225. [CrossRef]
20. Esmaeili, M.; Hammes, S.; Tosatto, S.; Geisler-Moroder, D.; Zech, P. Safe Reinforcement Learning for Buildings: Minimizing Energy Use While Maximizing Occupant Comfort. *Energies* **2025**, *18*, 5313. [CrossRef]
21. Sanchez, J.; Cai, J. Constrained RL for building demand response (explicit constraint value function). *Appl. Energy* **2025**, *in press*.

22. Vázquez-Canteli, J.R.; Nagy, Z. Reinforcement Learning for Demand Response: A Review. *Appl. Energy* **2019**, *235*, 1072–1089. [CrossRef]
23. Karniadakis, G.E.; Kevrekidis, I.G.; Lu, L.; Perdikaris, P.; Wang, S.; Yang, L. Physics-informed Machine Learning. *Nat. Rev. Phys.* **2021**, *3*, 422–440. [CrossRef]
24. Jiang, Z.; Wang, X.; Li, H.; Hong, T.; You, F.; Drgoňa, J.; Vrabie, D.; Dong, B. Physics-informed ML for building performance simulation—Review. *Patterns/Cell Press* **2025**, *18*, 100223.
25. Saeed, M.H.; Kazmi, H.; Deconinck, G. Dyna-PINN: Physics-informed Deep Dyna-Q for building heating control. *Energy Build.* **2025**, *324*, 114879. [CrossRef]
26. Jiang, Z.; Wang, X.; Dong, B. Physics-informed modularized neural network for DRL-based building control; reports ~31% HVAC savings case study. *Adv. Appl. Energy* **2025**, *19*, 100237. [CrossRef]
27. Drgoňa, J.; Arroyo, J.; Figueroa, I.C.; Blum, D.; Arendt, K.; Kim, D.; Perarnau, E.; Oravec, J.; Wetter, M.; Vrabie, D.L.; et al. All You Need to Know about Model Predictive Control for Buildings. *Annu. Rev. Control* **2020**, *50*, 190–232. [CrossRef]
28. Killian, M.; Kozek, M. Ten Questions Concerning Model Predictive Control for Energy Efficient Buildings. *Build. Environ.* **2016**, *105*, 403–412. [CrossRef]
29. Oldewurtel, F.; Parisio, A.; Jones, C.N.; Gyalistras, D.; Gwerder, M.; Stauch, V.; Lehmann, B.; Morari, M. Use of Model Predictive Control and Weather Forecasts for Energy Efficient Building Climate Control. *Energy Build.* **2012**, *45*, 15–27. [CrossRef]
30. Dobbs, J.R.; Hencey, B.M. Model Predictive HVAC Control with Online Occupancy Model. *Energy Build.* **2014**, *82*, 675–684. [CrossRef]
31. Privara, S.; Váňa, Z.; Široký, J.; Ferkl, L.; Cigler, J.; Oldewurtel, F. Building Modeling as a Crucial Part for Building Predictive Control. *Energy Build.* **2013**, *56*, 8–22. [CrossRef]
32. Zhang, Z.; Lam, K.P. Practical implementation and evaluation of deep reinforcement learning control for a radiant heating system. In Proceedings of the 5th Conference on Systems for Built Environments, Shenzen, China, 7–8 November 2018; pp. 148–157. [CrossRef]
33. Mnih, V.; Kavukcuoglu, K.; Silver, D.; Rusu, A.A.; Veness, J.; Bellemare, M.G.; Graves, A.; Riedmiller, M.; Fidjeland, A.K.; Ostrovski, G.; et al. Human-level Control through Deep Reinforcement Learning. *Nature* **2015**, *518*, 529–533. [CrossRef]
34. Lillicrap, T.P.; Hunt, J.J.; Pritzel, A.; Heess, N.; Erez, T.; Tassa, Y.; Silver, D.; Wierstra, D. Continuous Control with Deep Reinforcement Learning. *arXiv* **2015**, arXiv:1509.02971.
35. Fujimoto, S.; van Hoof, H.; Meger, D. Addressing Function Approximation Error in Actor–Critic Methods. In Proceedings of the 35th International Conference on Machine Learning, Stockholm, Sweden, 10–15 July 2018; PMLR: Cambridge, MA, USA, 2018; Volume 80, pp. 1587–1596.
36. Raissi, M.; Perdikaris, P.; Karniadakis, G.E. Physics-Informed Neural Networks: A Deep Learning Framework for Solving Forward and Inverse Problems Involving Nonlinear PDEs. *J. Comput. Phys.* **2019**, *378*, 686–707. [CrossRef]
37. Afram, A.; Janabi-Sharifi, F. Review of Modeling Methods for HVAC Systems. *Appl. Therm. Eng.* **2014**, *67*, 507–519. [CrossRef]
38. *ASHRAE Handbook—Fundamentals*; Chapter 1: Psychrometrics; ASHRAE: Atlanta, GA, USA, 2021.
39. Deru, M.; Field, K.; Studer, D.; Studer, D.; Benne, K.; Griffith, B.; Torcellini, P. *U.S. Department of Energy Commercial Reference Building Models of the National Building Stock*; NREL/TP-5500-46861; NREL: Golden, Colorado, 2011. Available online: https://www.nrel.gov/docs/fy11osti/46861.pdf (accessed on 4 October 2025).
40. ASHRAE Guideline 14-2014. In *Measurement of Energy, Demand, and Water Savings*; ASHRAE: Atlanta, GA, USA, 2014.
41. Schaul, T.; Quan, J.; Antonoglou, I.; Silver, D. Prioritized Experience Replay. *arXiv* **2016**, arXiv:1511.05952. [CrossRef]
42. Efron, B. Bootstrap Methods: Another Look at the Jackknife. *Ann. Stat.* **1979**, *7*, 1–26. [CrossRef]
43. Wilcox, S.; Marion, W. *Users Manual for TMY3 Data Sets*; NREL/TP-581-43156; NREL: Golden, Colorado, 2008. [CrossRef]
44. Fanger, P.O. *Thermal Comfort: Analysis and Applications in Environmental Engineering*; Danish Technical Press: Copenhagen, Denmark, 1970.
45. *ASHRAE Standard 55-2020*; Thermal Environmental Conditions for Human Occupancy. ASHRAE: Atlanta, GA, USA, 2020.
46. Paszke, A.; Gross, S.; Massa, F.; Lerer, A.; Bradbury, J.; Chanan, G.; Killeen, T.; Lin, Z.; Gimelshein, N.; Antiga, L. PyTorch: An Imperative Style, High-Performance Deep Learning Library. *Adv. Neural Inf. Process. Syst. 32 (NeurIPS)* **2019**, *32*, 8024–8035.
47. Kingma, D.P.; Ba, J. Adam: A Method for Stochastic Optimization. *arXiv* **2015**, arXiv:1412.6980.
48. Vaswani, A.; Shazeer, N.; Parmar, N.; Uszkoreit, J.; Jones, L.; Gomez, A.N.; Kaiser, L.; Polpsukhin, I. Attention Is All You Need. *Adv. Neural Inf. Process. Syst. (NeurIPS)* **2017**, *30*, 5998–6008.
49. Roijers, D.M.; Vamplew, P.; Whiteson, S.; Dazeley, R. A Survey of Multi-Objective Sequential Decision-Making. *J. Artif. Intell. Res.* **2013**, *48*, 67–113. [CrossRef]

*energies*

MDPI

*Article*

# Multi-Scale Predictive Modeling of RTPV Penetration in EU Urban Contexts and Energy Storage Optimization

**Vasileios Kapsalis *, Georgios Mitsopoulos, Dimitrios Stamatakis and Athanasios I. Tolis**

Sector of Industrial Management and Operational Research, School of Mechanical Engineering, National Technical University of Athens, 9, Heroon Polytechniou Str., Zografou Campus, 15780 Athens, Greece; gdmitsopoulos@mail.ntua.gr (G.M.); stamatakis92@mail.ntua.gr (D.S.)

* Correspondence: bkapsal@mail.ntua.gr

**Abstract**

Prosumer energy storage behavior alongside national rooftop photovoltaics (RTPV) penetration metrics is essential for decarbonization pathways in buildings. A research gap persists in quantitatively assessing storage strategies under varying regulatory frameworks that integrate both technical and financial dimensions while accounting for behavioral heterogeneity and policy feedback. This study introduces a novel degradation-aware, feedback-preserving framework that optimizes behind-the-meter storage design and operation, enabling realistic modeling of prosumer responses on large-scale RTPV adoption scenarios. Long Short-Term Memory (LSTM) and Compound Annual Growth (CAGR) models applied for the RTPV penetration rates projections in European urban contexts. The increasing rates in the Netherlands, Spain, and Italy respond to second-order regression behavior, with the former to emit signals of saturation and the latter to perform mixed anelastic and reverse elastic curves of elasticities. Accordingly, Germany, France, the United Kingdom (UK), and Greece remain in an inelastic area by 2030. The building RTPV energy storage arbitrage formulation is treated as a linear programming (LP) problem using a convex and piecewise linear cost function, a Model Predictive Control (MPC), Auto Regressive Moving Average (ARMA) and Auto Regressive Integrated Moving Average (ARIMA) statistical forecasts and rolling horizon in order to address the uncertainty of the load and the ratio $\kappa$ of the sold to purchased electricity price. Weekly arbitrage gains may drop by up to 9.1% due to stochasticity, with maximized gains achieved at battery capacities between 1C and 2C. The weekly gain per cycle performs elastic, anelastic, and reverse behavior of the prosumer across the range of $\kappa$ values responding to different regulatory mechanisms of pricing. The variability of economic incentives suggests the necessity of flexible energy management strategies.

**Keywords:** RTPV penetration; LSTM; CAGR; epigraph linearization and stochastic modeling; energy storage optimization; regulatory frameworks

## 1. Introduction

The penetration of rooftop photovoltaic (RTPV) systems in Europe is a key indicator of the energy transition towards clean energy in the European Union and an important part of decentralized electricity generation, especially in urban areas, while offering energy independence, $CO_2$ emission reduction, and cost savings for consumers [1]. The European Solar Rooftops Initiative provides for the mandatory installation of photovoltaics in new public and commercial buildings from 2026 and in new residential buildings from 2029 [2].

From an energy performance perspective, the current nearly Zero Energy Building (nZEB) standards for new buildings up to 2030 become mandatory, while the recast Energy Performance of Buildings Directive (EPBD) has set more ambitious goals for the period ahead. In this context, the energy performance and the contribution of renewables between countries significantly vary across European Union (EU) countries, and more efforts are required [3], considering the differences in support policies, schemes, capital subsidies, building density, land availability, sunshine, and energy market structure. In 2023, the EU installed 55.9 Giga Watt (GW) of new PV capacity, of which 37 GW (66%) was rooftop solar [4], and the trend remains in favor of the decentralized rooftop energy production [5].

### 1.1. The Penetration of Rooftop Photovoltaics in Europe

The distribution of Gross Value Added (GVA) at each stage of the photovoltaic (PV) value chain, as well as the comparative contribution of the EU to the rest of the world (ROW), is depicted in Figure 1.

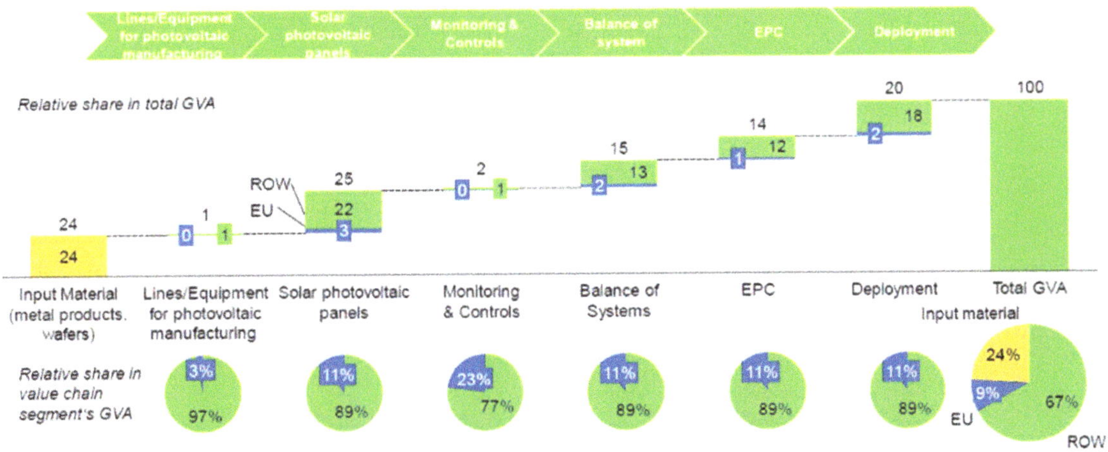


**Figure 1.** Distribution of the Gross Value Added (GVA) at each stage of the photovoltaic (PV) value chain and the contribution of the EU [2].

A major proportion of the economic value is produced outside the EU, due to the global dominance of Asian countries in solar panel production. This can substantiate the need to strengthen domestic production and EU technological dominance in the context of energy self-sufficiency.

In the era of high renewable energy penetration in the energy mix, the electricity flows present new challenges and needs [6]. Global warming mitigation efforts have focused on researching rooftop PV electricity generation and its innovative technologies as a tool to meet the increasing energy needs in cities [7,8]. The new role of the active consumer, also known as prosumer, is becoming demanded within a dynamic framework of flexibility, resilience, storage, and balancing [9,10]. The wholesale and retail organization of the markets is based on forecasting electricity prices and volume differences, under the technical constraints of the system, which offers opportunities to compensate for economic benefits through electricity pricing arbitrage. The policies used to manage technical and financial distortions aim to smooth the peak values and coincide with the fluctuations of the demand and supply curves, under different regulated frameworks. The optimal strategy may involve a progressive adjustment of the factor values depending on the level of Renewable Energy Systems (RES) penetration, technological developments, and grid needs [11,12]. The energy policy makers and regulators, through their assessment of the economic efficiency of existing, contribute to the evidence-based design of strategies that enhance the penetration of RES and the sustainability of the energy system [13–15].

Penetration rates and the value of storage assets in buildings have been investigated in [16]. In this study, they directly correlate the controlled load dispatched and the optimized storage to improve the wholesale value of RTPV and battery with the dynamic pricing framework included. In another study, low rates in PV penetration decreased volatility, and the existence of a feed-in model with no PV depicted lower storage revenues than a corresponding one with PV, under a share even greater than 10% [17]. A PV storage system for trading in the market has been developed in [18]. They analyze the sensitivity of households to market signals and the multi-scale parameters to achieve energy savings against the non-distributed energy generation. Finally, in [19], they consider the growing deployment of photovoltaic (PV) systems in EU member states and apply these insights to a comprehensive life cycle assessment (LCA).

It is worth mentioning that market penetration scenarios are related to the building-level arbitrage, while the increasing rate of PV penetration in urban areas results in significant changes in the electricity market dynamics. High rates of RTPV in the system increase price volatility and decrease the marginal value of self-consumption, creating opportunities for energy storage asset management. The latter profitability also depended on local volatility and forecasting quality [20].

Penetration Studies and Forecasting Methods

In a series of studies, refs. [21,22] simulate and explore the changing values of hybrid RTPV-plus-battery storage systems under different scenarios of increasing market penetration. In another study, they provide a Support Vector Machine (SVM) forecasting framework to upscale the local PV generation level into a regional market pricing signal [23]. Useful metric evaluations and forecast methods of PV production in high penetration rates, and simulation of price volatility can be found in [1]. Other machine learning techniques, such as LSTM and Convolution Neural Networks (CNN), are widely used, while newer methods emerge and highlight advances in deep learning and forecasting applications [24,25]. In [26], the optimal dispatching problem is addressed using LSTM or eXtreme Gradient Boosting (XGBoost) to enhance the forecast marginal prices of the system.

### *1.2. Energy Storage in RTPV Applied Systems*

RES reinforces the need for advanced storage technologies due to the stochastic nature and intermittent production of these sources. Storage systems allow the shift of generated energy to periods of higher demand, thus offering the possibility of improving the reliability, flexibility, and stability of the grid. Energy storage technologies present distinct technical characteristics, capabilities, and limitations, different from each other [27], for example, electrochemical [28], mechanical [29,30], thermal [31], or chemical storage [29], to name some. For instance, the RTPV system is utilized in [32] on an nZEB framework with an arbitrage mechanism, where photovoltaic (PV) systems installed on buildings decide whether to sell electricity to the grid or convert it into hydrogen for storage. The stored hydrogen is later utilized to generate electricity through a hydrogen-powered generator, enhancing energy autonomy and flexibility.

The critical characteristics of battery energy storage systems [33] mainly refer to the energy capacity and density, the round-trip efficiency (denoting the power discharge to charge ratio), the lifetime and degradation, the response time, and the installation and maintenance costs, under different projected PV levelized cost of electricity (LCOE) [34], besides the stochasticity and intermittent production [35–37].

As energy storage becomes a critical component of RTPV-designed and applied systems, it becomes necessary to define and understand the appropriate evaluation parameters, as well as the technical limits of the available technologies in space-constrained applications, such as building applications, towards a decarbonized energy transition [38]. The efficient cycling operation [39], the aging characteristics [40], the ramp rating, and the C-rate of the battery storage applications (denotes cascaded capacities) are important parameters to be considered [41]. The energy storage systems play an important and multifaceted role in the operation of distribution networks. They provide flexible loads [42], shave peaks, increase self-consumption, support ancillary services, and improve power quality. Their effective utilization requires addressing the relevant technical and institutional challenges, which necessitate the further development of control techniques and optimal energy management strategies for viable investments [43–47]. In parallel, reduced-order dynamic modeling approaches in other multi-energy domains, such as integrated electricity–gas systems, have been proposed to capture system coupling and ensure stable, computationally efficient operation [48], sharing conceptual similarities with control-oriented optimization in RTPV–storage systems. Similarly, resilient control and load frequency regulation strategies have been investigated in microgrids integrating distributed renewable energy sources, where Model Predictive Control and hierarchical architectures enhance stability under intermittent generation and communication uncertainty [49].

#### 1.2.1. Energy Arbitrage in Energy Storage Systems

The concept of energy arbitrage concerns the exploitation of the difference in electricity prices at various times to achieve economic benefit through the storage and shifting of energy consumption or production [50]. The basic principle of energy arbitrage is based on the volatility of electricity prices, which is observed in both the wholesale and retail markets, due to fluctuations in energy demand and supply. Energy prices tend to be lower during hours of reduced demand (e.g., evening hours), when there is a surplus of available energy, while they increase during hours of high demand (e.g., afternoon or evening peak hours), when generation may have difficulty meeting consumption requirements [51,52].

Arbitrage is distinguished into two main forms: the grid and the self-consumption. The former stores purchased energy from the grid at low prices and returned it at higher values [53]. The latter stores energy during hours of high local RES production and uses it to cover local consumption when grid prices are higher or the RES production is reduced. Each form is appropriate for different pricing models (e.g., net metering, net billing, Time of Use (ToU), and dynamic pricing) [54–56]. An approach that expands on local consumption is suggested by [57]. The suggested approach analyses multiple cases combining local storage, community storage, and internal trading, showing that community storage coupled with trading maximizes renewable consumption, reduces grid dependence, and achieves cost savings in residential communities.

In net metering schemes, the injected and absorbed volume of energy into and by the grid is offset at an almost equivalent price to that of the energy consumed; therefore, the economic incentives are missed due to the low grid parity [58,59]. In net billing schemes, the price of energy returned to the grid is compensated at a lower value than that purchased from it, which essentially strengthens the incentives for the use of storage systems and the increase in self-consumption [60]. In other dynamic pricing models, such as ToU, the billing of energy depends on the time of the day; thus, it is related to the demand and self-consumption, leading to a stronger economic incentive for energy storage arbitrage [61]. The stored energy at valley hours at low prices is consumed or sold at peak hours when the price is higher [62].

Besides the financial point of view, certain technical parameters are to be optimized for the best operational results in storage systems [63]. The energy storage arbitrage can balance the continuous demand and supply loads and manage critical events such as outages or sudden drifts of RES, providing resilience and reliability of the system [64], long-term ancillary services [65,66], and integrated value stacking for multimodal power systems [67], or frequency reserve [68]. The way it is used to control the process of energy storage units under arbitrage is a trade-off between scalable complexity and accuracy between the automation techniques levels and the specific goals of the predicted data [69–74].

#### 1.2.2. Optimization in Energy Storage Systems Applications

Linear programming (LP) is widely used in the field of energy optimization and in particular problems, such as generation and consumption scheduling, energy storage management, economic network management, and renewable energy (RES) integration, and generation unit operation scheduling [75], energy arbitrage and storage management, distributed generation management from RES [76], management of energy networks and infrastructures [77] and financial management and energy market planning [78]. To account for the stochastic nature of renewable energy output, a scenario-driven optimization method has been proposed. For example, ref. [79] uses a Deep Belief Network (DBN) to generate time-series PV scenarios and incorporates demand-side response and dynamic reconfiguration, improving PV absorption and reducing planning costs. In energy arbitrage and storage applications, they combined with advanced linearization techniques, Model Predictive Control (MPC) [80,81], and advanced econometric analysis [82] to manage the nonlinearities and the stochastic nature of the optimal control problems. Day-ahead optimization incorporating feedback correction via MPC is proposed in [83] to address uncertainty and predict the performance of energy storage systems in a radial distribution network. The need for a sufficient number of breakpoints and the increase in variables and constraints are also considered to increase the complexity of the process [84,85]. The implementation of the Model Predictive Control (MPC) toolbox, based on rolling horizon optimization, has also been used [86]. Mixed Integer Linear Programming-based and nature-inspired optimization approaches have been employed in [87,88], where Electric Vehicle (EV) battery storage is modeled as a flexible load from the system operator's perspective.

### *1.3. Motivation and Research Gaps*

The motivation is reinforced by the real need to make informed decisions in an ever-changing energy landscape, where the accuracy of the forecast and the sensitivity of the models play a decisive role. Analysis of the energy storage behavior of prosumers with national-level metrics on RTPV penetration is a necessity for advancing decarbonization strategies in the building sector.

There is a research gap in the quantitative analysis of storage strategies under various regulatory electricity market policies. This includes consideration of technical and financial modeling with both short-term optimization and forecasting, as well as long-term penetration prediction. Heterogeneity in prosumer storage attitudes in conjunction with feedback fidelity under evolving tariff frameworks is rarely represented while remaining computationally tractable for country-scale sensitivity analyses.

The main research questions posed here are:

1. What are the estimated penetration rates of rooftop PV capacity up to 2030 across different European countries? (4.1)
2. How are the rooftop PV storage dispatch, the capacity of storage, and the arbitrage gain in urban areas affected by the electricity pricing framework? (4.2)

3. How can uncertainty be effectively quantified regarding technical and economic optimization? (4.3)

#### 1.3.1. Structure of the Work

The structure of this work presents the theoretical understanding of the problems, the results, and the drawing of conclusions. The paper is organized into the following sections: Section 1 introduces a brief literature review and the research questions. Section 2 provides methodology, mathematical formulations, and the structure of the optimization models. Section 3 presents the results of the different model parameters and the discussion on the research questions. Section 4 sets forth an extensive discussion, and Section 5 summarizes the main conclusions.

#### 1.3.2. Contribution of This Study

The increasing penetration of PV creates challenges in production variability, impacts on the wholesale market volatilities, and flexible pricing policies, making optimization at the building level necessary.

The predicting models bridge the importance of integrating large-scale penetration forecasts with localized optimization strategies to capture the full spectrum of value creation and gain novel insights into the role of distributed storage assets in both grid stability and economic efficiency in urban energy systems.

## 2. Materials and Methods

The concept of multi-scale prediction in this study refers to the use of two distinct temporal horizons—short-term and long-term forecasting—to support energy management and policy assessment. The short-term prediction focuses on forecasting building load and the electricity market for both purchasing and selling prices. These forecasts serve as inputs to a stochastic arbitrage optimization framework, which enables several key insights: (a) it verifies the validity of the proposed behind-the-meter formulation that integrates building load, battery storage, and RTPV systems—since, if the formulation were suboptimal, the battery would remain inactive; (b) it identifies the most suitable battery type for the given operational context; and (c) it determines the profitability conditions under varying ratios $\kappa$.

On the other hand, the long-term prediction concerns the expected penetration of RTPV systems, providing a tool to evaluate the effectiveness of national energy policies, where positive trends indicate a supportive environment for such investments. Together, these two forecasting layers constitute a multi-scale prediction framework, encompassing both operational and policy-level analyses.

### *2.1. Penetration Forecast*

The forecasting of the penetration of photovoltaic systems on rooftops requires the use of appropriate methods, which can incorporate the uncertainty, the heterogeneity of the data, and the complexity of the development of RES at the national and regional levels [89,90].

LSTM-type Neural Networks and Mathematical Growth (CAGR) models are chosen to explore the penetration rates in certain EU countries. The former can incorporate complex and nonlinear relationships, memorize significant temporal patterns, and have proven success in related problems of energy demand, load, or renewable energy generation forecasting [91,92]. The latter provided interpretable prediction curves and are used as a point of comparison with the results of LSTMs, to assess the coherence between statistical and machine learning.

The implementation is conducted using MATLAB tools [93], adapting curves to historical data for each country based on the combined sources [94–98] and the tables in Appendix A.

2.1.1. LSTM Model

The internal structure of an LSTM block is shown in Appendix B and describes how LSTM can better manage long-term correlations in sequential data [99].

An LSTM memory layer with 50 units, a fully connected layer leading to the output via a regression layer, and a sequence input layer constitute the three main layers of the implemented LSTM model, using the Adam optimizer method for training in 300 epochs, with gradient clipping to avoid model instability.

After training, the next six years (2025–2030) are predicted to use a rolling forecasting system, with the next value predicted by the model using the last five observations of the normalized series as input. The process was repeated for each subsequent year, and after this added to the input, which was smoothed, using the inverse of the normalization to convert the predicted values back to the original power series.

The 50 LSTM units that make up the neural network are selected based on relevant research and experiments, with long-term dependencies in the time-series being preserved and learned with a sufficiently substantial number of LSTM units, but short-term correlations can still be learned. The Adam optimization technique is used for training, which is considered a fast and reliable method for converging neural networks, and a threshold of 300 epochs is chosen to ensure sufficient learning time without overfitting. The balance between learning speed and stability is achieved with an initial learning rate of 0.005, with the model gradients also subject to a constraint (Gradient Threshold) in order to avoid explosive levels that could lead to instability during training (Figure 2).

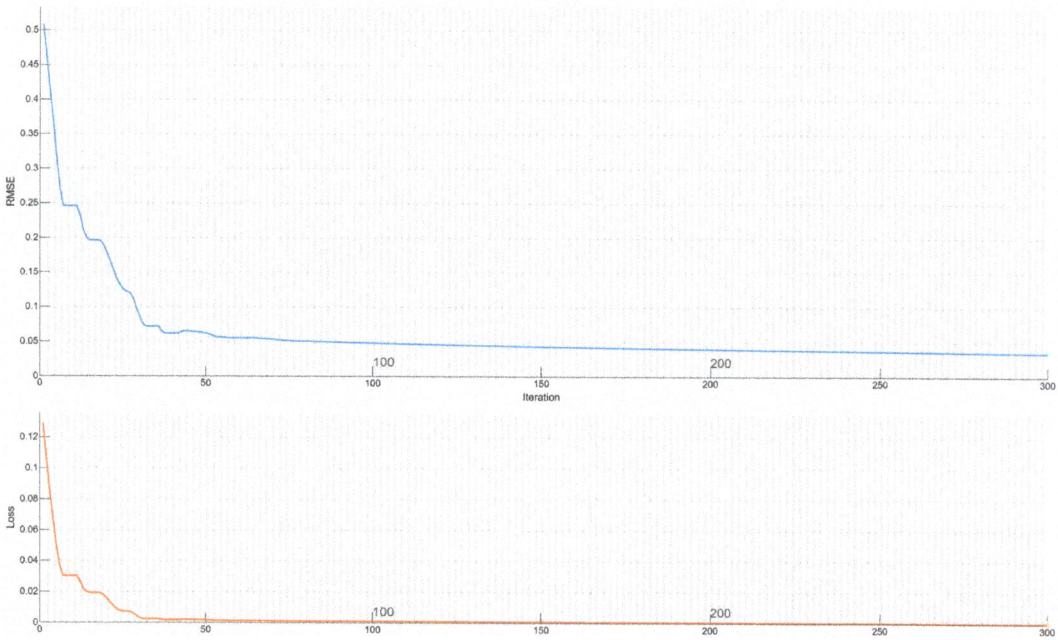


**Figure 2.** Root Mean Squared Error (RMSE) (blue) and Loss function (red) of the LSTM training and forecasting process.

### 2.1.2. Growth Model

The Mathematical Growth Models are based on simplified assumptions about the shape of the growth curve and use historical data to approximate the future evolution of the phenomenon. Although they do not consider complex socioeconomic or technological factors, they offer an interpretable and often sufficiently accurate approach, especially in environments with stable trends. They are simple but effective approaches for modeling the temporal evolution of phenomena such as the development of photovoltaic installations. They include exponential, logarithmic, and logistic models, which are widely used when the data shows a clear trend of increase with saturation. These models are widespread in technological forecasting and the study of dynamic markets. The Compound Annual Growth Rate (CAGR) model is used here in comparison with the LSTM model results.

## 2.2. *Building RTPV with Storage Formulation*

The energy storage arbitrage optimization problem is formulated and briefly described in Figure 3 in both deterministic and stochastic modes. The energy flows are referred to as the building behind the meter. This is where the energy storage arbitrage is optimized using only the electricity prices from the grid. More details are found in [12], where the original code and data for this work are motivated and validated for a two-step solution. Then, the formulation was extended to N steps with piecewise linearization and combined with stochastic consideration.

A linear mathematical model of the RTPV system is formulated through the epigraph linearization technique. Then, the Model Predictive Control (MPC) with rolling horizon is combined with ARMA/ARIMA to capture the uncertainty in prices and loads is implemented. The MATLAB software (Version: 9.14.0 (R2023a)) is used. The function *linprog* uses the simplex duality method, and the MPC controller solves an optimization problem at each time step, using information about the current state of the system and forecasts for its future behavior. For the detailed training and control of the model, econometrics of ARMA and ARIMA techniques have been inserted in each step, considering the compromise between the predictive and the control horizon related to energy management and arbitrage applications. The total time horizon or duration has been partitioned into N steps, $i \in (1, \ldots, N)$, with duration $h_i$, and the electricity price and correspond to consumption, purchasing price $p_b(i)$ when it is greater than or equal to zero, or selling price $p_s(i)$ less than zero. A particularly crucial factor for evaluating the economic efficiency of energy arbitrage is the ratio of sale to purchase price ($\kappa_i = p_s(i)/p_b(i)$).

### 2.2.1. Graphical Analysis

The cost function consists of four linear segments, each with a different slope, which change at the intersection points and depending on the sign of the net load (zi). The sign of the net load (zi) determines the form and position of the slope changes. The slopes are directly influenced by the buying/selling prices and the charging/discharging efficiencies. Each segment corresponds to a different operational state: buying/selling and charging/discharging. The piecewise linear and convex structure of the function ensures a unique optimum and allows for efficient resolution using linear programming. The intersection points define the boundaries, where the economic benefit from charging or discharging changes determines the operational strategy. In Figure 3b, the definition of the segments is depicted.

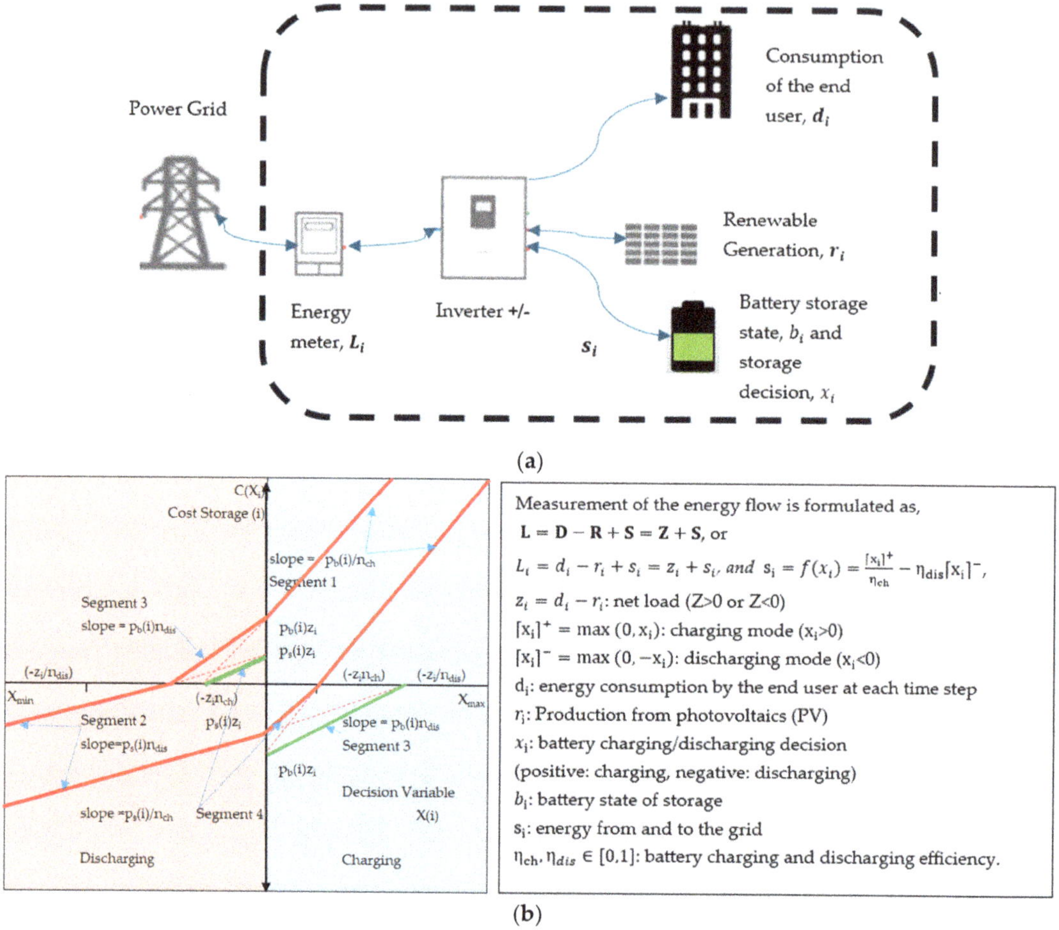


**Figure 3.** (**a**) Boundaries of the RTPV system and the energy flow. (**b**) Graphical analysis: The cost of storage is piecewise linear and convex and consists of two parts, discharging at negative values of xi and charging at positive values of $X_i$. Each region corresponds to a different physical function of the battery (buying/selling energy). Each part has a different slope due to charging/discharging efficiency. Interpretation of gradients for zi > 0 and zi < 0 modes. Segment 1 & slope $p_b(i)/n_{ch}$): Energy market & battery charging gradient, Segment 2 & slope $p_s(i)/n_{dis}$: Battery discharging & selling gradient: Segment 3 & slope ($p_b(i)/n_{dis}$): Switching between buying and selling gradient, Segment 4 & slope $p_s(i)/n_{ch}$: Selling with charging or buying with discharging gradient.

### 2.2.2. Intersection Points and Their Economic Interpretation

The diagram is characterized by two critical intersection points, referring to the intersections with the vertical (y-intercept) and the horizontal axis (x-intercept). The former represents the cost when the battery activity is zero, and it indicates the cost of covering the net load exclusively from the grid without the involvement of the battery. The latter shows the value of the change in charging ($X_i$), where the total economic interaction with the grid is zeroed out, meaning there is a balance between purchased and sold energy.

This is a critical point for the economic analysis of the system's operation. The objective function of the problem is, by definition, piecewise linear and convex, which are fundamental characteristics for the linear programming (LP) algorithms to solve it.

Convexity ensures the uniqueness of the optimal solution, eliminating the possibility of a local minimum. To leverage this convexity and enable the use of powerful LP solvers, the epigraph formulation is applied. This transformation introduces auxiliary variables ($t_i$), which envelop the actual cost at each time step, thus converting the original convex but nonlinear function into a set of linear constraints (segments) (Appendix C).

2.2.3. Optimization Problem and Epigraph Formulation

The optimization problem for the time horizon, N, in our graphical analysis, is related as

$$\min \sum_{i=1}^{N} C^i(X_i) = \max\{\text{segment1}, \text{segment2}, \text{segment3}, \text{segment4}\}, \tag{1}$$

where the decision variables vector, $X_i$, includes the energy flows from and to the battery in each time step. The equivalent epigraph problem is written as

$$\min_{\{x_i, t_i\}} \sum_{i=1}^{N} t_i, \tag{2}$$

Constraints

1. Epigraph linearization for each segment

$$\begin{aligned} \text{segment 1} &: \frac{p_b(i)}{\eta_{ch}} x_i + z_i p_b(i) \leq t_i \\ \text{segment 2} &: p_s(i)\eta_{dis} x_i + z_i p_s(i) \leq t_i \\ \text{segment 3} &: p_b(i)\eta_{dis} x_i + z_i p_b(i) \leq t_i \\ \text{segment 4} &: \frac{p_s(i)}{\eta_{ch}} x_i + z_i p_s(i) \leq t_i, \end{aligned} \tag{3}$$

The auxiliary variable $t_i$ defines the actual cost at each step, allowing for resolution with LP. The introduction of this constraint into the cost function replaces the nonlinear formulation with a linear piecewise approach.

2. Energy balance

$$l_i = z_i + f(x_i) = (d_i - r_i) + \left( \frac{\lceil x_i \rceil^+}{\eta_{ch}} - \eta_{dis} \lceil x_i \rceil^- \right), \forall i, \tag{4}$$

The energy balance requires the remaining energy after the usage of PV or the storage to be covered or injected into the grid.

3. Ramping rate

Replacing the quantity $f(x_i)$ by the $s_i = f(x_i)$ (to include the losses), we introduce the range for the battery ramp rate,

$$s_i \in \left[S^i_{min}, S^i_{max}\right], \forall i, \tag{5}$$

were

$$S^i_{min} = \eta_{dis} \delta_{min} h_i, \tag{6}$$

the maximum discharging to the grid and

$$S^i_{max} = \frac{\delta_{max} h_i}{\eta_{ch}}, \tag{7}$$

the maximum charging from the grid and

$$\delta \in [\delta_{min}, \delta_{max}] \ \forall i, \tag{8}$$

denotes the ramp rate.

The ramping rate defines the limitation of the changes in the battery and the inverter due to technical specifications.

4. Capacity

$$b_{min} - b_0 \leq \sum_{j=1}^{i} x_j \leq b_{max,} - b_0, \forall i, \tag{9}$$

$$b_i = b_{i-1} + x_i, \ \forall i = 1, \ldots, N, \tag{10}$$

$$b_i \in \left[b_{min}^i, b_{max}^i\right], \forall i, \tag{11}$$

with the battery energy storage state, $b_i$ to remain in the range 0–100% at every step.

The capacity constraints ensure realistic operation of the battery between the limits of the state of charge (SoC). In every step, battery energy is that of the previous state plus the change.

5. Energy prices come from the convexity condition of the system

$$p_b(i) \geq p_s(i), \ \forall i, \tag{12}$$

6. Non-negativity for auxiliary and decision variables

$$t_i \geq 0, x^+{}_i, x^-{}_i \geq 0 \, i, \tag{13}$$

2.2.4. Simulation Strategy

The algorithm implements an MPC (Model Predictive Control) model: at every moment in time, it recalculates the optimal strategy based on new forecasts. It combines stochastic forecasting and optimal linear programming. It allows for dynamic adjustment to real and future market/consumption conditions. The basic steps and a detailed flowchart are depicted in Figure 4.

2.2.5. Horizon Time Limits, and Basic Indices of Performance Evaluation

Parametric analysis of the coefficient κ is an important procedure for assessing its impact on the performance and sustainability of energy systems with storage [18]. The analysis of the parameter κ, which has already been denoted as the ratio of sale to purchase price, is studied under the impact of different pricing policies (net metering—net billing), battery C rates (maximum charging/discharging rate as a percentage of capacity) and the equivalent (normalized) cycle operation defined according to the full depth of discharge (DoD). The relationship between the prediction horizon, $Np$, and control horizon, $Nc$, directly affects the performance and behavior of the MPC and usually is $Nc \leq Np$. The use of a large prediction horizon combined with a small control horizon provides an ideal trade-off between adaptability, accuracy, and computational efficiency, as it allows the system to respond immediately to added information while having a broader view of its future behavior. Implementation of scenarios is made both with full knowledge of future prices (deterministic optimization) and with uncertain forecasting (forecast-based MPC), to capture the impact of forecast error on decision-making, an element that is the subject of increasing interest in relevant literature. The economic benefit from arbitrage expresses the net reduction in total energy costs achieved through the strategy of storing and transporting energy at different time periods, taking advantage of the volatility of electricity prices.

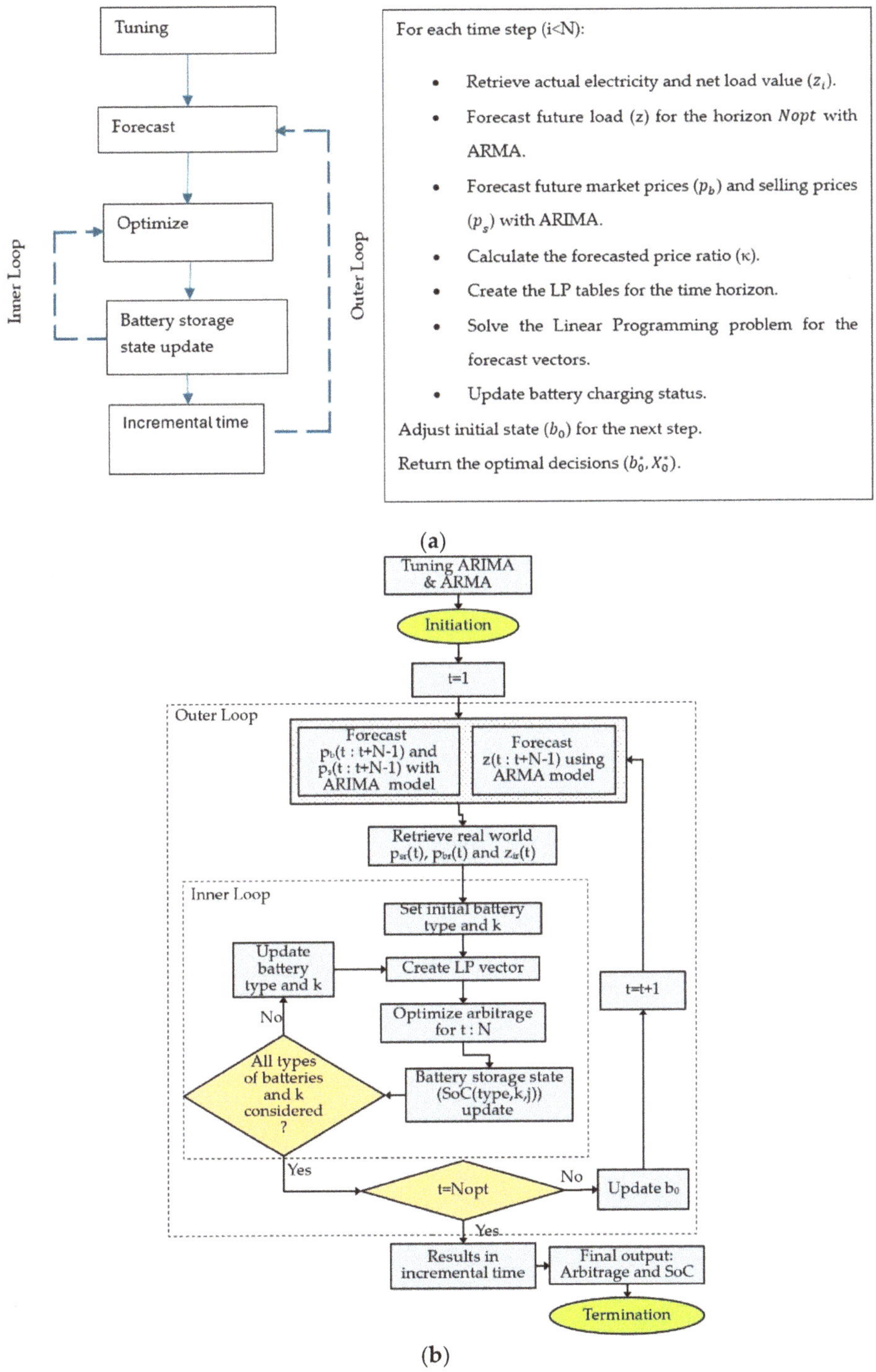


**Figure 4.** (**a**) Implementation of forecasting models via Epigraph LP—MPC—ARMA/ARIMA stochastic model (**b**) Detailed flowchart.

The basic parameters for the evaluation of prediction errors are:

$$\text{Mean Absolute Error—MAE}: \text{ MAE} = \frac{1}{n}\sum_{t=1}^{n}\left|Y_t - \hat{Y}_t\right|, \tag{14}$$

$$\text{Mean Squared Error—MSE}: \text{ MSE} = \frac{1}{n}\sum_{t=1}^{n}\left(Y_t - \hat{Y}_t\right)^2, \tag{15}$$

$$\text{Root Mean Squared Error—RMSE}: \text{ RMSE} = \sqrt{\frac{1}{n}\sum_{t=1}^{n}\left(Y_t - \hat{Y}_t\right)^2} \tag{16}$$

$$\text{Mean Absolute Percentage Error—MAPE}: \text{ MAPE} = \frac{100\%}{n}\sum_{t=1}^{n}\left|\frac{Y_t - \hat{Y}_t}{Y_t}\right|, \tag{17}$$

In Table 1, the basic originating conditions of the mode are summarized.

**Table 1.** Basic originating conditions for battery models.

| Parameters | Value |
| --- | --- |
| $b_{min}$ | 200 Wh |
| $b_{max}$ | 2000 Wh |
| b0 | 1000 Wh |
| $\eta_{ch} = \eta_{dis}$ | 0.95 |
| $\delta_{max} = -\delta_{min}$ | |
| C0.25–C2 | 500 W |
| C0.25–C2 | 1000 W |
| C0.25–C2 | 2000 W |
| C0.25–C2 | 4000 W |
| Cost of batteries at [C0.25, C2] | [0.2, 0.7] USD/W |
| Electricity price | [0.04, 0.14] USD/kWh |

The cost of batteries depicts the range from C0.25 to C2 storage rates of the simulations per nominal pack supply power. Accordingly, the electricity prices denote the data fluctuation range.

## 3. Results

### 3.1. Evolution of RTPV Penetration

The long short-term memory (LSTM) artificial neural network, trained using historical time-series data from 2015 to 2024, was used to predict the installed photovoltaic capacity on rooftops by country in the future, using the built-in MATLAB time-series processing and neural network training functions.

The future development of installed photovoltaic capacity on rooftops in selected European countries up to 2030 is depicted in Figure 5, through the application of LSTM-type neural networks and the CAGR model. The comparative evaluation of the two approaches is useful for interpretability and adaptability to understand the behavior of the models under conditions of uncertainty.

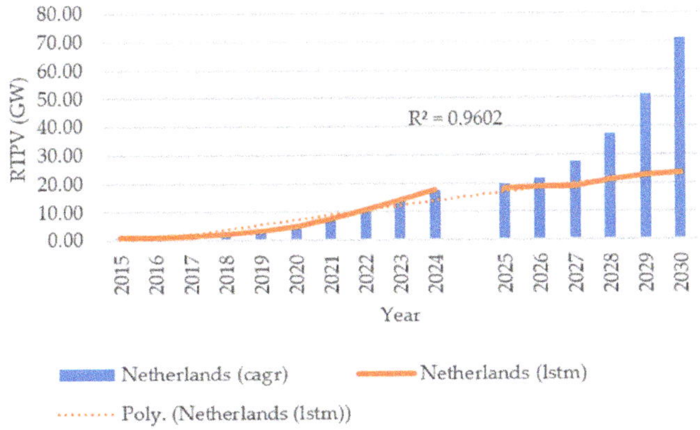


(**a**)

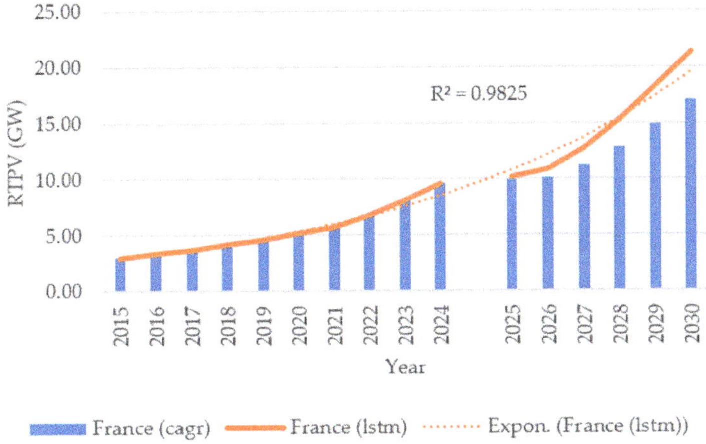


(**b**)

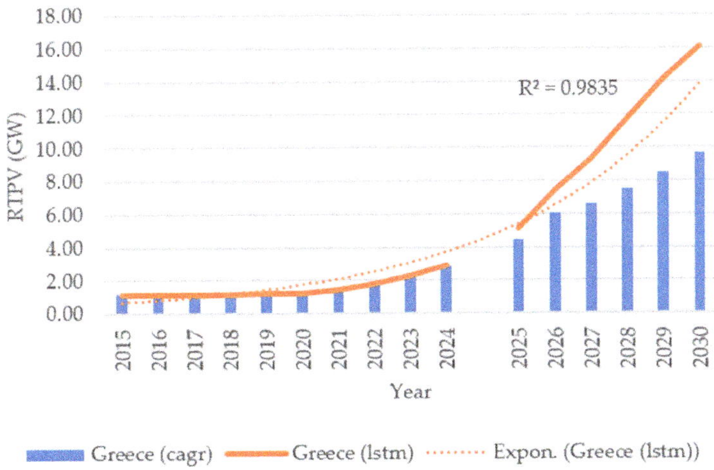


(**c**)

**Figure 5.** *Cont.*

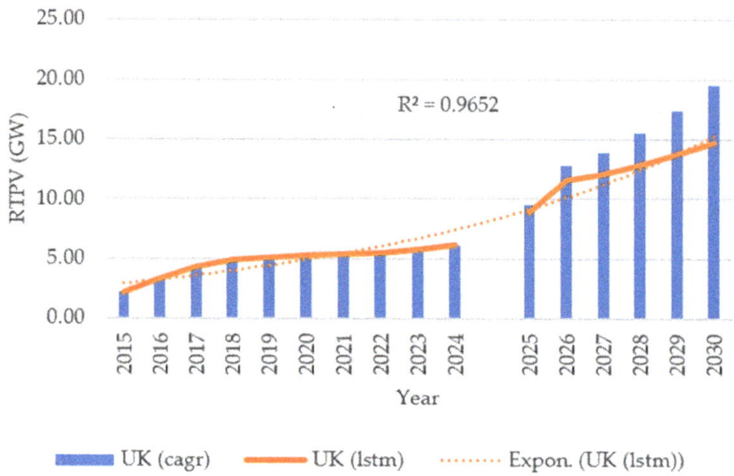


(**d**)

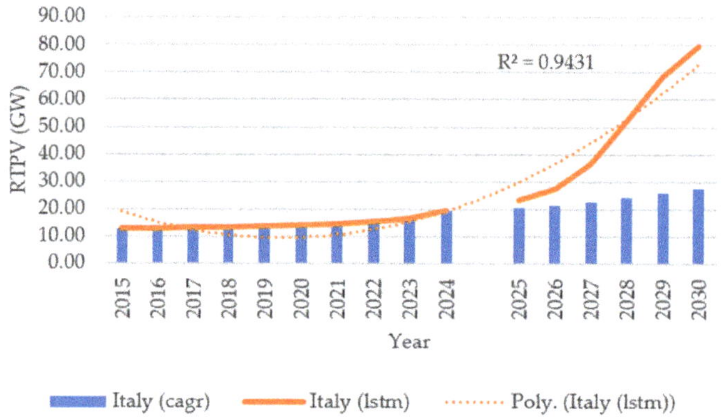


(**e**)

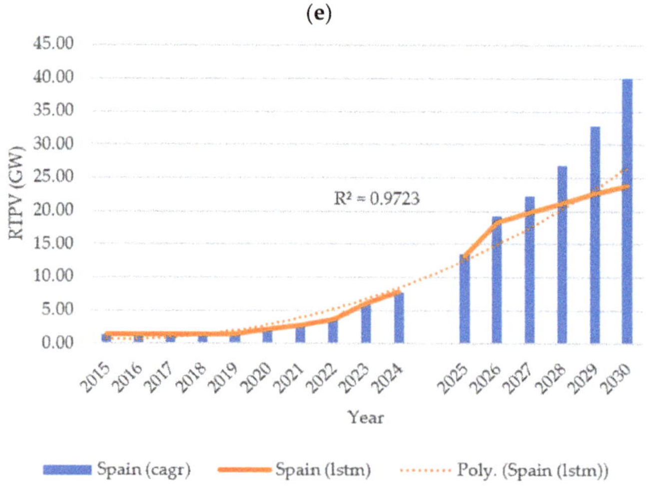


(**f**)

**Figure 5.** *Cont.*

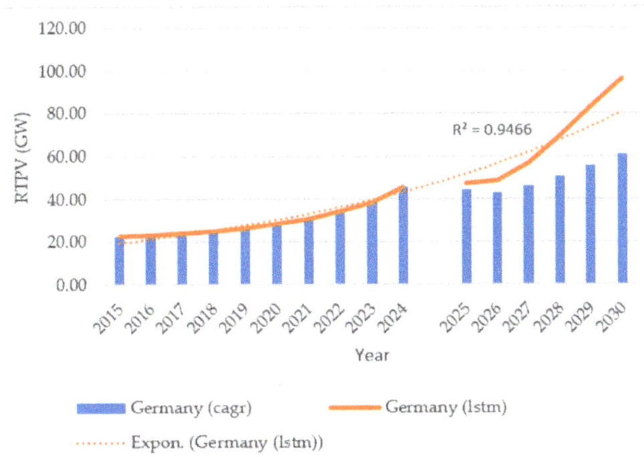


(**g**)

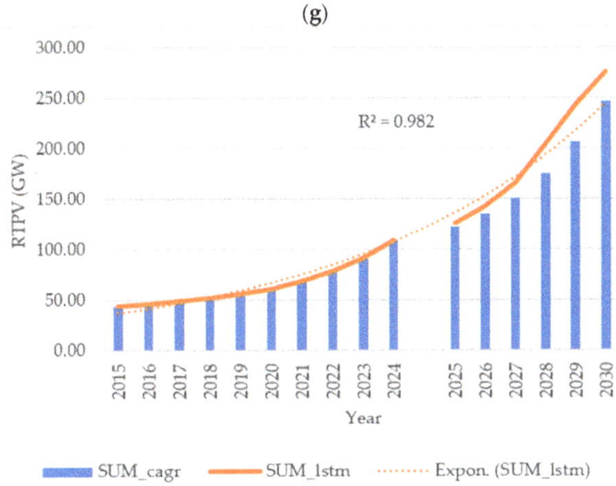


(**h**)

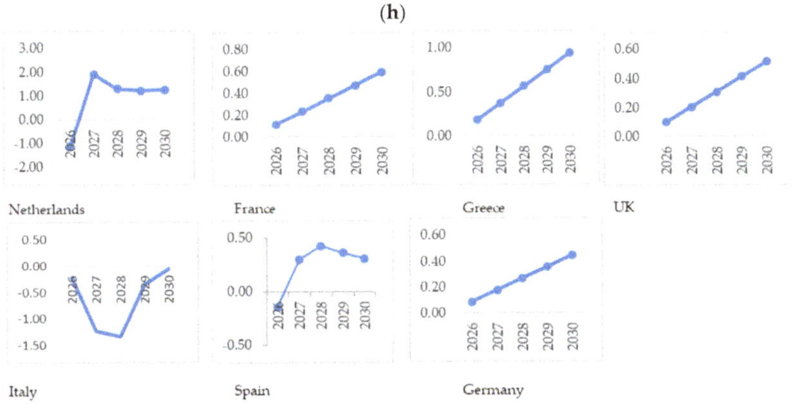


(**i**)

**Figure 5.** Predictive evolution of the RTPV in seven countries (**a**–**g**) and the cumulative trend of them (**h**). The trend with the best $R^2$ is depicted in each LSTM series of data and refers to an exponential or polyonymic equation rather than a linear one. The elasticity curves are depicted in (**i**).

In the Netherlands, there is a strong and increasing growth in the photovoltaic sector, due to the limited available space and the favorable policy for rooftop installations. The LSTM forecast shows steady growth in rooftop installations, reflecting a market that is gradually approaching saturation. The LSTM model accurately utilizes the characteristics of the country's development path, adapting the forecast curve to mature market conditions. The CAGR model predicts a rapid acceleration in installed capacity until 2030, reflecting both a mature market and increasing public and private investment in the sector. The exponential shape of the forecast curve reinforces the image of a country with robust growth potential in the decentralized RES sector.

In France, the curve shows a steady and gradually increasing trend in installed PV capacity in France, with the forecast showing an acceleration of growth after 2024. The LSTM forecast reflects the increasing support provided by the French state to self-generation and net metering programs, as well as the strategic target of total installed capacity by 2030. The model's response to historical data demonstrates its ability to accurately capture the underlying growth dynamics, predicting a significant increase from around 2025 to 2030 in the rooftop PV sector. The CAGR model shows a steady exponential growth path over the period 2025–2030. The curve follows the dynamics of previous years, confirming the accelerating trend of solar development in the country.

In Greece, both models show a steady acceleration trend towards 2030, reflecting the increase in self-generation applications and support through the net metering scheme, the European average more quickly. The LSTM exploits the underlying historical dynamics, considering institutional stability, investments in small-scale projects, and technological developments, and produces an optimistic but well-documented forecast for the penetration of rooftop PV in the Greek energy strategy. The CAGR model indicates institutional strengthening, the push for self-production systems, and the fall in installation costs. The model does not explicitly incorporate the effects of political or social variables; it adequately captures the need for the expected policy dynamics that affect the electricity sector, as previous periods have shown.

In the United Kingdom, the curves initially show a saturation phase between 2018 and 2021, reflecting the relative inertia of the market after the cessation of feed-in tariff subsidies. However, after 2020, 2021, a recovery trend is evident again, with an increase in installed capacity, which is attributed to innovative programs and the priorities of the Net Zero Strategy. The LSTM forecast captures this dynamic, with a continuous increase until 2030, reflecting the increase in self-consumption, the fall in installation costs, and the improvement of storage technologies. The projected trend records moderate but steady growth, suggesting a more mature and stable policy framework. The forecast based on the Compound Annual Growth Rate (CAGR) also shows a robust growth momentum until 2030. After a period of relative stagnation between 2016 and 2022, the trend accelerates from 2023 onwards. This is in line with government efforts to boost decentralized energy generation and initiatives. Although the CAGR model does not incorporate political or technological factors, it offers a clear picture of long-term market trends.

In Italy, after a mild increase until 2020, the Italian market shows obvious signs of acceleration from 2021 onwards, probably as a result of supportive tax incentives and subsidies (such as Super bonus 110%). The LSTM model identifies this changing pattern and predicts robust growth by 2030. The LSTM forecast highlights the momentum that self-generation has gained, as well as detecting changes in the growth trend related to external interventions or technological developments. The curve resulting from the CAGR model indicates steady future growth, which, although based on average annual growth and not on dynamic characteristics, is in line with observed trends.

In Spain, the curves demonstrate a clear growth phase, with a strong upward trend from 2020 onwards. The abolition of the sun tax and reforms in favor of self-generation with net metering have contributed to the rapid growth of rooftop PV. The LSTM network, trained on historical data, captures this transition and predicts continued growth in installed capacity at a slower pace than in other countries. The forecast suggests a steady strengthening of the market, in line with the policy focus on boosting decentralized generation. The use of the CAGR model demonstrates a growth rate that reflects the removal of political barriers (such as the sun tax) and the gradual strengthening of measures in favor of self-consumption. The projected trend is realistic and consistent with Spain's policy agenda to boost RES.

In Germany, a pioneer in solar energy, the LSTM model captures the continuous upward trend of the last decade and predicts a significant acceleration in growth after 2025. The projected capacity is almost triple by 2030. This response reflects the success of support policies through the Renewable Energy Sources Act law, feed-in-tariff, and net metering schemes that favor self-generation. The LSTM accurately exploits historical trends and time dependencies, offering a realistic projection for the German rooftop PV market, which is considered the most mature in Europe. The projection based on the CAGR model also shows a continued acceleration in growth during the period 2025–2030, reflecting the ongoing support policies, technological innovations, and increasing social acceptance.

Summarizing, it has a strong relationship of policy incentives with the acceleration or the decline of the penetration rates in each country and different states of saturation regarding the sensitivity of policy mechanisms adoption due to anelastic or elastic range of the curves. Netherlands, Spain, and Italy penetration rates respond to second-order regression behavior, with the former to emit signals of saturation and the latter to perform a mixed anelastic and reverse elastic curve. Germany, France, the UK, and Greece remain in an inelastic area by 2030.

### 3.2. *Evolution of Price and Net Load*

Before presenting the aggregated results of the energy storage arbitrage analysis, it is necessary to present the basic characteristics of the input signals used, as well as the key performance indicators of the system. In this way, a clear framework of reference is formed for the results that follow, in Figure 6, and the interpretation of the trends observed in the subsequent technical-economic analysis becomes possible.

#### 3.2.1. Arbitrage Gain per Battery Model and Coefficient $\kappa$

The difference in performance between the models is pronounced for high $\kappa$, highlighting the importance of battery C-rate in maximizing profit. In the stochastic scenario (MPC), arbitrage returns retain their dependence on the pricing factor and the technical parameters. The rapidly responding battery models (C2, C1) clearly perform better responses. The performance of all models, at low $\kappa$ values (<0.25), is nearly neutralized regardless of technical flexibility (Figure 7).

#### 3.2.2. Cycles of Operation per Battery Model and Coefficient $\kappa$

The difference between the models remains noticeable for high $\kappa$ but tends to disappear as the price spread decreases. The activity of all batteries is automatically filtered by the market environment: if there are not enough arbitrage incentives, battery degradation is limited regardless of its technical capabilities. The weekly analysis reveals how pricing policy can affect the long-term sustainability of the system. The Model Predictive Control (MPC) operation under uncertainty can lead to slightly increased activity, especially for the faster models. The actual price spread remains the key factor; when $\kappa$ is low, the number of cycles is significantly limited. The difference in the number of cycles between deterministic and stochastic becomes more pronounced only in favorable markets. Under favorable pricing policies (high κ), MPC can lead to slightly more cycles due to a more sensitive response to dynamic signals. The differences in the number of cycles are more evident in fast battery models and when arbitrage has a strong economic incentive. In restrictive environments with low κ, the results of the two approaches converge, with minimal cycles, and all models eliminate the incentives (Figure 8).

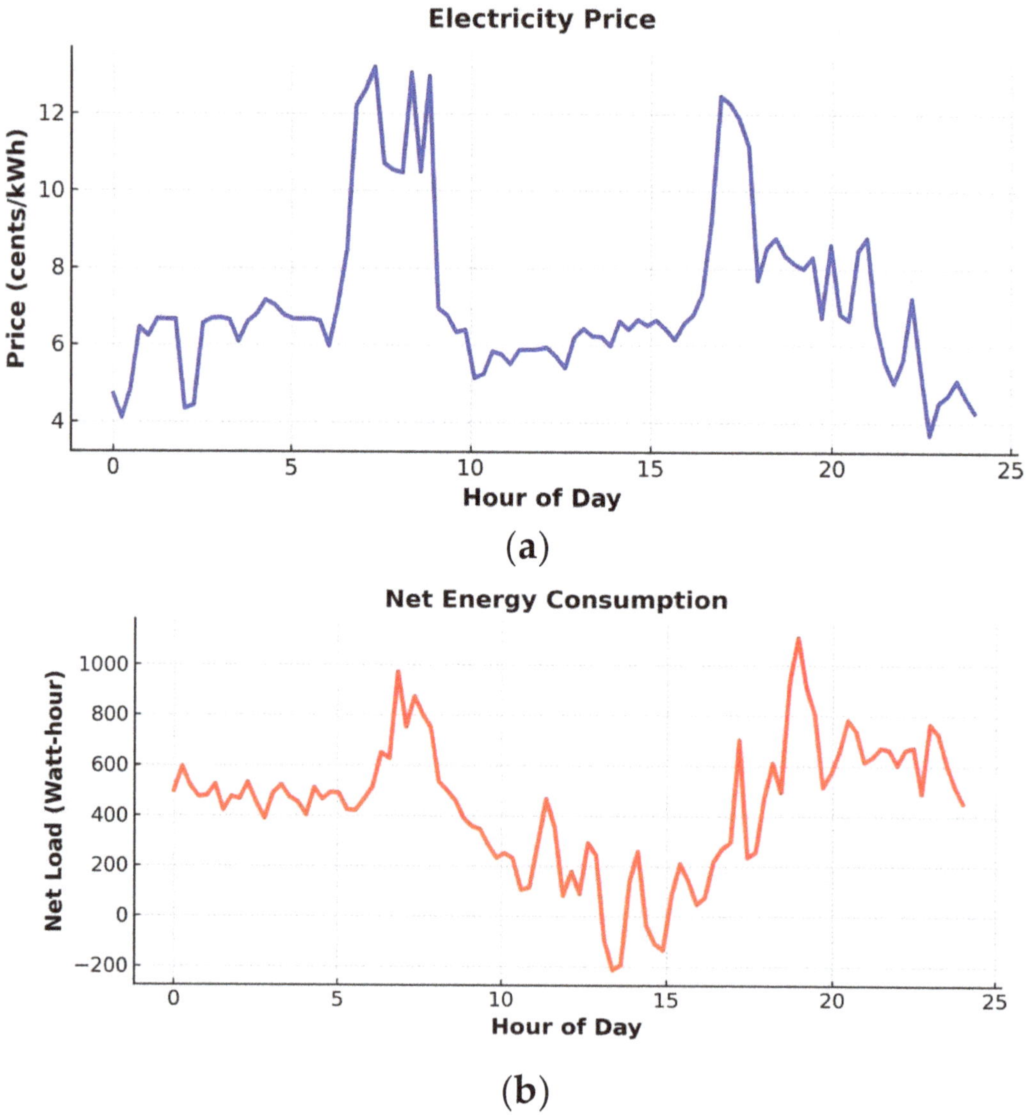


**Figure 6.** (**a**) Evolution of the price (in cents/kWh) per hour for one day. It shows fluctuations/peaks that affect arbitrage. (**b**) Hourly evolution of net load (consumption—PV production). Positive value: need to purchase from the grid. Negative value: surplus/injection.

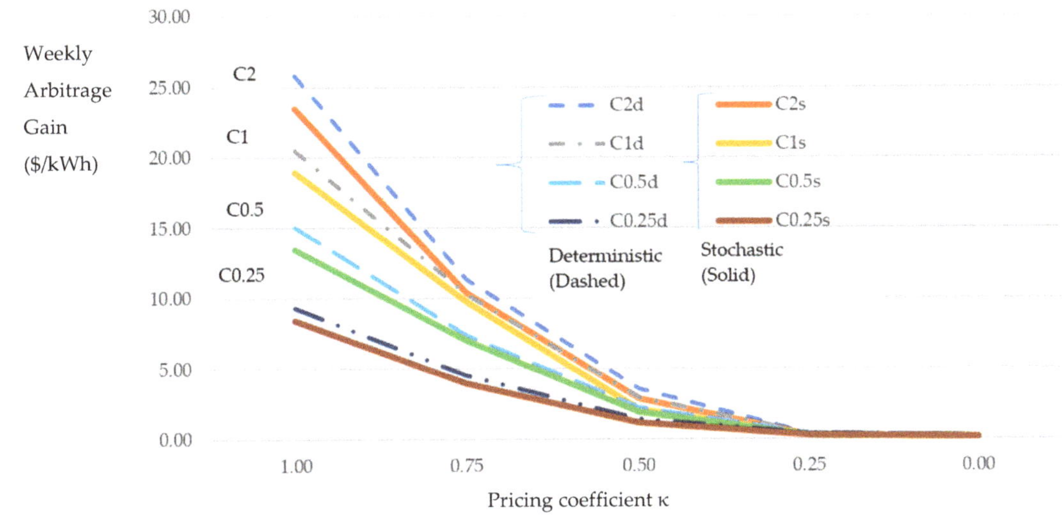


**Figure 7.** Deterministic vs. Stochastic weekly arbitrage gain.

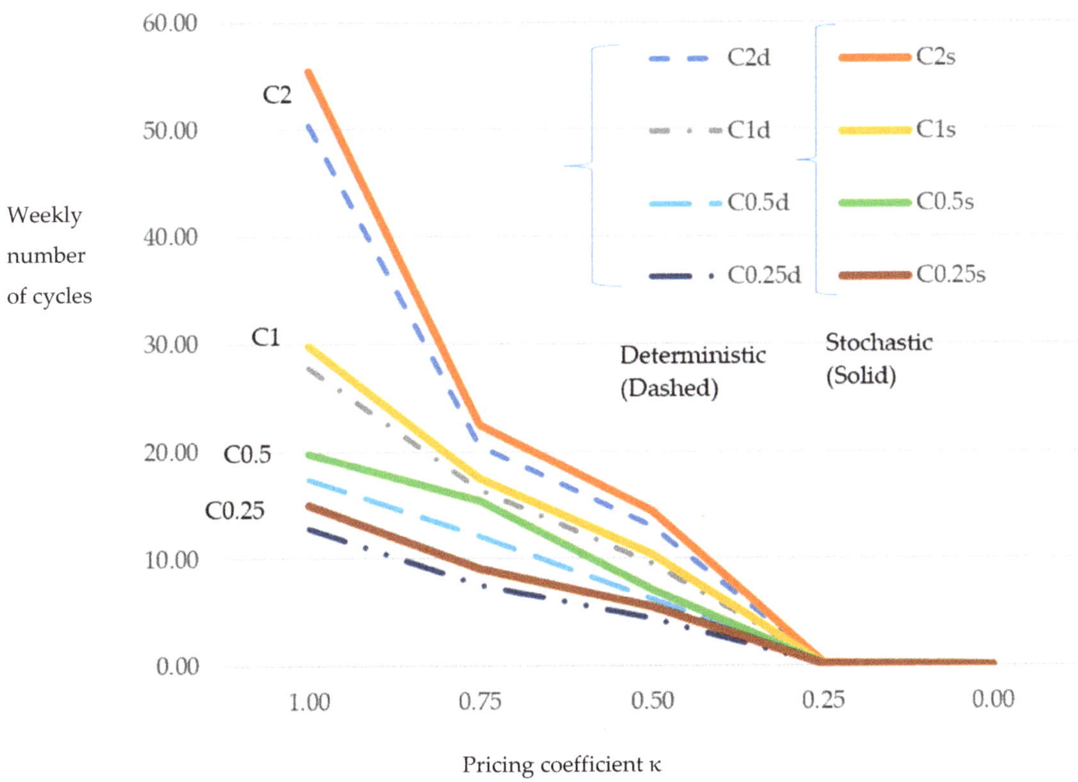


**Figure 8.** Deterministic vs. Stochastic weekly battery cycles.

### 3.2.3. Arbitrage Gain per Cycle per Battery Model and Coefficient κ

The gain per cycle increases sharply as the number of cycles decreases (low κ), because few arbitrage events accumulate all the profit. For high κ, the index values stabilize and diversify across battery models, reflecting the diverse ways of exploiting each technical configuration. The extreme high gain per cycle for exceptionally low κ reflects the rarity of opportunities. The deterministic strategy, with perfect knowledge of future prices and load, always yields slightly higher profits compared to stochastic MPC—especially for high κ and flexible batteries. The difference decreases as κ approaches zero, which means when the market does not provide real arbitrage opportunities, or the type of algorithm has little to no significant impact. Uncertainty affects fast battery models, which are based on dynamic responses, more intensely, while slow ones are affected less. (Figure 9).

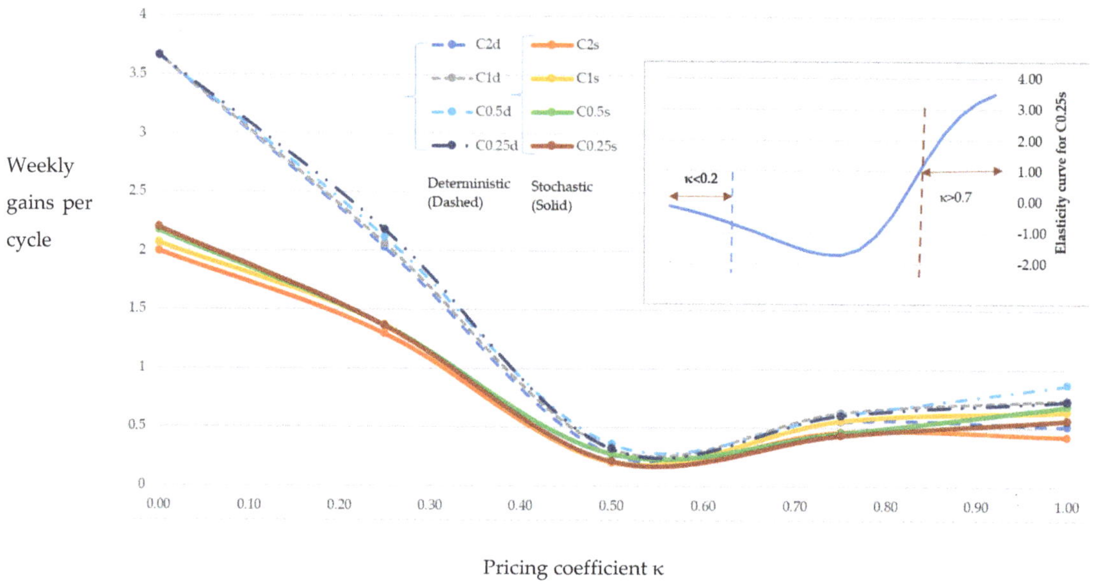


**Figure 9.** Deterministic vs. Stochastic weekly gains per battery cycle (USD/kWh.cycle).

### 3.3. *Overall Techno-Economic Interpretation of Uncertainty and Flexibility Response*

The synthesis of the results for the number of charge/discharge cycles, as presented in the previous sections for all battery models and pricing scenarios, highlights crucial aspects of the sustainability and resilience of energy storage systems to arbitrage management. Overall, this study demonstrates that the number of battery cycles operating is not just a technical quantity but fully incorporates the effects of economic incentives and market policy. The correct design of a storage system should therefore be based on a balance between maximizing arbitrage gain and limiting long-term degradation, guided by real market signals and the technical requirements of each application.

A comparison is made between the proposed stochastic approach and its deterministic counterpart, as illustrated in Figures 7–9. The weekly arbitrage gains for different batteries' C rates may reach up to about 26 USD/kWh for high κ values and neutralized for κ < 0.25. The stochasticity reduces the gains to 9.1%. The full cycles are increased up to 50 cycles and decreased to zero for κ values below 0.25. The gains per cycle are minimized by around 0.5 to 0.6 USD/kWh.cycle, increased in the range of 2 to 3.6 USD/kWh.cycle for κ approaching zero, and stabilized for high κ values in the range of 0.5 to 1 USD/kWh.cycle. The window curve of Figure 9 performs the sensitivity evolution for the C0.25s rate, as a representative result to explore the anelastic, the reversed, and the elastic zone of the weekly gains per cycle in the range of κ from zero to one. It indicates the response of prosumer flexibility to the market policy incentives at scale.

## 4. Discussion

The analysis of the results of the storage system clearly highlighted the decisive role of pricing policy in the actual exploitation of arbitrage opportunities. The pricing coefficient (κ) acts as a determining factor for the economic incentive, directly affecting both the frequency of operational cycles and the potential profit of the system.

In the short term, provide information relevant to decision-making for aggregators, building owners, businesses, and potential investors. These decisions concern the viability of the investment. In the long term, it offers a tool to assess the effectiveness of national energy policies, where positive trends reflect a favorable environment for such investments.

### *4.1. The Estimated Rooftop PV Capacity up to 2030 in Selected European Countries*

The curves of both models show a steady and gradually increasing trend in installed PV capacity in all countries, but with variable acceleration between models and countries. These variations reflect the varying rate of adopting the strategic targets capacity by 2030. Surprisingly, the underlying growth dynamics of each country are homogenized at an exponential rate when the total penetration of the seven countries is considered.

LSTM models captured the declining rates in each country due to a shortage of regulatory policies and the accelerating rates within the periods of policy changes. The CAGR gives a clear picture of the general trend; it does not consider dynamic variables (technological progress, regulatory changes, for example), which makes it limited compared to more flexible approaches such as LSTMs, and it is used for comparison.

The degree to which national markets respond to policy mechanisms varies significantly, depending on whether their respective adoption curves fall within elastic, anelastic, or inelastic ranges. For instance, the penetration rates in the Netherlands, Spain, and Italy display sensitivity characterized by second-order regression behavior. In the case of the Netherlands, the data suggest the emergence of saturation effects, indicating that further policy incentives may lead to diminishing returns in terms of increased penetration. Spain and Italy, on the other hand, exhibit a combination of anelastic and reverse elastic responses, reflecting a mix of limited responsiveness and potential for rebound effects as market conditions evolve.

In contrast, Germany, France, the United Kingdom, and Greece are projected to remain within inelastic regions of their adoption curves through 2030. This suggests that, for these countries, variations in policy mechanisms may have minimal influence on further increasing penetration rates during this period. The persistence of inelasticity in these markets highlights the need for alternative strategies or structural changes to stimulate further growth in adoption.

### *4.2. Electricity Pricing Framework Impacts RTPV–Storage Policies*

The value of the compensation factor directly affects the economic incentives to increase self-consumption and the use of storage. For values of $\kappa \approx 1$, users do not have a strong economic incentive to store and self-consume the energy they produce, as selling directly to the grid is economically equivalent to consuming it. For values of $\kappa < 1$, there is a clear economic incentive for users to increase self-consumption through storage, in order to maximize their economic efficiency and reduce their dependence on the grid. For exceptionally low k, extremely high values of gain per cycle are observed, due to the minimum number of cycles performed in a week.

In markets with high κ, the efficiency per cycle is uniform and stable, with small variations between technologies, making them more effective in the initial stages to encourage the initial deployment of RES systems. The convergence of values between battery models at extreme κ values indicates that the operating rate plays a smaller role when arbitrage does not have sufficient financial incentive. When κ approaches unity (ideal net metering policies), arbitrage becomes more attractive, resulting in the maximization of both the arbitrage gain and the operational load of the battery.

Low κ values are more suitable for promoting self-consumption and storage. In restrictive environments where the selling price is a small percentage of the purchasing price (low κ), the economic benefit from arbitrage is almost nullified, regardless of technical flexibility, control, the strategy of the structure of the market, and existing policies. For κ lower than 0.25, all configurations converge to zero returns. The buy-selling spread effectively acts as a natural filter, activating or deactivating the dynamics of the storage system.

Market policy incentives affect the degradation rates of battery storage, and the weekly gain per cycle performs elastic, anelastic, and reverse behavior of the prosumer across the range of κ values. According to the results, for $\kappa < 0.25$, there is no motivation for arbitrage. As the κ value becomes higher, we need a larger battery capacity. The increase in capacity from 0.25C to 2C increases the gains by 161%, while the corresponding change from 1C to 2C leads to 6% increase. The capacity between 1C and 2C maximizes profit.

### *4.3. Effect of Deterministic and Stochastic Applied Models on Technical and Economic Management and Efficiency*

#### 4.3.1. The Impact of Forecasting Uncertainty and Operating Cycles

Forecasting uncertainty in energy prices and load is one of the most significant challenges in the realistic operation of storage systems with arbitrage. Unlike theoretical deterministic scenarios, where the system has complete knowledge of the future, every real application relies on forecasting algorithms that exhibit error, both in the short term and the long term. This uncertainty has a direct and measurable impact on both the economic benefits and the operational wear of the battery.

In the context of the analysis, it is observed that the presence of uncertainty in forecasts leads to a slightly reduced overall arbitrage gain per cycle, compared to the ideal scenario. The MPC algorithm, attempting to offset the risk of forecasting errors, tends to have low or highly corrective behavior during the control horizon. This results, especially for the more dynamic battery models, in an increase in the number of charging/discharging cycles. However, the increased frequency of cycles does not necessarily imply greater profit for the system.

On the contrary, it may lead to increased wear of the battery, with direct consequences on maintenance costs and the lifespan of the equipment. This phenomenon highlights the importance of choosing an appropriate management strategy in order to achieve the optimal balance between capitalizing on trading opportunities and protecting the technical reliability of the system.

Of particular interest is the fact that, in environments with limited financial incentives (low price spread), the impact of uncertainty is minimized. The system itself, through optimization equations, filters the decisions, drastically reducing cycles and operational intensity, regardless of the accuracy of the forecast. This constitutes a form of automatic adjustment to the market structure, as the system learns to act only when the opportunity is hugely beneficial.

Overall, the analysis shows that uncertainty in forecasting is not just a mathematical or computational problem, but a practical limitation that directly affects both the economic viability and the long-term technical behavior of energy storage systems. Systematic monitoring and recording of operational cycles, combined with continuous improvement of forecasting algorithms, are necessary prerequisites for realistic design and successful operation of RTPV and battery storage in real environments.

4.3.2. Effects on Operating Cycles and Technical Battery Management

The choice of battery models (C2, C1, C0.5, C0.25) and, consequently, the C-rate measurably affect performance only when there is a real economic signal justifying it. Flexible battery models with high C-rate perform better when the market environment offers enough arbitrage opportunities—primarily due to their ability to react faster to fluctuations. In the opposite case, i.e., in markets with low spread or low κ, the choice of technical configuration does not result in significant differentiation in performance. This finding is of particular value for optimal sizing and investment choice, as oversizing a storage system without a corresponding economic signal leads to low utilization and excessive costs. Therefore, it is critical to adjust the technical parameterization (size, C-rate, depth of cycles) to the actual conditions of each market and the corresponding pricing framework. The weekly arbitrage gains can be reduced to 9.1%.

### *4.4. Limitations and Future Research*

The relationship with the macro RTPV penetration level and the micro storage optimization in nZEB buildings is of paramount multiparameter importance to further analyze, since it delves into the limits of the operational constraints of the electricity system. The present research work offers a first framework for a comparative quantification of the penetration of rooftop photovoltaics in European selective countries, and the building storage optimization effects on the electricity market policies, but leaves many fields open for further deepening. The development of more complex forecasting tools, the integration of additional parameters. Demographic data, climatic data, local support measures, regulatory changes, impacts on utilities or prosumer behavior data, may allow the construction of multifactorial forecasting models that better reflect the complexity of the market. Value stacking adds multiple revenue flows, even for low κ values, contributing to the higher rates of penetration and viability. However, a trade-off between them, the technical confinement of the system and the increased curtailment costs, needs awareness is limited.

## 5. Conclusions

The predicting model highlights the importance of integrating large-scale penetration forecasts with localized optimization strategies to capture the full spectrum of value creation. National markets differ in how they respond to policy mechanisms, with countries like the Netherlands, Spain, and Italy showing varying degrees of elasticity and sensitivity to incentives, including signs of saturation or rebound effects. In contrast, Germany, France, the UK, and Greece are expected to remain largely inelastic through 2030, indicating limited responsiveness to policy changes and the need for alternative strategies to drive further adoption.

By analyzing RTPV and storage optimization with the building arbitrage strategies, novel insights into the role of distributed storage assets in both grid stability and economic efficiency in urban energy systems are found. Weekly arbitrage gains are limited to 9.1%, with battery degradation and profitability showing elastic, anelastic, and reverse responses to policy incentives across different κ values. For $\kappa < 0.25$, arbitrage is unprofitable, while increasing battery capacity from 0.25C to 2C boosts gains by 161%, with optimal profits achieved between 1C and 2C.

Arbitrage revenues are overly sensitive to storage capacity sizing, cycles, and control strategies, and the market conditions, which, in turn, quantify expected impacts on electricity pricing and volatility.

The strategic choice and sizing of the battery model and the gains should not be based on theoretically maximized functionality but should consider the actual economic environment and the desired degree of utilization. RTPV and battery storage provide critical multimodal services, enhancing the flexibility and security of the system.

These challenges make the integration of storage at the building level crucial for maximizing assets and revenues, and at the same time, reduce the pressure on the grid depending on the corresponding pricing model.

**Author Contributions:** Conceptualization, A.I.T. and V.K.; methodology, A.I.T. and V.K.; software, D.S.; validation, A.I.T., V.K. and G.M.; formal analysis, V.K.; investigation, G.M. and D.S.; resources, V.K.; data curation, V.K. and D.S.; writing—original draft preparation, V.K.; writing—review and editing, V.K., A.I.T. and D.S.; visualization, G.M.; supervision, A.I.T.; project administration, A.I.T. and V.K.; funding acquisition, A.I.T. All authors have read and agreed to the published version of the manuscript.

**Funding:** This research project (no.: 14812, BIPV-City) is implemented in the framework of H.F.R.I call "Basic Research Financing (Horizontal support of all Sciences)" under the National Recovery and Resilience Plan "Greece 2.0" funded by the European Union—NextGenerationEU.

**Data Availability Statement:** The raw data supporting the conclusions of this article will be made available by the authors on request. The data are not publicly available due to [insert reason here].

**Conflicts of Interest:** The authors declare no conflicts of interest.

## Abbreviations

The following abbreviations are used in this manuscript:

| | |
|---|---|
| RTPV | Rooftop Photovoltaics |
| LSTM | Long Short-Term Memory |
| CAGR | Compound Annual Growth Rate |
| LP | Linear Programming |
| MPC | Model Predictive Control |
| ARIMA | Auto Regressive Integrated Moving Average |
| ARMA | Auto Regressive Moving Average |
| UK | United Kingdom |
| nZEB | nearly Zero Energy Building |
| EU | European Union |
| GW | Giga Watt |
| GVA | Gross Value Added |
| ROW | Rest of the World |
| RES | Renewable Energy Systems |
| LCA | Life Cycle Assessment |
| CNN | Convolution Neural Networks |
| XGBoost | eXtreme Gradient Boosting |
| LCOE | Leveraged Cost of Energy |
| MAE | Mean Absolute Error |
| MSE | Mean Squared Error |
| RMSE | Root Mean Squared Error |
| MAPE | Mean Absolute Percentage Error |
| SoC | State of Charge |
| DoD | Depth of Discharge |

## Appendix A

**Table A1.** Estimated installed PV and RTPV power (GW) and penetration percentage ratio in European Countries. Missing data are filled with verified estimations.

| Country | PV Installed Power (GW) | RTPV Installed Power (GW) | Percentage (%) |
|---|---|---|---|
| Germany | 82.4 | 51.0 | 62 |
| Spain | 39.5 | 8.4 | 21 |
| Italy | 30.3 | 21.1 | 70 |
| Netherlands | 23.9 | 19.0 | 80 |
| France | 24.6 | 12.2 | 50 |
| UK | 16.2 | 5.7 | 35 |
| Greece | ~5 (2022) | ~2.5 | 50 |

**Table A2.** Total installed RTPV power per country per year in MW (est.).

| | Country | 2015 | 2016 | 2017 | 2018 | 2019 | 2020 | 2021 | 2022 | 2023 | 2024 |
|---|---|---|---|---|---|---|---|---|---|---|---|
| 1 | Germany | 23,533 | 24,406 | 25,374 | 27,093 | 29,347 | 32,201 | 36,021 | 40,486 | 49,042 | 59,242 |
| 2 | Spain | 1411 | 1413 | 1416 | 1429 | 2642 | 3040 | 4114 | 6993 | 8613 | 10,530 |
| 3 | Italy | 13,230 | 13,498 | 13,777 | 14,075 | 14,605 | 15,155 | 15,815 | 17,188 | 20,852 | 25,872 |
| 4 | The Netherlands | 1220 | 1708 | 2328 | 3686 | 5782 | 8888 | 11,858 | 15,680 | 19,123 | 20,848 |
| 5 | France | 3569 | 3851 | 4305 | 4814 | 5364 | 5958 | 7301 | 8670 | 10,271 | 12,621 |
| 6 | UK | 3840 | 4765 | 5104 | 5224 | 5338 | 5420 | 5566 | 5860 | 6367 | 7048 |
| 7 | Greece | 1171 | 1171 | 1172 | 1192 | 1275 | 1479 | 19,245 | 2443 | 3163 | 4333 |

## Appendix B

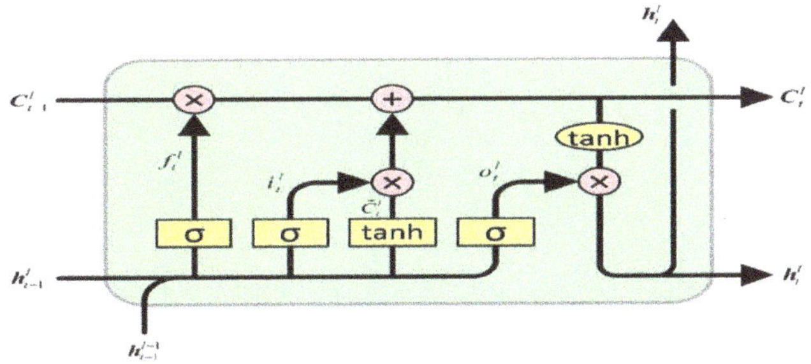


**Figure A1.** Structure of a network unit LSTM.

Each unit includes interactive layers $(1-1, 1)$ which consist of gates, $i_t$: input gate, $o_t$: output gate, $f_t$: forget gate, $C_t$: cell gate. They filter information from the previous step (t − 1) and transfer it to the next step (t + 1) according to appropriate functions $(i_t, o_t, f_t, \widetilde{C}_t)$ and weights $(W_i, W_o, W_f, W_c)$, able to modulate input and hidden states $(h_{t-1}, h_t)$, through adjustments within the training process. The forgetting gate decides what piece of information to keep or forget from the cell state.

$$f_t = \sigma(W_f[h_{t-1}, x_t] + b_f), \quad \text{(A1)}$$

$\sigma$: activation function (logistic)
$b_f$: bias of forget gate.
$x_t$: input at time t,

The input gate determines and decides the input values to be updated, and the cell gate is calculated as:

$$\mathrm{i_t} = \sigma(\mathrm{W_i}[\mathrm{h_{t-1}}, \mathrm{x_t}] + \mathrm{b_i}), \tag{A2}$$

$$\tilde{\mathrm{C}}_\mathrm{t} = \tanh(\mathrm{W_c}[\mathrm{h_{t-1}}, \mathrm{x_t}] + \mathrm{b_c}), \tag{A3}$$

*tanh*: activation function (hyperbolic tangent)

In the final state, the output gate decides the part of the cell to be output, such as:

$$\mathrm{o_t} = \sigma(\mathrm{W_o}[\mathrm{h_{t-1}}, \mathrm{x_t}] + \mathrm{b_o}), \tag{A4}$$

$$\mathrm{h_t} = \mathrm{o_t} + \tanh(\mathrm{C_t}), \tag{A5}$$

## Appendix C

For any optimization problem of the form

$$\min_x C(x), \tag{A6}$$

$$\text{s.t.} \quad \mathrm{h(x)} \le 0, \tag{A7}$$

The epigraph form of the above is the equivalent optimization problem.

$$\min_{x,z} t, \tag{A8}$$

$$\text{s.t.} \quad \mathrm{h(x)} \le 0, \tag{A9}$$

$$\mathrm{C(x)} \le \mathrm{t}, \tag{A10}$$

where the auxiliary optimization variable $t \in R$. The combination of the above results into the following

$$\min_{x,z} t, \tag{A11}$$

$$\text{s.t.} \min_z \max\{\, h(x),\ C(x) - t\} \le 0, \tag{A12}$$

## References


1. IEA-PVPS. *State of the Art Reports on PV Forecast. Task 14*; IEA-PVPS: Paris, France, 2013.
2. *European Commission EU Solar Energy Strategy*; EU: Spain, Italy, 2022.
3. Maduta, C.; D'Agostino, D.; Tsemekidi-Tzeiranaki, S.; Castellazzi, L. From Nearly Zero-Energy Buildings (NZEBs) to Zero-Emission Buildings (ZEBs): Current Status and Future Perspectives. *Energy Build.* **2025**, *328*, 115133. [CrossRef]
4. *Solar Power Europe EU Market Outlook for Solar Power 2023–2027*; Solar Power Europe: Brussels, Belgium, 2023.
5. *Solar Power Europe EU Market Outlook for Solar Power 2024–2028*; Solar Power Europe: Brussels, Belgium, 2024.
6. Abbas, M.; Zhang, Y.; Koura, Y.H.; Su, Y.; Iqbal, W. The Dynamics of Renewable Energy Diffusion Considering Adoption Delay. *Sustain. Prod. Consum.* **2022**, *30*, 387–395. [CrossRef]
7. Mitsopoulos, G.; Kapsalis, V.; Tolis, A.; Karamanis, D. Innovative Photovoltaic Technologies Aiming to Design Zero-Energy Buildings in Different Climate Conditions. *Appl. Sci.* **2024**, *14*, 8950. [CrossRef]
8. Zhang, Z.; Qian, Z.; Chen, M.; Zhu, R.; Zhang, F.; Zhong, T.; Lin, J.; Ning, L.; Xie, W.; Creutzig, F.; et al. Worldwide Rooftop Photovoltaic Electricity Generation May Mitigate Global Warming. *Nat. Clim. Change* **2025**, *15*, 393–402. [CrossRef]
9. Díaz-González, F.; Sumper, A.; Gomis-Bellmunt, O.; Villafáfila-Robles, R. A Review of Energy Storage Technologies for Wind Power Applications. *Renew. Sustain. Energy Rev.* **2012**, *16*, 2154–2171. [CrossRef]
10. Hashmi, M.U.; Mukhopadhyay, A.; Busic, A.; Elias, J. Optimal Control of Storage under Time Varying Electricity Prices. In Proceedings of the 2017 IEEE International Conference on Smart Grid Communications (SmartGridComm), Dresden, Germany, 23–27 October 2017; IEEE: New York, NY, USA, 2017; pp. 134–140.

11. Talavera, D.L.; Muñoz-Cerón, E.; Ferrer-Rodríguez, J.P.; Nofuentes, G. Evolution of the Cost and Economic Profitability of Grid-Connected PV Investments in Spain: Long-Term Review According to the Different Regulatory Frameworks Approved. *Renew. Sustain. Energy Rev.* **2016**, *66*, 233–247. [CrossRef]
12. Hashmi, M.U.; Mukhopadhyay, A.; Busic, A.; Elias, J.; Kiedanski, D. Optimal Storage Arbitrage under Net Metering Using Linear Programming. In Proceedings of the 2019 IEEE International Conference on Communications, Control, and Computing Technologies for Smart Grids (SmartGridComm), Beijing, China, 21–23 October 2019; IEEE: New York, NY, USA, 2019; pp. 1–7.
13. Box, G.E.P.; Jenkins, G.M.; Reinsel, G.C. *Time Series Analysis*; Wiley: Hoboken, NJ, USA, 2008; ISBN 9780470272848.
14. Luo, X.; Wang, J.; Dooner, M.; Clarke, J. Overview of Current Development in Electrical Energy Storage Technologies and the Application Potential in Power System Operation. *Appl. Energy* **2015**, *137*, 511–536. [CrossRef]
15. Acosta-Banda, A.; Aguilar-Esteva, V.; Hechavarría Difur, L.; Campos-Mercado, E.; Cortés-Martínez, B.; Patiño-Ortiz, M. Grid-Connected Photovoltaic Systems as an Alternative for Sustainable Urbanization in Southeastern Mexico. *Urban Sci.* **2025**, *9*, 329. [CrossRef]
16. O'Shaughnessy, E.; Cutler, D.; Ardani, K.; Margolis, R. Solar plus: Optimization of Distributed Solar PV through Battery Storage and Dispatchable Load in Residential Buildings. *Appl. Energy* **2018**, *213*, 11–21. [CrossRef]
17. Hartner, M.; Permoser, A. Through the Valley: The Impact of PV Penetration Levels on Price Volatility and Resulting Revenues for Storage Plants. *Renew. Energy* **2018**, *115*, 1184–1195. [CrossRef]
18. Nguyen, S.; Peng, W.; Sokolowski, P.; Alahakoon, D.; Yu, X. Optimizing Rooftop Photovoltaic Distributed Generation with Battery Storage for Peer-to-Peer Energy Trading. *Appl. Energy* **2018**, *228*, 2567–2580. [CrossRef]
19. Dimnik, J.; Topić Božič, J.; Čikić, A.; Muhič, S. Impacts of High PV Penetration on Slovenia's Electricity Grid: Energy Modeling and Life Cycle Assessment. *Energies* **2024**, *17*, 3170. [CrossRef]
20. Antweiler, W. Microeconomic Models of Electricity Storage: Price Forecasting, Arbitrage Limits, Curtailment Insurance, and Transmission Line Utilization. *Energy Econ.* **2021**, *101*, 105390. [CrossRef]
21. Schleifer, A.H.; Murphy, C.A.; Cole, W.J.; Denholm, P.L. The Evolving Energy and Capacity Values of Utility-Scale PV-plus-Battery Hybrid System Architectures. *Adv. Appl. Energy* **2021**, *2*, 100015. [CrossRef]
22. Schleifer, A.H.; Murphy, C.A.; Cole, W.J.; Denholm, P. Exploring the Design Space of PV-plus-Battery System Configurations under Evolving Grid Conditions. *Appl. Energy* **2022**, *308*, 118339. [CrossRef]
23. Kim, T.; Kim, J. A Regional Day-Ahead Rooftop Photovoltaic Generation Forecasting Model Considering Unauthorized Photovoltaic Installation. *Energies* **2021**, *14*, 4256. [CrossRef]
24. Miraftabzadeh, S.M.; Di Martino, A.; Longo, M.; Zaninelli, D. Deep Learning in Power Systems: A Bibliometric Analysis and Future Trends. *IEEE Access* **2024**, *12*, 163172–163196. [CrossRef]
25. Singh, U.; Singh, S.; Gupta, S.; Alotaibi, M.A.; Malik, H. Forecasting Rooftop Photovoltaic Solar Power Using Machine Learning Techniques. *Energy Rep.* **2025**, *13*, 3616–3630. [CrossRef]
26. Wu, W.; Chou, S.-C.; Viswanathan, K. Optimal Dispatching of Smart Hybrid Energy Systems for Addressing a Low-Carbon Community. *Energies* **2023**, *16*, 3698. [CrossRef]
27. Kapsalis, V.C. *Multi Energy Systems of the Future*; Elsevier: Amsterdam, The Netherlands, 2021; ISBN 9780128228975.
28. Schimpe, M.; Naumann, M.; Truong, N.; Hesse, H.C.; Santhanagopalan, S.; Saxon, A.; Jossen, A. Energy Efficiency Evaluation of a Stationary Lithium-Ion Battery Container Storage System via Electro-Thermal Modeling and Detailed Component Analysis. *Appl. Energy* **2018**, *210*, 211–229. [CrossRef]
29. Beaudin, M.; Zareipour, H.; Schellenberglabe, A.; Rosehart, W. Energy Storage for Mitigating the Variability of Renewable Electricity Sources: An Updated Review. *Energy Sustain. Dev.* **2010**, *14*, 302–314. [CrossRef]
30. Hannan, M.A.; Lipu, M.S.H.; Hussain, A.; Mohamed, A. A Review of Lithium-Ion Battery State of Charge Estimation and Management System in Electric Vehicle Applications: Challenges and Recommendations. *Renew. Sustain. Energy Rev.* **2017**, *78*, 834–854. [CrossRef]
31. Denholm, P.; Jorgenson, J.; Hummon, M.; Jenkin, T.; Palchak, D. *The Value of Energy Storage for Grid Applications (Report Summary) (Presentation), NREL (National Renewable Energy Laboratory)*; National Renewable Energy Lab: Golden, CO, USA, 2013.
32. Stamatellos, G.; Stamatellou, A.-M. The Interaction between Short- and Long-Term Energy Storage in an NZEB Office Building. *Energies* **2024**, *17*, 1441. [CrossRef]
33. Chen, X.; Li, C.; Tang, Y.; Li, L.; Du, Y.; Li, L. Integrated Optimization of Cutting Tool and Cutting Parameters in Face Milling for Minimizing Energy Footprint and Production Time. *Energy* **2019**, *175*, 1021–1037. [CrossRef]
34. Lebedeva, K.; Borodinecs, A.; Palcikovskis, A.; Wawerka, R.; Skandalos, N. Estimation of LCOE for PV Electricity Production in the Baltic States—Latvia, Lithuania and Estonia until 2050. *Renew. Sustain. Energy Transit.* **2025**, *7*, 100110. [CrossRef]
35. Gungor, V.C.; Sahin, D.; Kocak, T.; Ergut, S.; Buccella, C.; Cecati, C.; Hancke, G.P. Smart Grid Technologies: Communication Technologies and Standards. *IEEE Trans. Ind. Inform.* **2011**, *7*, 529–539. [CrossRef]
36. Ghiani, E.; Galici, M.; Mureddu, M.; Pilo, F. Impact on Electricity Consumption and Market Pricing of Energy and Ancillary Services during Pandemic of COVID-19 in Italy. *Energies* **2020**, *13*, 3357. [CrossRef]

37. Najafi, M.; Mobarez, M.D.; Vahidi, B.; Hosseinian, S.H.; Gharehpetian, G.B. Stochastic Techno-Economic Assessment of a Grid-Connected Rooftop Photovoltaic in Iran. *Renew. Sustain. Energy Rev.* **2025**, *219*, 115808. [CrossRef]
38. Kapsalis, V.; Maduta, C.; Skandalos, N.; Bhuvad, S.S.; D'Agostino, D.; Yang, R.J.; Udayraj; Parker, D.; Karamanis, D. Bottom-up Energy Transition through Rooftop PV Upscaling: Remaining Issues and Emerging Upgrades towards NZEBs at Different Climatic Conditions. *Renew. Sustain. Energy Transit.* **2024**, *5*, 100083. [CrossRef]
39. Lee, M.; Park, J.; Na, S.-I.; Choi, H.S.; Bu, B.-S.; Kim, J. An Analysis of Battery Degradation in the Integrated Energy Storage System with Solar Photovoltaic Generation. *Electronics* **2020**, *9*, 701. [CrossRef]
40. Schmalstieg, J.; Käbitz, S.; Ecker, M.; Sauer, D.U. A Holistic Aging Model for Li(NiMnCo)O2 Based 18650 Lithium-Ion Batteries. *J. Power Sources* **2014**, *257*, 325–334. [CrossRef]
41. Berrueta, A.; Heck, M.; Jantsch, M.; Ursúa, A.; Sanchis, P. Combined Dynamic Programming and Region-Elimination Technique Algorithm for Optimal Sizing and Management of Lithium-Ion Batteries for Photovoltaic Plants. *Appl. Energy* **2018**, *228*, 1–11. [CrossRef]
42. Mears, A.; Martin, J. Fully Flexible Loads in Distributed Energy Management: PV, Batteries, Loads, and Value Stacking in Virtual Power Plants. *Engineering* **2020**, *6*, 736–738. [CrossRef]
43. Couture, T.; Gagnon, Y. An Analysis of Feed-in Tariff Remuneration Models: Implications for Renewable Energy Investment. *Energy Policy* **2010**, *38*, 955–965. [CrossRef]
44. Campoccia, A.; Dusonchet, L.; Telaretti, E.; Zizzo, G. Comparative Analysis of Different Supporting Measures for the Production of Electrical Energy by Solar PV and Wind Systems: Four Representative European Cases. *Solar Energy* **2009**, *83*, 287–297. [CrossRef]
45. Burns, J.E.; Kang, J.S. Comparative Economic Analysis of Supporting Policies for Residential Solar PV in the United States: Solar Renewable Energy Credit (SREC) Potential. *Energy Policy* **2012**, *44*, 217–225. [CrossRef]
46. Kazempour, S.J.; Conejo, A.J.; Ruiz, C. Strategic Generation Investment Using a Complementarity Approach. *IEEE Trans. Power Syst.* **2011**, *26*, 940–948. [CrossRef]
47. Long, P.D.; Tram, N.H.M.; Ngoc, P.T.B. Financial Mechanisms for Energy Transitions: A Review Article. *Fulbright Rev. Econ. Policy* **2024**, *4*, 126–153. [CrossRef]
48. Huang, Y.; Sun, Q.; Chen, Z.; Wenzhong Gao, D.; Bach Pedersen, T.; Guldstrand Larsen, K.; Li, Y. Dynamic Modeling and Analysis for Electricity-Gas Systems with Electric-Driven Compressors. *IEEE Trans. Smart Grid* **2025**, *16*, 2144–2155. [CrossRef]
49. Hu, Z.; Su, R.; Veerasamy, V.; Huang, L.; Ma, R. Resilient Frequency Regulation for Microgrids Under Phasor Measurement Unit Faults and Communication Intermittency. *IEEE Trans. Ind. Inform.* **2025**, *21*, 1941–1949. [CrossRef]
50. Sioshansi, R.; Denholm, P.; Jenkin, T.; Weiss, J. Estimating the Value of Electricity Storage in PJM: Arbitrage and Some Welfare Effects. *Energy Econ.* **2009**, *31*, 269–277. [CrossRef]
51. Conejo, A.J.; Carrión, M.; Morales, J.M. *Decision Making Under Uncertainty in Electricity Markets*; Springer: Boston, MA, USA, 2010; Volume 153, ISBN 978-1-4419-7420-4.
52. Weron, R. Electricity Price Forecasting: A Review of the State-of-the-Art with a Look into the Future. *Int. J. Forecast.* **2014**, *30*, 1030–1081. [CrossRef]
53. Darghouth, N.R.; Barbose, G.; Wiser, R. The Impact of Rate Design and Net Metering on the Bill Savings from Distributed PV for Residential Customers in California. *Energy Policy* **2011**, *39*, 5243–5253. [CrossRef]
54. Luthander, R.; Widén, J.; Nilsson, D.; Palm, J. Photovoltaic Self-Consumption in Buildings: A Review. *Appl. Energy* **2015**, *142*, 80–94. [CrossRef]
55. Sebestyén, V. Renewable and Sustainable Energy Reviews: Environmental Impact Networks of Renewable Energy Power Plants. *Renew. Sustain. Energy Rev.* **2021**, *151*, 111626. [CrossRef]
56. Bai, Y.; Wang, J.; He, W. Energy Arbitrage Optimization of Lithium-Ion Battery Considering Short-Term Revenue and Long-Term Battery Life Loss. *Energy Rep.* **2022**, *8*, 364–371. [CrossRef]
57. Hussain, A.; Kim, H.-M. Enhancing Renewable Energy Use in Residential Communities: Analyzing Storage, Trading, and Combinations. *Sustainability* **2024**, *16*, 891. [CrossRef]
58. Eid, C.; Reneses Guillén, J.; Frías Marín, P.; Hakvoort, R. The Economic Effect of Electricity Net-Metering with Solar PV: Consequences for Network Cost Recovery, Cross Subsidies and Policy Objectives. *Energy Policy* **2014**, *75*, 244–254. [CrossRef]
59. García-Suso, F.; Molina-García, A.; Fernández-Guillamón, A.; Bueso, M.C. Alternative Non-Optimal Orientations in Highly PV Self-Consumption Integration: Exploring Spanish Prosumers as a Case Study. *Renew. Energy* **2026**, *256*, 123987. [CrossRef]
60. Hashemi, M.; Jenkins, G.; Milne, F. Rooftop Solar with Net Metering: An Integrated Investment Appraisal. *Renew. Sustain. Energy Rev.* **2023**, *188*, 113803. [CrossRef]
61. Sriyono; Aditya Pramana, P.A.; Kusumadinata, A. Mitigating the Impact of Photovoltaic Penetration on the Electricity System Using Advanced Metering Infrastructure. In Proceedings of the 2024 International Seminar on Intelligent Technology and Its Applications (ISITIA), Mataram, Indonesia, 10–12 July 2024; IEEE: New York, NY, USA, 2024; pp. 82–86.

62. Hoppmann, J.; Volland, J.; Schmidt, T.S.; Hoffmann, V.H. The Economic Viability of Battery Storage for Residential Solar Photovoltaic Systems—A Review and a Simulation Model. *Renew. Sustain. Energy Rev.* **2014**, *39*, 1101–1118. [CrossRef]
63. Obi, M.; Jensen, S.M.; Ferris, J.B.; Bass, R.B. Calculation of Levelized Costs of Electricity for Various Electrical Energy Storage Systems. *Renew. Sustain. Energy Rev.* **2017**, *67*, 908–920. [CrossRef]
64. Denholm, P.; Margolis, R.M. Evaluating the Limits of Solar Photovoltaics (PV) in Traditional Electric Power Systems. *Energy Policy* **2007**, *35*, 2852–2861. [CrossRef]
65. Hashmi, M.U.; Labidi, W.; Busic, A.; Elayoubi, S.-E.; Chahed, T. Long-Term Revenue Estimation for Battery Performing Arbitrage and Ancillary Services. In Proceedings of the 2018 IEEE International Conference on Communications, Control, and Computing Technologies for Smart Grids (SmartGridComm), Aalborg, Denmark, 29–31 October 2018; IEEE: New York, NY, USA, 2018; pp. 1–7.
66. Biggins, F.A.V.; Homan, S.; Ejeh, J.O.; Brown, S. To Trade or Not to Trade: Simultaneously Optimising Battery Storage for Arbitrage and Ancillary Services. *J. Energy Storage* **2022**, *50*, 104234. [CrossRef]
67. Schwidtal, J.; Agostini, M.; Coppo, M.; Bignucolo, F.; Lorenzoni, A. Integrating Distributed Energy: Value Stacking for PV with Power-to-Gas 2022. *Energy Proc.* **2021**, *20*, 765.
68. Dufo-López, R.; Lujano-Rojas, J.M.; Artal-Sevil, J.S.; Bernal-Agustín, J.L. Optimising Grid-Connected PV-Battery Systems for Energy Arbitrage and Frequency Containment Reserve. *Batteries* **2024**, *10*, 427. [CrossRef]
69. Oudalov, A.; Chartouni, D.; Ohler, C. Optimizing a Battery Energy Storage System for Primary Frequency Control. *IEEE Trans. Power Syst.* **2007**, *22*, 1259–1266. [CrossRef]
70. Xu, Y.; Tong, L. Optimal Operation and Economic Value of Energy Storage at Consumer Locations. *IEEE Trans. Autom. Control.* **2017**, *62*, 792–807. [CrossRef]
71. Castillo-Cagigal, M.; Caamaño-Martín, E.; Matallanas, E.; Masa-Bote, D.; Gutiérrez, A.; Monasterio-Huelin, F.; Jiménez-Leube, J. PV Self-Consumption Optimization with Storage and Active DSM for the Residential Sector. *Solar Energy* **2011**, *85*, 2338–2348. [CrossRef]
72. Khalilpour, R.; Vassallo, A. Planning and Operation Scheduling of PV-Battery Systems: A Novel Methodology. *Renew. Sustain. Energy Rev.* **2016**, *53*, 194–208. [CrossRef]
73. Higgins, A.; Grozev, G.; Ren, Z.; Garner, S.; Walden, G.; Taylor, M. Modelling Future Uptake of Distributed Energy Resources under Alternative Tariff Structures. *Energy* **2014**, *74*, 455–463. [CrossRef]
74. Seward, W.; Qadrdan, M.; Jenkins, N. Revenue Stacking for behind the Meter Battery Storage in Energy and Ancillary Services Markets. *Electr. Power Syst. Res.* **2022**, *211*, 108292. [CrossRef]
75. Kou, P.; Liang, D.; Gao, L. Distributed EMPC of Multiple Microgrids for Coordinated Stochastic Energy Management. *Appl. Energy* **2017**, *185*, 939–952. [CrossRef]
76. Fang, X.; Yang, Q.; Wang, J.; Yan, W. Coordinated Dispatch in Multiple Cooperative Autonomous Islanded Microgrids. *Appl. Energy* **2016**, *162*, 40–48. [CrossRef]
77. Zakaria, A.; Ismail, F.B.; Lipu, M.S.H.; Hannan, M.A. Uncertainty Models for Stochastic Optimization in Renewable Energy Applications. *Renew. Energy* **2020**, *145*, 1543–1571. [CrossRef]
78. Hu, J.; Shan, Y.; Guerrero, J.M.; Ioinovici, A.; Chan, K.W.; Rodriguez, J. Model Predictive Control of Microgrids—An Overview. *Renew. Sustain. Energy Rev.* **2021**, *136*, 110422. [CrossRef]
79. Yang, H.; Liu, Q.; Xiao, K.; Guo, L.; Yang, L.; Zou, H. Scenario-Driven Optimization Strategy for Energy Storage Configuration in High-Proportion Renewable Energy Power Systems. *Processes* **2024**, *12*, 1721. [CrossRef]
80. Gomez-Exposito, A.; Conejo, A.J.; Cañizares, C. (Eds.) *Electric Energy Systems: Analysis and Operation*, 2nd ed.; CRC Press: Boca Raton, FL, USA, 2018; ISBN 9781315192246.
81. Birge, J.R.; Louveaux, F. *Introduction to Stochastic Programming*; Springer: New York, NY, USA, 2011; ISBN 978-1-4614-0236-7.
82. Shobande, O.A.; Ogbeifun, L.; Tiwari, A.K. Net-Zero Transitions: Advancing Dynamic Econometric Analysis of Carbon Tax, Renewable Energy, and Circular Economy on Government Actions. *J. Environ. Manag.* **2025**, *378*, 124761. [CrossRef]
83. Pan, P.; Wang, Z.; Chen, G.; Shi, H.; Zha, X. Multi-Time Scale Optimal Dispatch of Distribution Network with Pumped Storage Station Based on Model Predictive Control. *Appl. Sci.* **2024**, *14*, 11122. [CrossRef]
84. Parisio, A.; Rikos, E.; Glielmo, L. Stochastic Model Predictive Control for Economic/Environmental Operation Management of Microgrids: An Experimental Case Study. *J. Process Control.* **2016**, *43*, 24–37. [CrossRef]
85. Hashmi, M.U.; Van Hertem, D.; van der Meer, A.; Keane, A. Linear Energy Storage and Flexibility Model with Ramp Rate, Ramping, Deadline and Capacity Constraints. In Proceedings of the 2024 IEEE PES Innovative Smart Grid Technologies Europe (ISGT EUROPE), Dubrovnik, Croatia, 14–17 October 2024; IEEE: New York, NY, USA, 2024; pp. 1–6.
86. Camacho, E.F.; Bordons Alba, C. *Model Predictive Control*; Springer: London, UK, 2013; ISBN 9780857293985.
87. Stamatakis, D.; Tolis, A.I. Coordinated Electric Vehicle Demand Management in the Unit Commitment Problem Integrated with Transmission Constraints. *Energies* **2025**, *18*, 4293. [CrossRef]

88. Semertzidis, G.; Stamatakis, D.; Tsalavoutis, V.; Tolis, A.I. Optimized Electric Vehicle Charging Integrated in the Unit Commitment Problem. *Oper. Res.* **2022**, *22*, 5137–5204. [CrossRef]
89. Antonanzas, J.; Osorio, N.; Escobar, R.; Urraca, R.; Martinez-de-Pison, F.J.; Antonanzas-Torres, F. Review of Photovoltaic Power Forecasting. *Solar Energy* **2016**, *136*, 78–111. [CrossRef]
90. Sulaiman, M.H.; Jadin, M.S.; Mustaffa, Z.; Daniyal, H.; Mohd Azlan, M.N. Short-Term Forecasting of Rooftop Retrofitted Photovoltaic Power Generation Using Machine Learning. *J. Build. Eng.* **2024**, *94*, 109948. [CrossRef]
91. Gulkowski, S.; Krawczak, E. Long-Term Energy Yield Analysis of the Rooftop PV System in Climate Conditions of Poland. *Sustainability* **2024**, *16*, 3348. [CrossRef]
92. Liu, C.-H.; Gu, J.-C.; Yang, M.-T. A Simplified LSTM Neural Networks for One Day-Ahead Solar Power Forecasting. *IEEE Access* **2021**, *9*, 17174–17195. [CrossRef]
93. The MathWorks Inc. *MATLAB*, version: 9.14.0 (R2023a); The MathWorks Inc.: Natick, MA, USA, 2023.
94. Fraunhofer Institute. Available online: www.Ise.Fraunhofer.de/Dam/Ise/de/Documents/Publications/Studies/Photovoltaics-Report.Pdf (accessed on 12 September 2025).
95. CBS. Available online: https://www.cbs.nl/ (accessed on 10 October 2025).
96. UK Statistics. Available online: www.gov.uk/search/research-and-statistics/ (accessed on 12 September 2025).
97. IRENA Renewable Energy Statistics. Available online: https://www.irena.org/-/media/Files/IRENA/Agency/Publication/2025/Mar/IRENA_DAT_RE_Capacity_Statistics_2025.pdf (accessed on 10 October 2025).
98. IEA-PVPS. *Snapshot of Global PV Markets 2023. International Energy Agency—Photovoltaic Power Systems Programme*; IEA-PVPS: Paris, France, 2023.
99. Nguyen, M.H.; Braeye, T.; Hens, N.; Faes, C. Multivariate Phenomenological Models for Real-Time Short-Term Forecasts of Hospital Capacity for COVID-19 in Belgium from March to June 2020. *Epidemiol. Infect.* **2022**, *150*, e12. [CrossRef]

energies

MDPI

*Article*

# Hydraulic Balancing of District Heating Systems and Improving Thermal Comfort in Buildings

**Stanislav Chicherin** [1,2]

[1] Thermo and Fluid Dynamics (FLOW), Faculty of Engineering, Vrije Universiteit Brussel (VUB), Pleinlaan 2, 1050 Brussels, Belgium; stanislav.chicherin@vub.be

[2] Brussels Institute for Thermal-Fluid Systems and Clean Energy (BRITE), Vrije Universiteit Brussel (VUB) and Université Libre de Bruxelles (ULB), 1050 Brussels, Belgium

**Abstract:** The relevance is introducing fourth generation district heating (4GDH), which decreases operation and maintenance costs by utilizing the efficiency of low temperature district heating (LTDH). The aim is to develop a methodology allowing for a more flexible heat demand model and accurate function describing the relationship between outdoor temperature and heat demand. It is represented by a black-box model based on historical data collected from heating, ventilation, and air conditioning (HVAC) systems. Energy delivery/consumption is analyzed with the help of a set of statistical and regression formulas. The analysis of operational data is then transformed to methodology to regulate heat supply with combined heat-and-power (CHP) generation. The key features are that the model takes into account thermal capacity and type of substation; the district heating (DH) plant is not assumed to have a fixed return temperature and generation profile. The novelty is an emphasis on DH operation and introduction of statistics into a dynamic simulation model. With no abnormal buildings, higher accuracy of modeling is achieved. Most of the consumers are pretty similar in thermal response, even though specific energy demand and heated volume may differ. Heat demand of an old building is better simulated with discrete regression, while those with pump-equipped substations are modeled with linear regression.

**Keywords:** renewables; energy integration; efficiency; environmental sustainability; district energy; simulation

## 1. Introduction

Compared to traditional district heating (DH) systems, the benefits of the fourth generation district heating (4GDH) concept are the reduced operation and maintenance (O&M) costs of central plants, increase in efficiency of heat distribution, reduction in capital expenditure, and great potential to optimize installed capacity [1]. The efficiency of low-temperature district heating (LTDH) systems is higher if the share of directly supplied energy is high [2,3]. LTDH systems are globally feasible by implementing the principle of heat carrier cascading and providing users situated in the newly developed area with service by integrating them into a high-temperature district heating (HTDH) system. However, with increasing space heating (SH) share and heated volume/area, along with high energy consumption and poor efficiency of old buildings connected to an HTDH system, the seasonal coefficient of performance of a new subnetwork is also low. The present paper shows that the primary measure before conversion to the 4GDH system is not only a reduction in energy consumption on the demand side but also a better understanding of its distribution with an emphasis on not only design data. Researchers are now looking for new options, such as booster heat pumps partially powered by solar energy [4]. We should know the design and actual heat demand to understand how much energy they may cover. Design values should be adjusted if a new neighborhood is under construction or operational data are unavailable. That correlates well with the development of more energy-efficient buildings with low SH consumption, thus constituting the relevance of the present research.

In practice, mass flow rates (and thus circulation pump energies) are largest in winter (for heating) and in summer (for cooling) and smallest in the intermediate seasons. If the DH system is large, the return temperature is high, issues of hydraulic balancing are expected, and energy consumption for pumping is also high.

### 1.1. Literature Review

Relative to the current study, Sommer et al. [5] assert that the energy required by circulation pumps is minor compared to the energy used by heat pumps, thereby having a negligible impact on the seasonal variation in overall energy consumption. However, in modeling scenarios, electricity usage for pump operation is typically treated as constant across all DH systems throughout the year, except in summer. This assumption stems from distribution networks' uniform and constant mass flow rate control while supply temperatures are adjusted.

Heat demand plays a critical role in any DH system and is a fundamental element of operational data. For example, findings in [6] are attributed to the distribution and composition of heat demand. District energy systems are particularly viable in highly populated French cities—280 cities with populations exceeding 30,000. To facilitate a transition to LTDH or 4GDH frameworks, it is essential to lower heat demand through demand-side measures (DSMs). Another approach involves enhancing the heat transfer capacity or area of radiators to maintain the same level of heat output with reduced supply temperatures. Since SH systems are often slightly oversized and poor regulation is a primary cause of elevated return temperatures, a 10–15% reduction in heat demand for new buildings may suffice. In comparison, as noted in [7], this reduction aligns with a potential decrease in residential energy consumption of 16–25%.

DH supply and return temperatures follow the energy demand profile, with the daily resolution showing the amount of energy at a DH plant that should be generated. Overconsumption means the supply temperature is excessive, while its lowest threshold is set to 70 °C. It coincides with periods of low energy consumption by the building inhabitants or when there is no heat demand for SH purposes, as at the beginning/end of the SH season. The maximum value for a representative winter day is 7.38 MWh.

In contrast to our approach, Quirosa et al. [8] emphasize that the instantaneous heat demand met by the DH system corresponds directly to the building's thermal response. They link this demand to the building's requirements and align it with a similar profile for the compressor power curve. Additionally, they maximize the system's performance based on the DH supply temperature and the heat pump's COP. However, on the analyzed winter day, the maximum observed value is significantly lower, reaching only 44 kW.

Harney et al. [9] examine a correlation between SH energy consumption and outdoor temperature. This obvious connection implies that as temperature decreases, the SH demand of the building increases. That clarifies why energy demand is higher in colder winter months, although no numerical assessments are given. When fuel consumption is related to warmer weather and consumer behavior decreases, it is high time modernization starts the process of conversion into LTDH and 4GDH concepts.

For HTDH, the main strategies include reducing distribution losses by lowering network temperatures and minimizing SH demand [10]. Building on this premise, further research could focus on reducing return temperatures in the network and improving modeling accuracy. Unlike earlier studies, this approach highlights low return temperatures as a critical factor for maintaining high performance and reducing primary energy consumption. Additionally, achieving a greater temperature difference at the system level facilitates DH system expansion without compromising the indoor comfort of existing users.

However, the reason for high return temperature is typically not a DH network but substation tuning, the thermal response of buildings, and occupants' behavior. During periods of low outdoor temperature, all the substations of a DH system are running while the valves in the secondary circuits are fully open. As outdoor temperature increases, the temperature sensors in front of each SH system close their valves, restricting the flow through radiators. Nonetheless, the development of accurate simulation models able to cover such dynamic behavior entails considerable effort [11]. According to Brandi et al. [11], detailed and sophisticated models cannot always simulate the complexity of heating, ventilation, and air conditioning (HVAC) systems operation and the effect of occupant behavior. Therefore, we developed an alternative, represented by a black-box model based on historical data collected from HVAC systems. After design, heat demand is adjusted to lower values, and a lower flow rate is also expected. The flow sensor installed in front of an SH system detects the flow reduction from a DH plant, and restores flow by opening the bypass valve to maintain the desired temperature in the return line.

Unlike earlier studies such as [12–15], energy demand in this research is modeled with greater flexibility, accounting for variations in outdoor temperatures, specific building characteristics, and regional factors, such as whether it is a weekday or a weekend. A new framework is introduced for simulating steady-state demand in branched DH networks, addressing the limitations of previous methods. Previously, models were based on a fixed-point iteration approach, with constraints and the Jacobian matrix configured manually [16].

Kurek et al. [17] focused on DH substations equipped with additional meters to measure only SH heat demand. In contrast, this study utilized data from shared DH substations to develop a forecast model applicable across winter, intermediate, and summer seasons, with design heat demand determined using a specific equation. This approach enhances the model's versatility.

Regarding methodologies, Ivanko et al. [18] examined control point temperature, defined as the final day of the SH season. They applied the least squares method to develop a regression function for two model segments by establishing a correlation between outdoor temperature and heat demand. Unlike the present study, they focused on identifying the control point temperature using piecewise regression analysis.

We developed distinct models for various outdoor temperature ranges using the same method. A similar approach is demonstrated in [18], confirming the broad applicability of regression methods for analyzing operational data. While top-down approaches rely on highly detailed consumption data—often unavailable at DH substations or certain groups of substations—these methods are closely tied to the DH system's hydraulic configuration due to their aggregation level.

In contrast to the present study, the model by Kristensen et al. [19] cannot be directly applied to predict heat consumption in specific subsections of an existing DH system. Furthermore, it is unsuitable for forecasting energy demand in newly constructed districts, such as those arising from urban expansion. We observed subtle temperature differences between supply and return lines by analyzing operational data from substations farther from a DH plant.

Barone et al. [20] supported this concept, identifying significant temperature drops in flow branches relative to the outlet temperature of the power station alongside minor drops in return branches. Similarly to our findings, they attributed this to abnormal thermal responses influenced by the minimal energy consumption of a small number of DH consumers. As noted in [21], two key reasons for high error rates were identified: (1) excessively high return temperatures after domestic hot water (DHW) or SH heat exchangers and (2) flawed design approaches when calculating the surface area of SH system radiators. For buildings with hydronic SH systems, prevalent in Russia and Europe, the design secondary supply temperature often results in return temperatures that fail to reach low values, though indoor comfort is maintained. To address low-temperature differentials, Ref. [21] recommends checking and upgrading the building's HVAC system operation and associated heat demand.

To analyze heat demand, Jebamalai et al. [22] utilized open-source street-level gas consumption data provided by local gas suppliers. These profiles were derived from extensive datasets and accounted for typical building types in Belgium. In contrast, this study assessed the system's total load by aggregating residential energy consumption, incorporating operational data, and using detailed profiles. Another distinction lies in the DH plant modeling: unlike fixed generation curves assumed throughout the year in [22], this study adopts a more dynamic approach.

In [23], the heat demand of all buildings in a neighborhood is aggregated into a single substation, representing the district's average demand. Their model of a district energy system comprises only a DH plant and a single aggregated district substation, which cannot capture temperature variations within supply and return lines. By comparison, this study developed a more detailed model, limiting aggregation to exclude abnormal heat demand profiles. This approach enhances accuracy while reducing the number of demand peaks.

DHW was not included in Luc et al.'s [24] model. However, DHW is typically a reason for hourly peaks in flow rates and return temperature. Since their model contains SH only, the baseline scenario is characterized by decreased energy consumption during the daytime, according to solar insolation. The difference is that the simulation model used here includes both SH and DHW consumption; we performed a more detailed study of some buildings.

Contribution to the pool of knowledge is the analysis of operational data, which is then transformed into developing a methodology to regulate heat supply into a branched network with CHP energy generation. The key feature is that the model considers thermal capacity and the type of substation. They both strongly influence the DH operation and imply that smart regulation strategies may be implemented to regulate the network according to actual heat demand rather than design one. Compared to the present paper, Guelpa et al. [25] report that the amount of thermal energy stored in heat accumulators is a considerable share of energy used to heat the various masses in the system. They conclude that this result is extremely significant, but no attempts have been made to study the effects of thermal capacities within the pool of operational data.

Farouq et al.'s [26] limitation is caused by the DH system. For instance, a reference-group-based approach cannot be directly applied to the building substations to adjust any automatic control mechanism when a faulty operation is detected. The first aspect of the novelty here is that we emphasize DH operation and cover the energy consumption, following solely individual consumers' requirements. The difference is that implementing centralized control achieves better results. The pressure level at a CHP plant may also change via a centralized supervisory control and data actuation (SCADA) system to deliver ample flow to DH substations.

Based on the surveyed literature, the second aspect of novelty entails deducing a relationship between outdoor and supply temperatures (i.e., fuel consumption) for any DH plant at any time. The next step is establishing a new regulation method for the possible energy demand ranges presented in the Results and Discussion. This method should be accurate for any season, including winter and moderate cold days. Unlike now, we used to assume that most substantial deviations happened during average outdoor temperatures of 0–20 °C, i.e., in the intermediate season or the end and the beginning of winter when the network temperatures and energy consumption experience peaks in volatility [27].

We declare that validated operational data are essential for such suggestions, although data treatment and smart data applications (big data, artificial neural networks) still require manual adjustment in most cases. The difference between the approach of Kleinertz et al. [28] and the one presented here is the necessary actions mentioned in the transformation strategy guideline. They include replacing missing consumption data and derivation of data determination and integrating cross-sectoral data in the study. However, some statements are similar, as we conclude that data from the network itself are crucial, as well as the type of buildings required to assess their flexibility and ability to follow the predetermined control strategy.

### 1.2. Summary

This paper introduces a novel approach to the transition from HTDH to 4GDH systems by emphasizing the critical role of operational data in developing a flexible heat demand model. Unlike previous studies, which often rely on design data alone or simplified fixed-point models, our research highlights the necessity of combining design and real-time operational data to understand energy distribution dynamics better. The primary innovation of this study lies in creating a new regulation methodology that establishes a dynamic relationship between outdoor temperature and heat demand, ensuring more accurate predictions for various consumer types and seasonal conditions.

Our model integrates statistical and regression formulas with operational data collected from an HTDH network during 2020–2021. This dual approach allows us to address the limitations of traditional methods by adapting to fluctuating energy demands and capturing the effects of thermal capacity, building characteristics, and consumer behavior. Furthermore, the study proposes a centralized control strategy using validated operational data, moving beyond the conventional decentralized methods that often fail to account for the network's dynamic behavior.

Another key contribution is introducing a new simulation framework for branched DH networks. Unlike earlier models that relied on aggregated heat demand at a single district substation, our method limits aggregation to reduce inaccuracies from demand peaks and incorporates CHP generation. Additionally, this framework incorporates smart regulation strategies, enabling real-time adjustments based on actual heat demand rather than design assumptions.

In summary, the innovations of this research are twofold: (1) the development of a flexible and dynamic heat demand model that leverages both design and operational data, and (2) the establishment of a new regulation methodology for DH systems that optimizes supply temperatures and flow rates across varying seasonal and consumer conditions. These advancements offer a robust foundation for transitioning to 4GDH systems and improving the overall efficiency and adaptability of DH networks.

The aim is to develop a methodology that allows a more flexible heat demand model and an accurate function that describes the relationship between outdoor temperature and heat demand for various types of consumers. It should be sufficiently generalized to require only substation operational data for simulation regardless of the network type. To the best of the authors' knowledge, to achieve this aim, energy delivery/consumption from the supply and return pipes has never been analyzed with the help of (a) the set of statistical and regression formulas and (b) operational data from an HTDH network.

## 2. Materials and Methods

This study aims to evaluate the performance of DH substations with a focus on return temperature efficiency, hydraulic balancing, and energy losses. The primary objectives include assessing the impact of substation configurations on heating performance, analyzing seasonal variations in supply and return temperatures, and identifying key inefficiencies in DH networks. The study employs a combination of operational data analysis, system modeling, and comparative evaluation against established performance benchmarks.

Operational data were sourced from a central SCADA system, covering key parameters such as supply and return temperatures, flow rates, and outdoor temperature fluctuations. The dataset spans multiple seasons to capture variations in heating demand and efficiency. Additional data, including building characteristics and heat consumption patterns, were gathered to support the analysis of substation performance across different building types.

The study employs a computational model to simulate DH system performance under various conditions. The model integrates real-world data with predefined system parameters, including substation configurations, mass flow rates, and thermal losses. To enhance accuracy, the model accounts for known inefficiencies such as poor mixing, heat losses due to insulation deficiencies, and delayed temperature adjustments in remote substations. Comparative modeling was performed using existing methodologies, such as Vandermeulen et al. [23] and Kurek et al. [17], to validate findings.

Model validation was conducted by comparing simulated results with recorded SCADA data. Key performance indicators (KPIs) such as return temperature deviation, energy efficiency, and heat distribution accuracy were analyzed. A sensitivity analysis was performed to assess the impact of varying input parameters, including supply temperature adjustments, mass flow regulation, and seasonal variations. Findings were cross-validated using industry benchmarks and previous studies to ensure consistency and reliability.

The traditional modeling approach assumes that heat demand for any given conditions (outdoor temperature) is stated as

$$Q^{T} = Q^{T}_{P} \cdot \frac{t_{in} - t_{out}}{t_{in} - DOT} \tag{1}$$

where $Q^{T}{}_{p}$ is design heat demand (at design outdoor temperature) [MW];

$t_{in}$ is the indoor temperature [°C];

$t_{out}$ is the outdoor temperature [°C];

DOT is the design outdoor temperature [°C].

Supply and return temperatures, respectively,

$$t^{sup}_{w} = t_{in} + \Delta t' \overline{Q}^{0.8} + \frac{\overline{Q} \cdot \theta'}{2} \tag{2}$$

$$t^{ret}_{w} = t_{in} + \Delta t' \overline{Q}^{0.8} - \frac{\overline{Q} \cdot \theta'}{2} \tag{3}$$

where $\Delta t'$ is the annual average temperature difference between the mean radiator temperature $\frac{t^{sup}_{wp} + t^{ret}_{wp}}{2}$ and indoor temperature $t_{in}$ [°C]

$$\Delta t' = \frac{t^{sup}_{wp} + t^{ret}_{wp}}{2} - t_{in} \tag{4}$$

where $t^{sup}_{wp}$ and $t^{ret}_{wp}$ are the design supply and return temperatures within an SH system.

The relative heat demand $\overline{Q}$ [-] is

$$\overline{Q} = \frac{Q^T}{Q_P^T} = \frac{(t_{in} - t_{out})}{(t_{in} - DOT)} \tag{5}$$

$\theta'$ temperature difference within an SH system [°C]:

$$\theta' = t_{wp}^{sup} - t_{wp}^{ret} \tag{6}$$

Mass flow rate is

$$G_w^K = \frac{Q_P^T}{C_w \cdot (t_{wp}^{sup} + t_{wp}^{ret})} \tag{7}$$

where $C_w = 4.187$ is the specific heat capacity of water [kJ/kgK].

True value of heat demand for a particular day is better described as

$$\widetilde{X}_i = X_i + \nabla_i \tag{8}$$

where $X_i$ is recorded (raw) heat demand for hour i; $\nabla_i$ error for hour i.

The average heat demand is

$$\overline{X} = \frac{\sum_{i=1}^{n} \widetilde{X}_i}{n} \tag{9}$$

where n is the number of the data points (hours, days, etc.).

The standard deviation of heat demand is

$$S_X = \sqrt{\frac{\sum_{i=1}^{n} \left(\widetilde{X}_i - \overline{X}\right)^2}{n-1}} \tag{10}$$

while sampling variance is

$$S_{\overline{X}} = \frac{S_X}{\sqrt{n}} = \sqrt{\frac{\sum_{i=1}^{n} \left(\widetilde{X}_i - \overline{X}\right)^2}{n(n-1)}} \tag{11}$$

To calculate the size of a bin, Equation (12) is suggested.

$$\Delta X = \frac{\widetilde{X}_{max} - \widetilde{X}_{min}}{r} \tag{12}$$

The (non-central) chi-squared distribution is

$$\chi_i^2 = \frac{(m_i - nP_i)^2}{nP_i} \tag{13}$$

Degrees of freedom k depend on the number of bins r

$$k = r - 3 \tag{14}$$

If $\chi^2_{k;\frac{1}{2}q} < \chi^2_k \le \chi^2_{k;1-\frac{1}{2}q}$, then the distribution is symmetric and unimodal. That means that Student's factor might be applied.

Frequency of heat demand with a value within the corresponding bin

$$P_i = \int_{\widetilde{X}_i}^{\widetilde{X}_{i+1}} p_x(x)dx = \Phi\left(\frac{\widetilde{X}_{i+1} - \overline{X}}{S_X}\right) - \Phi\left(\frac{\widetilde{X}_i - \overline{X}}{S_X}\right) \tag{15}$$

where $\widetilde{X}_i$, $\widetilde{X}_{i+1}$ are left and right bin boundaries, respectively; $\Phi(z)$ is the probability density for the normal distribution,

$$\Phi(z) = \frac{1}{\sqrt{2\pi}} \int_{-\infty}^{z} e^{-\frac{1}{2}t^2} dt \tag{16}$$

Abnormal values of heat demand were found as

$$\nu_1 = \frac{\widetilde{X}_{max} - \overline{X}}{S_X} \text{ and } \nu_2 = \frac{\overline{X} - \widetilde{X}_{min}}{S_X}, \tag{17}$$

Calculated $\nu_1$ and $\nu_2$ values are then compared to the threshold values of $\nu$. For every acceptable level of error (e.g., 3%, 7%, 10%), these threshold values can be extrapolated using Table 1.

**Table 1.** Threshold $\nu$-values.

| Number of Data Points (Hours, Days) n | Acceptable Accuracy of Modeling 10% | Acceptable Accuracy of Modeling 5% |
|---|---|---|
| 3 | 1.414 | 1.414 |
| 4 | 1.710 | 1.728 |
| 5 | 1.917 | 1.972 |
| <...> | <...> | <...> |
| 10 | 2.414 | 2.616 |
| 11 | 2.470 | 2.689 |
| <...> | <...> | <...> |
| 28 | 2.929 | 3.258 |
| <...> | <...> | <...> |
| 39 | 3.065 | 3.413 |
| 40 | 3.075 | 3.424 |
| <...> | <...> | <...> |
| 52 | 3.175 | 3.534 |
| <...> | <...> | <...> |

With the help of Table 1, the threshold values can be extrapolated for any number of the data points n. If $\nu_1$ or $\nu_2$ are higher than $\nu$, then the corresponding $\widetilde{X}_{max}$ or $\widetilde{X}_{min}$ are called abnormal values of specific heat demand.

Accuracy is defined as

$$\delta = t_p S_{\overline{X}} \tag{18}$$

where $t_p$ is the Student factor for k set as

$$k = n - 1, \tag{19}$$

Systematic error is

$$\theta = k\sqrt{\sum_{i=1}^{m} \theta_i^2}, \tag{20}$$

where $\theta_i$ are bin boundaries; k is the adjusting factor, set to 1.1 if an acceptable modeling error is 10% or to 1.4 for an error of 5%.

If

$$\frac{\theta}{S_{\overline{X}}} < 0.8 \tag{21}$$

then

$$\Delta = \delta \tag{22}$$

If

$$\frac{\theta}{S_{\overline{X}}} > 8, \tag{23}$$

then

$$\Delta = \theta, \tag{24}$$

otherwise

$$\Delta = KS_{\Sigma}, \tag{25}$$

where K is the ratio of systematic error and standard deviation of heat demand:

$$K = \frac{\delta + \theta}{S_{\overline{X}} + \sqrt{\sum_{i=1}^{m} \frac{\theta_i^2}{3}}}, \tag{26}$$

$S_{\Sigma}$ standard score:

$$S_{\Sigma} = \sqrt{\sum_{i=1}^{m} \frac{\theta_i^2}{3} + S_{\overline{X}}^2}. \tag{27}$$

Heat demand for every building for any given temperature range is then defined as

$$X = \overline{X} \pm \Delta, \tag{28}$$

To compare, Harney et al. [9] calculated heat demand using the EnergyPlus model.

To sum up, input parameters applied for modeling are outdoor, supply/return temperatures, and flow rates at both local (substation) and central (DH plant) levels.

## 3. Case Study

This paper presents two case studies to uncover the advantages of using a novel modeling approach. Due to the similar solutions encountered in low-temperature district heating or cooling worldwide, case studies are presumed quite indicative and equally acceptable for any individual application worldwide. Both are DH systems in Omsk, Russia. The first is the set of districts connected to a large CHP plant (CHPP#5) with a higher-than-supposed return temperature, exacerbated by its high variation throughout the building stock. After measuring network temperatures and flow rates, the highest return temperature and the lowest modeling accuracy were recorded in sections that mainly serve residential buildings. Therefore, the second case study is a district comprising a section of a traditional second generation HTDH system in Omsk. Its design heat demand is set to 250 MW and reasoned to circa 15,000 inhabitants living with a designed indoor temperature of 20 °C and a primary network temperature regime of 150/70, which has never been met. The geographical arrangement of the buildings, together with the configuration of a transmission network, is illustrated in Figure 1.

**Figure 1.** Scheme of the studied DH system in the Omsk city (Russia). Residential neighborhoods are yellow. Industrial zones and business parks are green. The DH network is presented as blue lines. The considered section of the DH network starting from CHPP#5 ('ТЭЦ-5') is highlighted in red.

This case study complies well with the current international development direction to shift towards higher energy efficiency in northern cold climates for rural heating in line with 4GDH and LTDH concepts. The building stock comprises 190 buildings, including grocery and hardware shops, office buildings, and three malls. The district's choice stems from it being a typical one with a significant share of residential buildings. They are responsible for more than 72% of the total heat deliveries of the DH network studied in this paper.

The buildings analyzed in this case study represent a range of typologies, from high-rise buildings to single-family houses, with varying insulation levels and heating systems. These buildings are located in several regions across Eastern Europe and Russia, where DH is prevalent.

The buildings in this study span a wide range of ages and types. The oldest buildings, constructed before the 1980s, comprise about 30% of the case study population. These buildings typically have poor insulation and are heated by centralized DH systems. Many buildings are inefficient due to outdated heating systems, leading to high heat demand and greater reliance on DH. The second group includes buildings constructed between the 1980s and 2000s, representing 40% of the case study buildings. These buildings generally have moderate insulation levels and less efficient heating systems than newer models. The remaining 30% of the buildings are modern, having been constructed after 2010. These newer buildings feature energy-efficient designs with high-quality insulation, modern heating systems, and, in some cases, renewable energy sources, such as solar panels for water heating. These buildings have lower heat demands due to their higher thermal efficiency.

Insulation levels are a critical factor influencing heating demand. About 50% of the buildings in the study have low insulation, typically those built before 1980, which results in higher heat losses and greater energy consumption. Around 40% of the buildings have moderate insulation, mostly those built in the 1980s to 2000s. In comparison, the remaining 10% of the buildings are highly insulated, which reduces their heating demand and improves energy efficiency.

The heating systems in these buildings vary significantly. Most older and mid-aged buildings rely on centralized DH systems, which distribute heat from a central source to multiple buildings through a network. Modern buildings, however, are more likely to use individual heating systems, such as gas-fired boilers or heat pumps, which are more energy-efficient and allow for better control of heating levels. These systems are generally more adaptable to the specific needs of each building, resulting in lower energy consumption and cost.

### 3.1. Data Collection

A comprehensive data collection approach was employed to analyze the energy performance of these buildings and their respective DH systems. With hourly resolution for heating demand and environmental parameters, data were gathered from utility companies, local heating providers, and building energy management systems. This dataset contains mean daily temperatures and flow rates recorded automatically at each line of a DH plant for more than 2 years, from 1 January 2020 until 31 December 2021. Unlike now, we used to assume the processing to bring together the makeup rate of all lines for identical periods [29]. Furthermore, that analysis was based on the dataset with hourly measured data. The heat supply and makeup water demand are analyzed for these periods separately. To compare, Volkova et al.'s [30] input parameters used for simulation were outdoor, supply, and return temperatures.

### 3.2. Modeling Framework

The modeling framework used for this study integrates building energy performance models with DH system simulations. These models consider building characteristics such as insulation levels, heating system types, and occupancy patterns. The simulation framework was designed to predict heating demand under different operational scenarios, including variations in outdoor temperature, energy prices, and the adoption of renewable energy technologies.

### 3.3. Validation Metrics

A range of metrics was used to validate the models, including the standard deviation and the total error, comparing the predicted heating demands with actual consumption data from the buildings over the study period. These metrics provide insight into the accuracy and reliability of the simulations, helping to refine the models and ensure that the predictions align with real-world performance. The validation process also includes sensitivity analysis to determine how different building characteristics, such as insulation levels and heating systems, impact the overall heating demand and system efficiency.

## 4. Results

It is relevant to note the potential to utilize all the delivered energy efficiently, which shows the ratio of exergy used to exergy delivered from the distribution system and depends on the substation configuration. Bearing in mind substations equipped with a pump, we start analyses in this section of a DH system with the simplified type of substation. They include throttle devices only (such as an orifice flange or a hydroelevator [31]) and are widespread in Russia, China, post-Soviet countries, and Eastern Europe. A substation with a hydroelevator and a mixing shunt is illustrated in Figure 2.

Figure 2 illustrates a substation with a hydroelevator and a mixing shunt, critical components in DH systems, particularly in regions like Russia, China, and Eastern Europe. These components regulate temperature and flow, influencing heating performance and occupant comfort. The efficiency of such configurations directly impacts the delivery of heat energy, making it essential to assess their role in ensuring optimal thermal conditions within buildings.

Most of the year, supply temperature varies between 70 and 100 °C, while the total flow rate experiences less significant fluctuations. Figure 3 shows that the total flow rate is higher for the days with higher DHW use (i.e., weekdays) and remains almost the same for others.

**Figure 2.** An example of key components of a typical substation: the hydroelevator and a mixing shunt are red.

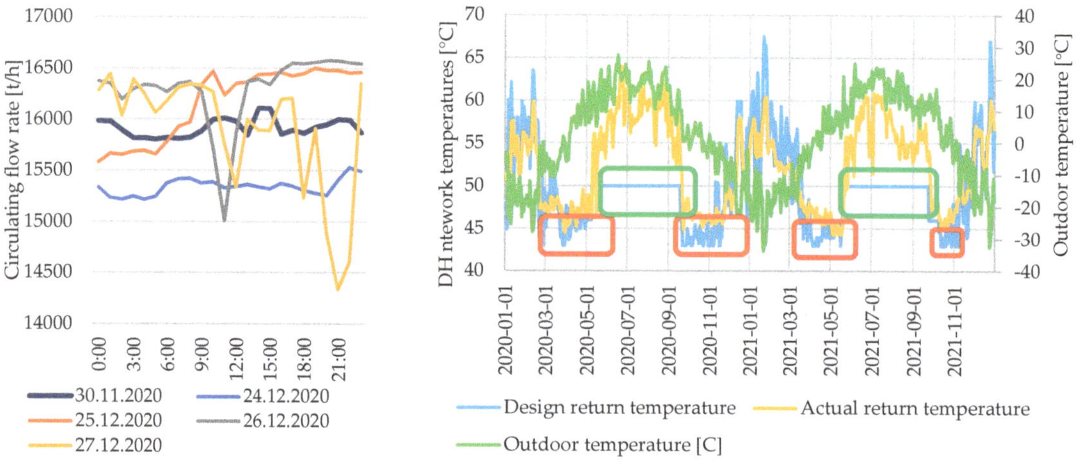


(**a**) (**b**)

**Figure 3.** Operational data obtained from a central SCADA system: (**a**) CHP flow rate and (**b**) return/outdoor temperature profiles.

Most of the year, supply temperatures fluctuate between 70 and 100 °C, while total flow rates vary based on demand. Figure 3 presents operational data from a central SCADA system, showing (a) CHP flow rates and (b) return and outdoor temperature profiles.

The data highlight the significant impact of outdoor temperatures on heating system performance. For instance, return temperatures are higher on warmer days, reducing heating efficiency. This effect is particularly noticeable during seasonal transitions (fall and spring), where excessive indoor temperatures result in energy wastage due to open windows. The hourly variation profiles demonstrate that heating loads are not evenly distributed throughout the day, influencing supply and return temperature stability.

It is worth noting that the hourly variation profile is not similar for all the days. For instance, peaks and valleys were more distinct on the 26th and 27th of December 2020 (refer to Figure 3a), i.e., for weekdays. However, the daily difference is attributed chiefly to the days with high outdoor temperatures (−5~+5 °C) when CHP production and system capacity are lower because of higher circulating flow and higher return temperature. During winter, the outdoor temperature ranges mostly between −20 and 0 °C according to the dataset recorded at the SCADA system values. There are several zones describing the correlation between actual and design return temperatures. The first zone is defined by the outdoor temperature ranging from −30 to −15 °C when the supply temperature is about 90–100 °C and SH-related consumption is high. This supply temperature is not enough—it should be 120 °C and above. Moreover, due to high heat losses, a temperature drop of up to 15 °C is registered on the way to the most remote substation. That means the return temperature also depends on the location of a substation. The second factor is irregular hourly temperature fluctuations due to consumers' uneven heating loads throughout the day. The second period occurs during falls and springs, when the supply temperature exceeds 70 °C due to DHW consumption, with excessive heat being released from the open windows to prevent indoor temperatures from rising. The latter time period is mostly responsible for inefficient energy use.

As shown in Figure 3b, high return temperatures lead to inefficiencies in DH systems, affecting energy distribution and electricity generation at CHP plants. The correlation between return temperature levels and substation efficiency has been well documented, including by Farouq et al. [26], who emphasize the importance of monitoring return temperature as a control metric in Sweden. The elevated return temperatures observed in this study suggest inefficiencies in mixing and flow regulation, particularly in older or poorly maintained substations.

High return temperatures (circa 11% higher) heavily affect the overall operational efficiency of the DH system and deteriorate electricity generation at a CHP plant. Furthermore, the return temperature is a decent parameter to check a substation for poor operation. In addition, according to Farouq et al. [26], the analysis of return temperature levels is a commonly applied control check on the operational efficiency of substations in Sweden.

To compare, Vandermeulen et al. [23] established a model where an average substation represents a whole residential area containing n buildings. Hence, the requested mass flow rate of the aggregated substation is multiplied by n, while the supply and return temperatures are assumed constant throughout the year. In [32], a cooling return temperature of 30 °C is assumed for the district energy substations of data centers. Moreover, all supply and return temperatures are fixed during the year again, which is hardly true.

However, Figure 3 shows a higher-than-should-be return temperature, which is reasoned to the fact that the optimal supply temperature reaches the substation a few hours after leaving a central plant. In developing countries, this ordinary consideration is further exacerbated by the poor condition of the DH network (Figure 4), which is responsible for high-temperature drops and delays.

**Figure 4.** An example of high distribution heat losses due to poor insolation.

Figure 4 demonstrates the significant heat losses in DH networks due to poor insulation. These losses contribute to increased energy demand and reduced system efficiency, ultimately affecting heating performance and thermal comfort. In developing countries, where DH networks often suffer from infrastructure degradation, temperature drops and delays exacerbate the problem. A temperature drop of up to 15 °C has been recorded on the way to remote substations, meaning that buildings in different locations within the network experience varying levels of heating performance. Such disparities underscore the need for targeted interventions, such as improved insulation and better hydraulic balancing strategies.

For its part, that results in high return temperature, hydraulic balancing issues, and low modeling accuracy. However, a 1-degree decrease is beneficial since it results in 22% lower heat losses (regarding $W/m^3$ unit) [33]. A possible reason is poor mixing at DH substations (mainly in urban districts), as throttle devices are increased in diameter, and substations are tuned during cold winter days.

However, performance adjustments in a new substation can be achieved without damaging the hydraulic balancing of the network. The problem is aggravated when the supply temperature or mass flow increases. Prosumers, plant, and a distribution network can adjust their mass flow rates without affecting the hydraulics of other prosumers. The highest difference between actual and design return temperature is registered during the coldest days (refer to Figure 3b). Moreover, the overall trend is negative because the monthly average difference increases, cf. 5.3 °C in January 2020 and 6.4 °C in January 2021.

Figure 5 visualizes the contribution of different buildings in total heat demand distributed by their total heated area.

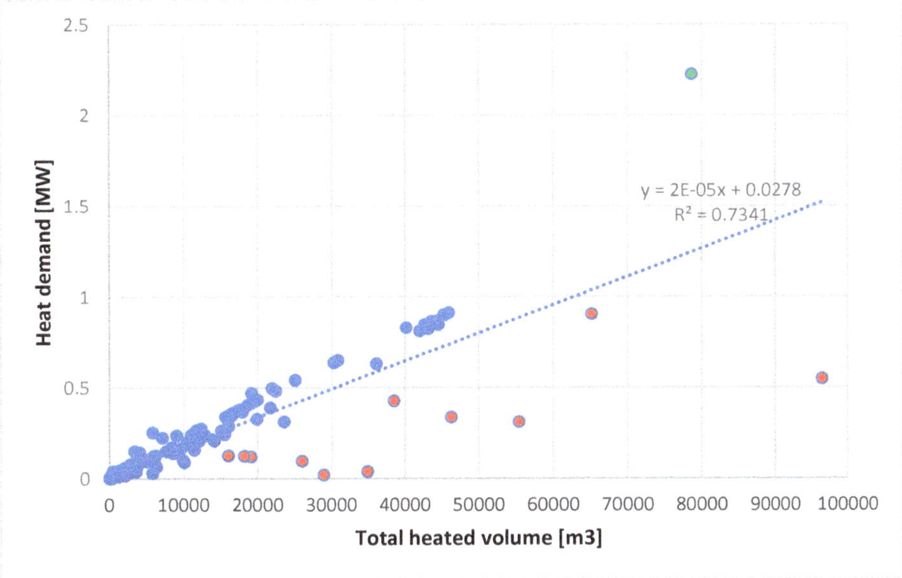


**Figure 5.** Abnormal values are colored in red. In particular, dynamically deployed substations achieve higher energy savings in the DH system than their linearly deployed counterparts.

Most abnormal consumers have a total heated volume of 15,000 $m^3$ and above. The possible hypothesis is that the higher the building volume, the lower the concentration of inhabitants/employees is. The best linear model, the partially linear autoregression with R2 of 0.7341, has formula as $Q = 2 \times 10^{-5} t_{out} + 0.0278$. To compare, Kurek et al. [17] obtained input features with the forward feature selection algorithm. However, similar correlations and selected weather features are also listed in [17].

Figure 5 highlights the contribution of different building types to total heat demand, with abnormal values indicated in red. The data reveal that larger buildings (above 15,000 $m^3$ in heated volume) tend to have lower inhabitant densities, which affects heating patterns. The figure shows that dynamically deployed substations achieve higher energy savings than linearly deployed ones. These findings align with prior studies [17], indicating that system design and deployment strategies are crucial in optimizing energy efficiency.

The final conclusion is that 12 consumers (6%) are responsible for low accuracy, while 11 seem energy-efficient. However, seven are office buildings, shops, and car parks with indoor temperatures of 12–18 °C, which is the main reason for low energy consumption. Only four buildings are new energy-efficient residential buildings, all equipped with pumps, and the rest with throttle devices. Excluding these four buildings, all the others have almost similar energy density—specific energy consumption per surface area. Linear heat density is one of the key indicators.

To compare, Patureau et al. [6] use linear heat density as a threshold to divide consumers into those with sufficient and insufficient values. For one of the major cities, the former is 8.3 km from the city center, while the latter is farther than 8.3 km but closer than 28.7 km away. As with us, locations favorable for a DH system contain mostly old buildings and feature a high share of heat demand from the residential sector.

At the same time, the newer buildings in the residential sector are much more energy efficient but exhibit a more fluctuating and stochastic heating pattern. The reason is that large buildings do not behave as a single zone with steady-state mixing and uniform indoor temperature distribution. In actual buildings, the solar gains are not distributed evenly across all the indoor spaces.

Figure 6 clarifies the large difference in the heat demand levels, which is typical among consumers.

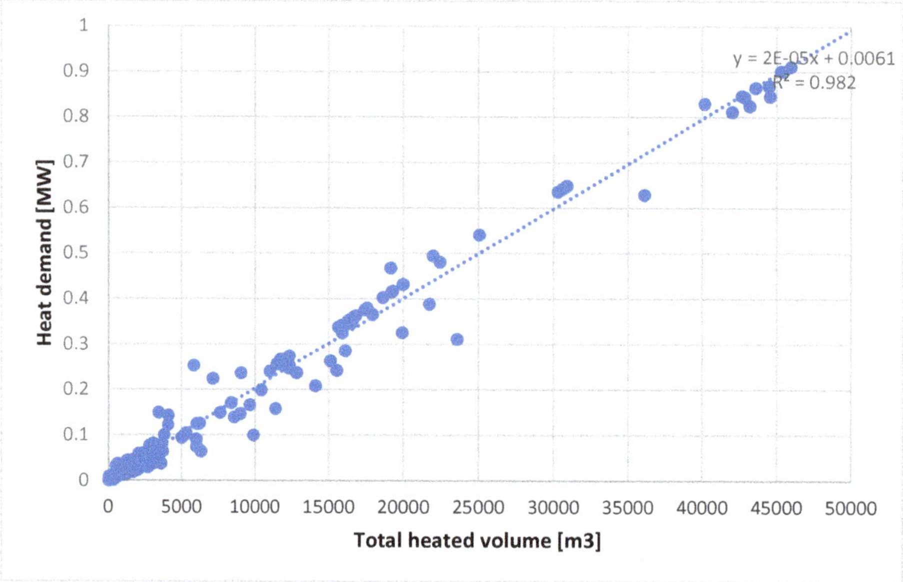


**Figure 6.** Abnormal values of heat demand have been cropped. That leads to higher accuracy and more evenly distributed SH demand in the modeled districts.

At the same time, most of the buildings in the case study area have an overall high but steady specific energy consumption. Excluding abnormal heat demand profile increases the accuracy and decreases the number of heat demand peaks. That takes any deviations from design values into account and makes it meaningful to calculate the accuracy of a novel or an error of a traditional method. At the same time, that does not cause any fundamental changes to the substation thermal response; therefore, this form of aggregation is considered viable for this research. Furthermore, DHW consumption and solar radiation do not influence the indoor comfort of all spaces in the same manner over the day, as the space normally has differences in the number of inhabitants/occupants and windows, e.g., with one or two orientations. With no abnormal buildings, high modeling accuracy (98.2%) is easily achieved using regression $y = 2 \times 10^{-5} + 0.0061$. If so, that results in a substantial improvement of temperature control performance compared to the traditional scenario, when the temperature control is performed with higher error due to dynamically deployed factors. Hence, it was possible to compare the recorded energy consumption for SH and DHW with the building footprint and height. For instance, 50% of consumers have a heated volume of 5000 $m^3$ and below, while 85% have a heated volume below 25,000 $m^3$. With no abnormal values, the total heat consumption of the DH system decreases by 7.3%.

The analysis showed that a quite high coefficient of determination ensures more efficient SH heat use. Modernizing calculations according to the heat demand of regular consumers may ease the design guidelines. Other upsides are establishing a benchmark for efficient energy use and easing decision-making when considering network expansion. The regulation system trained with such a dataset can adapt to the different requirements in different conditions, achieving higher accuracy than following the reference method. The control strategy becomes more solid and sound while the issue of high return temperature is alleviated.

To compare, Ivanko et al. [18] considered the data for the energy consumption in a hotel in Norway, which has a substation with two separate heat meters for the SH and DHW. Their model had R2 of 0.97 and 0.76 for SH and DHW heat use, respectively. The difference with [6] is that each district comprises an equivalent number of occupants (for the residential areas) or staff (for the commercial ones).

The operational heat demand of Omsk, Russia, for 2021 (frequency histogram shown in Figure 7) is input for each building type to the average building demand profile.

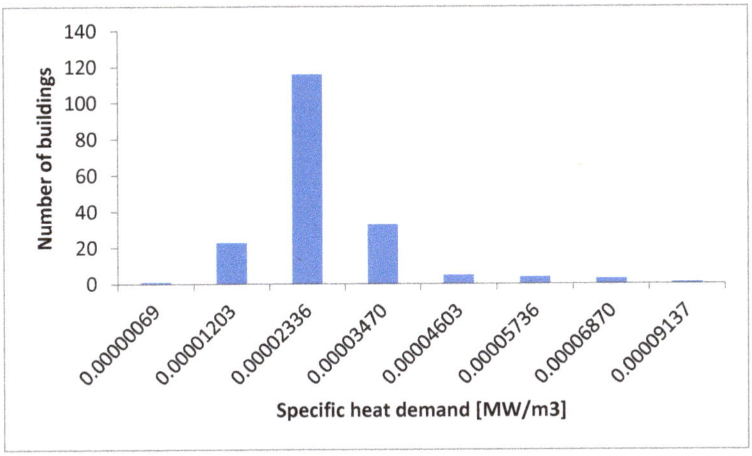


**Figure 7.** Frequency distribution of specific heat demand.

In total, 62% of consumers have specific values ranging between $1.2 \times 10^{-5}$ and $2.3 \times 10^{-5}$ MW/m$^3$. The total energy consumption of the 20–30 old buildings remains almost unchanged, resulting in the same specific heat demand value. However, poor network conditions increase distribution losses (and thus fuel consumption) in the heating network's branch to the oldest buildings. We see a reduction here, which is well represented by 0.00002336 $\pm$ 8.2% according to Equation (28). Due to further demand-side measures and new buildings constructed in the DH area, average specific heat demand is expected to decrease by 2–3% yearly.

The results suggest two possible directions. One is to use the specific load analyses and the nascent network analyses against the districts of other cities when modeling. For instance, for neighborhoods that may switch to such configurations, the district-level energy demand data are mapped to a GIS map using the total volume [m$^3$] or surface area [m$^2$] ratio for each archetype. Another way is to configure an in-district heat distribution system 'as is' but fix current malfunctions on the demand side, where the difference is the largest.

Now, all the buildings are divided into the following reference groups (archetypes): old traditional ones (>0.00003470 MW/m$^3$), new energy efficient (>0.00005736, but <0.00003470 MW/m$^3$), and industrial (<0.00005736 MW/m$^3$). To compare, in [22], all the buildings are divided into similar building types: residential, commercial (<0.15 GWh/year), and industrial (>0.15 GWh/year). Specific heat demand in energy use per unit of heated volume is also more accurate than traditional indicators such as linear heat density. Based on linear heat density, Patureau et al. [6] determine the area acceptable for placing a DH system—its average surface is 3.7 km$^2$, while an area of 14.2 km$^2$ has insufficient linear heat density. These conclusions may become incorrect once the MW/m$^3$ indicator is applied. Another relevant comparison is the potential for improvement due to energy efficiency measures on the consumers' side (in the LHCD area); in [4], a value of 2.4% is reported.

When studying the archetype of old traditional buildings, the most evident trend in Figure 8 is a significant increase in energy consumption on the coldest winter days.

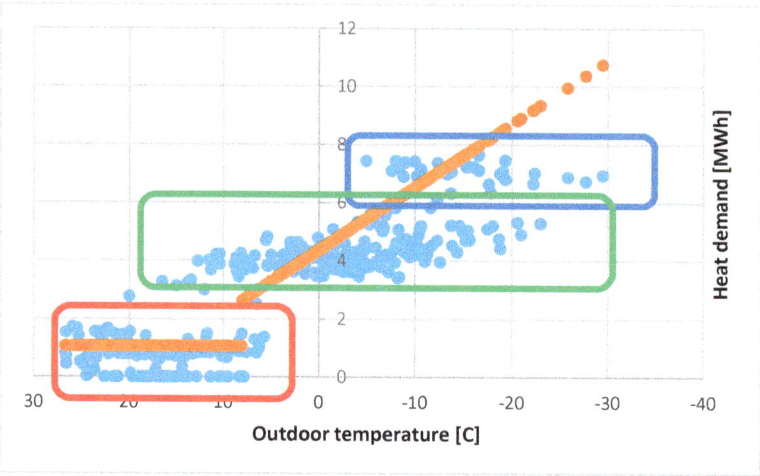


**Figure 8.** Correlation between actual (blue points) and design (orange ones) heat demand and outdoor temperature. Zones outlined in blue and green represent periods when the flow rate could be cut since the actual heat demand for old buildings is much lower. The days with high average outdoor temperatures ranging from 5 to 30 °C are the days when DHW demand was the most important factor (red).

The annual energy consumption is 1.063 GWh with a peak level of 7.64 MW. The annual average demand is 3.19 MW, with a dispersion of 4.26 MW. In a residential building, heat is consumed from the DH system to cover both SH and DHW demands. Different systems' operation modes are also noticed, as indicated by various energy demand levels. There are three temperature ranges when heat demand does not depend on outdoor temperature but could be related to the building's substation tuning or occupancy scheduling. Therefore, most of the time, the temperature of return water fed back into the network is larger than it should be. Moreover, heat demand is usually larger than the design curve (orange), while accuracy is pretty low.

That means the traditional modeling approach (orange dots) for an old building is incorrect. The heat demand of a building is better simulated with a discrete regression like

$$\textit{if 3-day average } t_{out} < 0.85DOT\textit{, then } Q = 0.75Q_{des},$$

$$\textit{otherwise } Q = 0.5Q_{des}$$

The time period with the highest energy demand (blue zone) is winter, with a peak SH energy demand of 7.8 kWh, accounting for 163.4% of average demand at outdoor temperatures of 0~−30 °C. Therefore, during the coldest winter days of peak heat consumption, the appropriate regulation strategy is expected to reduce the difference between energy delivery and consumption until it becomes negative at any specific time step. It means that this zone depends on thermal inertia as the peak demand correlates with maximum heat delivery and energy accumulated within building envelopes for the normal consumption profile (green zone). It may also vary from district to district depending on its location with respect to a DH plant and flow velocity. However, in addition to distance, heat losses, and flow velocity, many more factors can affect energy efficiency annually. There are even more factors that may have an impact on heat demand during several years of operation [34].

Moderate-cold days represent a time range where enough heat is supplied for SH purposes, and substations could be set back to normal regulation. That is due to external physical factors such as higher outdoor temperatures (−10~+5 °C mostly), more intensive solar radiation, and no winds, which may naturally increase the building's indoor temperature and thus decrease the SH consumption during this time. The SH demand is much lower in the fall and spring months. However, heat demand does not reduce to zero in March, April, and September, even for outdoor temperatures above 0 °C.

Outlined with red, the range indicates the time period when heat is supplied for DHW only and the SH systems are forcibly turned off. To compare, Guelpa [25] points out that the excessive amount of heat energy required over the peak demand (0.4 GWh) is mainly consumed to heat the system's thermal masses (0.33 GWh). In [9], the period with the highest heat consumption is winter, with a total SH energy demand of 276 kWh, accounting for 26.4% of annual demand. To sum up, in [9,25], the SH demand is much higher in the winter months while—the same as here—heat consumption plummets to zero in summertime.

The novel model outperforms the traditional approach in all the simulated periods with the sampling variance Sx′ (Equation (11)) of 0.68 MW for the winter season, 0.23 MW for the summer season, and 0.51 MW for the intermediate season. At the same time, the MAPE is 8.4% for the winter season, 4.3% for the summer season, and 6.9% for the intermediate season. To compare, in [17], MAPE was 2.6/3.2% for the winter season, 4/5.8% for the summer season, and 7.4/8.5% for the intermediate one for an ANN with autoregressive input forecast/Ridge Regression, respectively. Much higher MAPE in winter is due to the inability to keep proper supply temperature according to outdoor temperature due to high heat losses, hydraulic balancing issues, and worn-out equipment. Wirtz et al. [32] report an annual heating demand of 6.36 GWh with a peak demand of 2.01 MW. The annual cooling demand is 10.04 GWh and shows a peak of 2.42 MW. Omsk has no district cooling network, even though building stock comprises university laboratories, malls, office buildings, and data centers—the same with [32].

Buildings with pump-equipped substations can be modeled with the help of simple regression, like y = kx + m (Figure 9).

The maximum energy consumption for the new building is 11.66 MWh, which is only 31.2% higher than the winter average. The deviation (Equation (10)) is also much lower, accounting for only 2.7 MWh. The difference with the previous Figure 8 means that there is almost no requirement to provide heat to the system components during peak conditions by increasing the SH supply temperature (Equation (2)) before the coldest days. We found a correlation between outdoor temperature and heat demand. Based on the identified correlation coefficient (0.89), it is relevant to conclude that outdoor temperature has a main role in supply and return temperatures in the outdoor temperature range of −30~+8 °C. External conditions also affect the system but to a lesser extent. It is also relevant to compare to our previous paper, covering regulation of supply temperature at the outlets of a CHP plant over 2 years where trend lines were $-799.04t_{out} + 26{,}581$ and $-820.23t_{out} + 26{,}630$ curves. There was no harmful effect of overheating due to supply and return temperature levels being lower. The overheating effect is checked against the design heat demand for the same outdoor temperature. Although the impact of other factors is limited, the difference between design and actual heat demand is higher in the range of low outdoor temperatures.

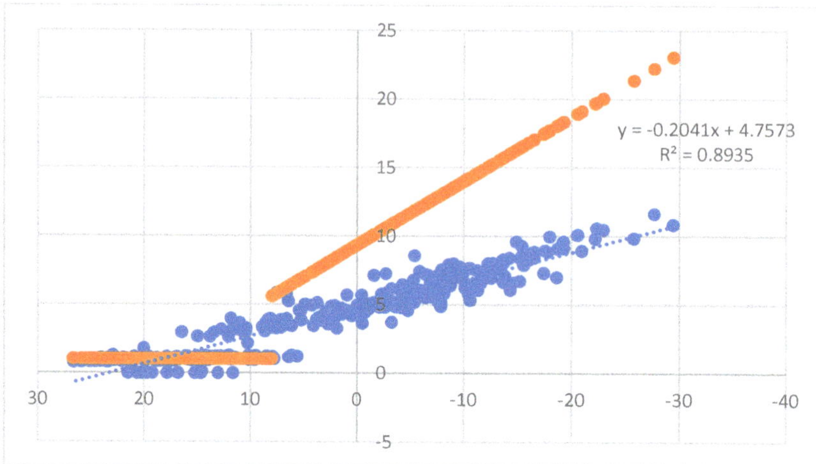


**Figure 9.** Heat demand given mainly in the new buildings. Please pay attention to the days with low outdoor temperatures ranging from −30 to 0 °C. Operational data points are presented in blue, while regression-based points for the same values of outdoor temperature are in orange.

There is an even clearer deviation between operational and design heat demand in new buildings, cf. Figure 8 for old buildings. This discrepancy is attributed to the fact that the SH systems have difficulties keeping indoor temperatures due to the aging of building envelopes. Therefore, the secondary temperatures are generally increased by manually tuning the SH system set points. In modern buildings, heat demand above the buckling point of +0.3 °C is set much lower due to DSMs. It also makes it much closer to design values within this range. DSMs is a term consolidating efforts taken to manipulate the consumption profiles by enhancing the performance of building envelopes, load management, and adapting the flexible demand profiles [35].

Decreasing the difference between operational and design heat demand in buildings is believed to promote the LTDH concept. If a high-temperature difference is achieved, the flow rates within the distribution are lower; therefore, the investments and operational costs to cover additional electricity consumption for pumping are also lower. Like the present paper, Romanov et al. [7] declare better potential for the lower temperature charts', once a better understanding of heat demand and higher modeling accuracy is achieved.

When the SH and DHW systems work simultaneously, the DH system is considered to follow heat mode. It consumes only the energy required to ensure indoor comfort from the DHC, provided that the supply and return temperatures are enough to cover the DHW requirement of buildings. This diminishing evolution is gradual and observed over the daytime steps.

To compare, in [8], the maximum peak demand of the combined district cooling and heating system was 64.2 kW for cooling and 12.4 kW for heating. That brings the idea of applying the method in detail to district cooling systems in the first place. It should also be emphasized that the actual demand profile and the simulated one are never totally similar due to the impact of uneven factors. For instance, the DH system inevitably accumulates some heat energy, thereby leveling heat demand fluctuations [36–38].

The aggregation is suggested as follows: as the heat demand profiles of each archetype became known (both old and new buildings), the average heat demand for each outdoor temperature is calculated according to y = f(x) regression. This average heat demand is then applied to the substation model, representing a single building substation for each temperature range. Another contribution to the pool of knowledge made is that the return temperature is no longer fixed. There is a generalized methodology that considers that return temperature varies with the supply temperatures and the overall heat demand.

## 5. Discussion

This research addresses several critical gaps in the existing literature regarding the performance and energy efficiency of DH systems. Previous studies often focus on isolated aspects of building energy use or heating systems without thoroughly integrating the diverse characteristics of buildings or the operational dynamics of DH systems. By including a comprehensive dataset of case study buildings spanning a range of typologies, insulation levels, and heating systems, this study provides a more holistic understanding of how various factors influence the efficiency of DH.

One of the key findings from this research is the significant impact of building age and insulation levels on heating demand. Older buildings, especially those constructed before the 1980s, have considerably higher energy consumption due to poor insulation and outdated heating systems. This finding is consistent with the existing literature, which shows that thermal inefficiency in older buildings leads to higher heating demands, often requiring higher energy input to maintain comfortable indoor temperatures. However, this study further quantifies these effects, showing that buildings with low insulation levels consume up to 30% more energy than highly insulated buildings of similar typology. This is an important contribution as it offers specific data to help inform strategies for retrofitting and improving the energy efficiency of older building stock.

Another significant contribution of this study is its exploration of the varying performance of DH systems based on building typologies and heating systems. While much of the existing research focuses on centralized DH systems, this study highlights the growing role of individual heating systems in newer buildings. These modern systems, such as gas boilers or heat pumps, offer greater energy efficiency and more precise control over heating demand, which is not often captured in traditional DH research. By integrating centralized and individual systems into the modeling framework, this research addresses the gap in understanding how mixed heating systems perform in different building contexts.

The validation of the modeling framework also represents a significant advancement. Many prior studies have relied on theoretical models or simplified assumptions, leading to less reliable conclusions. By validating its predictions against actual heating data over a four-year period, this research provides a more accurate and trustworthy model for future studies. The use of metrics like mean absolute error (MAE) and root mean square error (RMSE) allows for a robust assessment of the model's accuracy, and the sensitivity analysis helps to pinpoint which factors (e.g., insulation levels, heating systems) most significantly affect the performance of DH systems. These methods provide a clearer picture of the uncertainties inherent in energy modeling, often limitations in previous works.

Furthermore, this study's inclusion of data over a two-year period, spanning from 2019 to 2021, addresses the gap in temporal analysis seen in earlier studies, which often focus on shorter timeframes. Including multiple years of data enables a more nuanced understanding of seasonal and long-term variations in heating demand, improving the reliability of the findings and making them more applicable to future energy policy decisions.

Regarding policy implications, the study provides valuable insights into the need for targeted interventions in renovating older buildings. It emphasizes that without addressing insulation deficiencies and improving heating systems in these buildings, the potential for reducing energy consumption in DH networks will remain limited. Additionally, the findings point to the importance of promoting more efficient heating systems in new constructions, which could further optimize energy use across DH networks.

Compared to existing studies, which may focus on either the building or the DH system in isolation, this research offers a more integrated view, considering the individual building characteristics and the broader DH infrastructure. It provides a more detailed and comprehensive approach to understanding how these systems can improve energy efficiency and offers a clear roadmap for future research.

Thermal comfort is a state of mind where occupants feel thermally satisfied with their surroundings. This involves a complex interplay of temperature, humidity (though less critical in this dry climate), air velocity, and personal factors. While individual preferences vary, this study aims to maintain indoor temperatures within a range that satisfies the majority of occupants. Specifically, a target temperature range of 20–24 °C is considered, aligning with recommendations from ASHRAE EN standards for typical office and residential buildings. These standards also guide acceptable humidity levels, which should ideally be monitored and controlled, although these data are not currently available for this study.

The regional climate of Omsk, Russia, presents significant seasonal challenges. Winter temperatures can plummet to −30 °C, requiring robust heating to maintain comfortable indoor conditions. Conversely, summer temperatures can exceed 30 °C, though district cooling is not currently implemented, making managing internal heat gains (solar radiation, equipment, occupants) crucial for preventing overheating. The transitional seasons (spring and fall) present unique challenges, as highlighted by the observation of excessive heat being released from open windows. This suggests inadequate control strategies to adapt to milder outdoor temperatures, leading to energy waste and potential discomfort due to temperature swings.

The existing data from the SCADA system (Figure 3) show a concerning trend of high return temperatures, indicating inefficiencies in heat delivery and potentially impacting thermal comfort. As mentioned in the text, poor mixing at DH substations can exacerbate this issue, leading to uneven temperature distribution within buildings. This can result in some areas being overheated while others remain underheated, compromising overall comfort. Furthermore, hydraulic balancing issues can also contribute to inconsistent heating, affecting energy efficiency and thermal comfort. Addressing these distribution network deficiencies is crucial not only for energy savings but also for ensuring occupant well-being.

Different building types within the district have varying thermal comfort requirements. Office buildings typically require tighter temperature control and higher ventilation rates due to higher occupancy densities and internal heat gains. On the other hand, residential buildings may have more flexibility in temperature settings, but occupant preferences and individual control become more important. Industrial buildings may have specific temperature and humidity requirements depending on the processes. While identifying different building archetypes (old, new, industrial), the current analysis does not explicitly address how the heating system caters to these diverse needs. Future work should include a more detailed analysis of thermal comfort requirements for each building type and how the heating system can be optimized to meet those needs.

While the current study focuses on energy consumption data, the research acknowledges the lack of direct occupant feedback regarding thermal comfort. Surveys or other methods of gathering occupant perception data should be incorporated into future studies. This would provide valuable insights into the actual thermal comfort experienced by building occupants and allow for data-driven optimization of the heating system. Without such data, the analysis assumes that indoor temperatures within the target range (20–24 °C) will generally provide acceptable thermal comfort. However, it is recognized that this is a simplification and that individual preferences may vary.

The analysis may also explore the relationship between energy efficiency measures and thermal comfort. For instance, reducing supply temperatures to improve efficiency may impact the ability of the system to maintain comfortable indoor temperatures during peak demand periods. Similarly, implementing DSMs, while beneficial for energy savings, should be carefully evaluated to ensure they do not negatively affect thermal comfort. A balanced approach is needed that considers both energy performance and occupant well-being.

To evaluate the effectiveness of the proposed model, a comparison is made against existing methods in the literature. Table 2 highlights key differences in temperature assumptions, heating efficiency, outdoor temperature impact, substation modeling, and energy loss considerations. Unlike previous models that rely on fixed temperature assumptions or simplified substation representations, the proposed approach incorporates real-world operational data, addressing key inefficiencies in DH systems.

**Table 2.** Comparison of the proposed model with existing methods regarding temperature assumptions, heating efficiency, outdoor temperature impact, substation modeling, and energy loss considerations.

| Criterion | Proposed Model | Vandermeulen et al. [23] | Farouq et al. [26] | Kurek et al. [17] |
|---|---|---|---|---|
| Temperature Assumptions | Dynamic return/supply temperature variations based on real SCADA data | Fixed supply and return temperatures year-round | Return temperature analysis used as an efficiency check | Weather-based feature selection for return temperature predictions |
| Heating Efficiency | Adjusts based on substation performance and hydraulic balancing | Aggregated substation approach for residential areas | Applied control check for operational efficiency | Uses selected weather features to predict heat demand |
| Impact of Outdoor Temperature | Correlates outdoor temperature (−30 to +5 °C) with system performance | Assumes constant supply/return temperatures | Evaluates return temperature fluctuations in Sweden | Uses a forward feature selection algorithm to choose relevant weather parameters |
| Substation Modeling | Incorporates real-world delays in optimal supply temperature reaching substations | Assumes ideal flow rate scaling for residential areas | Focuses on identifying poor-performing substations | Uses autoregression model to estimate heat demand |
| Energy Loss Consideration | Identifies high heat losses due to poor insulation and hydraulic imbalances | Does not consider real-world heat losses | Uses return temperature as an indirect efficiency indicator | Identifies correlations between temperature and heat demand |

## 6. Conclusions

This research exposes a new formulation for the operational process of an HTDH with a high share of residential consumers and buildings requiring renovation. The novelty is the introduction of statistics into a CAD-aided dynamic simulation model. Its effectiveness is demonstrated on three representative periods: summer, cold (winter), and moderate cold (fall and spring) days. The methodology yields the following key findings:

1. A minority of consumers are responsible for low modeling accuracy and, therefore, hydraulic balancing issues.
2. Office buildings, shops, and car parks with low indoor temperatures are buildings with low specific energy demand. Encountering new energy-efficient residential buildings using low specific energy demand as a criterion is highly unlikely.
3. With no abnormal buildings, higher modeling accuracy is achieved.
4. Most consumers are similar in thermal response. However, specific energy demands may differ.
5. The traditional modeling approach is incorrect. The heat demand of a building is better simulated with a discrete regression.
6. Heat demand of buildings with pump-equipped substations can be modeled with the help of simple linear regression with acceptable error.

While this study provides valuable insights into the modeling and analysis DH systems, several limitations should be acknowledged. Firstly, the modeling framework relies on the availability and accuracy of building-level data, such as insulation levels, heating system types, and occupancy patterns. Any inaccuracies in these inputs may affect the reliability of the predictions. Additionally, the study focuses primarily on a physics-based approach, which, although transparent and interpretable, may not fully capture complex nonlinear relationships present in some DH dynamics—an area where data-driven models could offer complementary value.

The temporal resolution of the data, while sufficient for the scope of this research, may also limit the ability to capture short-term fluctuations in heating demand, particularly during extreme weather events or abrupt changes in occupancy. Moreover, the case study buildings, although varied in age, typology, and heating systems, may not fully represent the broader diversity of building stocks in other regions, potentially affecting the generalizability of the findings.

Lastly, while the validation process included historical data over multiple years, further testing across different climatic zones and DH configurations would strengthen the robustness of the proposed model. Future research could also explore hybrid modeling approaches, combining physics-based and machine-learning methods to enhance predictive accuracy and adaptability.

In addition, several remarks on return temperature have been made. Accordingly, a particular advantage of the present paper is assessing heat demand within LTDH and 4GDH concepts, intending to spread this technique to other regions and support DH transformation. Additional advantages of the novel method over the traditional one are its greater support in planning, its simplicity regarding further spatial extension of the network, and its robust operation. This includes evaluating transformation measures with an easy dataset, which any DH system may provide. In future work, the possible emphasis on a more sophisticated real-life check of the applicability of the approaches and guidelines is to be presented.

**Funding:** This project received funding from VLAIO in Belgium, ICON project OPTIMESH (VLAFLX7, & FLUX50 ICON Project Collaboration Agreement—HBC.2021.0395).

**Data Availability Statement:** The data presented in this study are available on request from the corresponding author. The data are not publicly available due to confidentiality reasons.

**Conflicts of Interest:** The author declares no conflicts of interest.

## References

1. Hering, D.; Cansev, M.E.; Tamassia, E.; Xhonneux, A.; Müller, D. Temperature control of a low-temperature district heating network with Model Predictive Control and Mixed-Integer Quadratically Constrained Programming. *Energy* **2021**, *224*, 120140. [CrossRef]
2. Kilkis, B. An exergy-based minimum carbon footprint model for optimum equipment oversizing and temperature peaking in low-temperature district heating systems. *Energy* **2021**, *236*, 121339. [CrossRef]
3. Meesenburg, W.; Ommen, T.; Thorsen, J.E.; Elmegaard, B. Economic feasibility of ultra-low temperature district heating systems in newly built areas supplied by renewable energy. *Energy* **2020**, *191*, 116496. [CrossRef]
4. Ziemele, J.; Talcis, N.; Osis, U.; Dace, E. A methodology for selecting a sustainable development strategy for connecting low heat density consumers to a district heating system by cascading of heat carriers. *Energy* **2021**, *230*, 120776. [CrossRef]
5. Sommer, T.; Sulzer, M.; Wetter, M.; Sotnikov, A.; Mennel, S.; Stettler, C. The reservoir network: A new network topology for district heating and cooling. *Energy* **2020**, *199*, 117418. [CrossRef]
6. Patureau, R.; Tran, C.T.; Gavan, V.; Stabat, P. The new generation of District heating & cooling networks and their potential development in France. *Energy* **2021**, *236*, 121477.
7. Romanov, D.; Pelda, J.; Holler, S. Technical, economic and ecological effects of lowering temperatures in the Moscow district heating system. *Energy* **2020**, *211*, 118680. [CrossRef]

8. Quirosa, G.; Torres, M.; Soltero, V.M.; Chacartegui, R. Energetic and economic analysis of decoupled strategy for heating and cooling production with $CO_2$ booster heat pumps for ultra-low temperature district network. *J. Build. Eng.* **2022**, *45*, 103538. [CrossRef]
9. Harney, P.; Gartland, D.; Murphy, F. Determining the optimum low-temperature district heating network design for a secondary network supplying a low-energy-use apartment block in Ireland. *Energy* **2020**, *192*, 116595. [CrossRef]
10. Chicherin, S. Methodology for analyzing operation data for optimum district heating (DH) system design: Ten-year data of Omsk, Russia. *Energy* **2020**, *211*, 118603. [CrossRef]
11. Brandi, S.; Piscitelli, M.S.; Martellacci, M.; Capozzoli, A. Deep reinforcement learning to optimise indoor temperature control and heating energy consumption in buildings. *Energy Build.* **2020**, *224*, 110225. [CrossRef]
12. Wojdyga, K. Predicting Heat Demand for a District Heating Systems. *Int. J. Energy Power Eng.* **2014**, *3*, 237–244. [CrossRef]
13. Morteza, A.; Yahyaeian, A.A.; Mirzaeibonehkhater, M.; Sadeghi, S.; Mohaimeni, A.; Taheri, S. Deep learning hyperparameter optimization: Application to electricity and heat demand prediction for buildings. *Energy Build.* **2023**, *289*, 113036. [CrossRef]
14. Terry, N.; Galvin, R. How do heat demand and energy consumption change when households transition from gas boilers to heat pumps in the UK. *Energy Build.* **2023**, *292*, 113183. [CrossRef]
15. Sasso, F.; Chambers, J.; Patel, M.K. Space heating demand in the office building stock: Element-based bottom-up archetype model. *Energy Build.* **2023**, *295*, 113264. [CrossRef]
16. Gonzalez-Salazar, M.; Klossek, J.; Dubucq, P.; Punde, T. Portfolio optimization in district heating: Merit order or mixed integer linear programming? *Energy* **2023**, *265*, 126277. [CrossRef]
17. Kurek, T.; Bielecki, A.; Świrski, K.; Wojdan, K.; Guzek, M.; Białek, J.; Brzozowski, R.; Serafin, R. Heat demand forecasting algorithm for a Warsaw district heating network. *Energy* **2021**, *217*, 119347. [CrossRef]
18. Ivanko, D.; Sørensen, Å.L.; Nord, N. Splitting measurements of the total heat demand in a hotel into domestic hot water and space heating heat use. *Energy* **2021**, *219*, 119685. [CrossRef]
19. Kristensen, M.H.; Hedegaard, R.E.; Petersen, S. Long-term forecasting of hourly district heating loads in urban areas using hierarchical archetype modeling. *Energy* **2020**, *201*, 117687. [CrossRef]
20. Barone, G.; Buonomano, A.; Forzano, C.; Palombo, A. A novel dynamic simulation model for the thermo-economic analysis and optimisation of district heating systems. *Energy Convers Manag.* **2020**, *220*, 113052. [CrossRef]
21. Jangsten, M.; Lindholm, T.; Dalenbäck, J.O. Analysis of operational data from a district cooling system and its connected buildings. *Energy* **2020**, *203*, 117844. [CrossRef]
22. Jebamalai, J.M.; Marlein, K.; Laverge, J. Influence of centralized and distributed thermal energy storage on district heating network design. *Energy* **2020**, *202*, 117689. [CrossRef]
23. Vandermeulen, A.; Van Oevelen, T.; van der Heijde, B.; Helsen, L. A simulation-based evaluation of substation models for network flexibility characterisation in district heating networks. *Energy* **2020**, *201*, 117650. [CrossRef]
24. Luc, K.M.; Li, R.; Xu, L.; Nielsen, T.R.; Hensen, J.L.M. Energy flexibility potential of a small district connected to a district heating system. *Energy Build.* **2020**, *225*, 110074. [CrossRef]
25. Guelpa, E. Impact of thermal masses on the peak load in district heating systems. *Energy* **2021**, *214*, 118849. [CrossRef]
26. Farouq, S.; Byttner, S.; Bouguelia, M.-R.; Nord, N.; Gadd, H. Large-scale monitoring of operationally diverse district heating substations: A reference-group based approach. *Eng. Appl. Artif. Intell.* **2020**, *90*, 103492. [CrossRef]
27. Dehghan, S.; Aunedi, M.; Ameli, H.; Strbac, G. Whole-system value of electrified district heating networks in decarbonising heat sector in the UK. *Electr. Power Syst. Res.* **2024**, *229*, 110144. [CrossRef]
28. Kleinertz, B.; Gruber, K. District heating supply transformation—Strategies, measures, and status quo of network operators' transformation phase. *Energy* **2022**, *239*, 122059. [CrossRef]
29. Junussova, L.R.; Abildinova, S.K.; Aliyarova, M.B.; Chicherin, S.V.; Junussov, T.J. The means to improve water treatment and to enhance power engineering performance of the water source heat pump. *Energetika. Proc. CIS High. Educ. Inst. Power Eng. Assoc.* **2018**, *61*, 372–380. [CrossRef]
30. Volkova, A.; Krupenski, I.; Ledvanov, A.; Hlebnikov, A.; Lepiksaar, K.; Latõšov, E.; Mašatin, V. Energy cascade connection of a low-temperature district heating network to the return line of a high-temperature district heating network. *Energy* **2020**, *198*, 117304. [CrossRef]
31. Chicherin, S.V. Unlocking a potential of district heating network efficient operation and maintenance by minimizing the depth of a trench system. *Bull. Tomsk. Polytech. Univ. Geo Assets Eng.* **2017**, *328*, 49–56.
32. Wirtz, M.; Kivilip, L.; Remmen, P.; Müller, D. 5th Generation District Heating: A novel design approach based on mathematical optimization. *Appl. Energy* **2020**, *260*, 114158. [CrossRef]
33. Jakubek, D.; Ocłoń, P.; Nowak-Ocłoń, M.; Sułowicz, M.; Varbanov, P.S.; Klemeš, J.J. Mathematical modelling and model validation of the heat losses in district heating networks. *Energy* **2023**, *267*, 126460. [CrossRef]
34. Massidda, L.; Marrocu, M. Total and thermal load forecasting in residential communities through probabilistic methods and causal machine learning. *Appl. Energy* **2023**, *351*, 121783. [CrossRef]

35. Coss, S.; Verda, V.; Le-Corre, O. Multi-objective optimization of District Heating Network model and assessment of Demand Side Measures using the load deviation index. *J. Clean. Prod.* **2018**, *182*, 338–351. [CrossRef]
36. Dénarié, A.; Aprile, M.; Motta, M. Dynamical modelling and experimental validation of a fast and accurate district heating thermo-hydraulic modular simulation tool. *Energy* **2023**, *282*, 128397. [CrossRef]
37. Rämä, M.; Pursiheimo, E.; Sundell, D.; Abdurafikov, R. Dynamically distributed district heating for an existing system. *Renew. Sustain. Energy Rev.* **2024**, *189*, 113947. [CrossRef]
38. Steinegger, J.; Wallner, S.; Greiml, M.; Kienberger, T. A new quasi-dynamic load flow calculation for district heating networks. *Energy* **2023**, *266*, 126410. [CrossRef]

*energies*

MDPI

*Article*

# Sky Temperature Forecasting in Djibouti: An Integrated Approach Using Measured Climate Data and Artificial Neural Networks

**Hamda Abdi [1], Abdou Idris [1,*] and Anh Dung Tran Le [2]**

[1] Faculté d'Ingénieurs, Université de Djibouti, CEALT, Djibouti 1904, Djibouti; hamdaabdi1998@gmail.com
[2] Laboratoire des Technologies Innovantes (LTI), Université de Picardie Jules Verne, Avenue des Facultés, 80025 Amiens, Cedex 1, France; anh.dung.tran.le@u-picardie.fr
* Correspondence: abdou_idriss_omar@univ.edu.dj; Tel.: +253-77823879

**Abstract:** Buildings exchange heat with different environmental elements: the sun, the outside air, the sky, and outside surfaces (including the walls of environmental buildings and the ground). To correctly account for building energy performance, radiative cooling potential, and other technical considerations, it is essential to evaluate sky temperature. It is an important parameter for the weather files used by energy building simulation software for calculating the longwave radiation heat exchange between exterior surfaces and the sky. In the literature, there are several models to estimate sky temperature. However, these models have not been completely satisfactory for the hot and humid climate in which the sky temperature remains overestimated. The purpose of this paper is to provide a comprehensive analysis of the sky temperature measurement conducted, for the first time, in Djibouti, with a pyrgeometer, a tool designed to measure longwave radiation as a component of thermal radiation, and an artificial neural network (ANN) model for improved sky temperature forecasting. A systematic comparison of known correlations for sky temperature estimation under various climatic conditions revealed their limited accuracy in the region, as indicated by low $R^2$ values and root mean square errors (RMSEs). To address these limitations, an ANN model was trained, validated, and tested on the collected data to capture complex patterns and relationships in the data. The ANN model demonstrated superior performance over existing empirical correlations, providing more accurate and reliable sky temperature predictions for Djibouti's hot and humid climate. This study showcases the effectiveness of an integrated approach using pyrgeometer-based sky temperature measurements and advanced machine learning techniques ANNs for sky temperature forecasting in Djibouti to overcome the limitations of existing correlations and improve the accuracy of sky temperature predictions, particularly in hot and humid climates.

**Keywords:** sky temperature; climate data; artificial neural networks

## 1. Introduction

The sky vault acts as an eventual heat sink for our planet, thereby providing a significant cooling resource for buildings. The morning dew is the direct consequence of this effect. This principle is particularly used for the passive cooling of buildings in hot regions like Djibouti. In fact, heat transfer between the exterior surface of a building envelope and the environment is based on radiative and convective phenomena. Radiative cooling, for instance, comes from surface heat loss due to longwave radiation emissions toward the sky. It is important to note that such cooling is most effective during the night when the sky is clear, and the humidity is low [1]. These observations are strictly related to the current challenge of reducing building energy consumption. In Africa, the building sector represents approximately 56% of the total energy use [2], with nearly half of this demand attributable to the air-conditioning of living spaces. To illustrate, in Djibouti, the energy consumption profile is primarily dominated by cooling and ventilation requirements, together accounting for over 70% of the total energy demand [3]. Furthermore, there is an upward trend in energy demand within the building sector in such African countries, which further underscores the importance of implementing passive cooling techniques for buildings. The sky temperature is a critical parameter for determining building energy performance. Sky temperature is a key input for calculating the heating and cooling energy needs of buildings to maintain thermal comfort. Accurate sky temperature values are essential for reducing errors and improving building energy simulations, which is crucial for optimizing energy consumption in heating and cooling systems. Choosing incorrect sky temperatures, even by a margin of ±10 °C, can significantly increase simulation errors, affecting the reliability of energy predictions [4].

Numerous studies concerning sky temperature models have been conducted since the early 20th century, resulting in a variety of correlations proposed by researchers [5–20] over the world. These correlations originate from various methods devised for the estimation of infrared radiation. Figure 1 provides a categorical representation of empirical models, distinguishing between the conditions of clear and cloudy skies. Within these two categories, we can identify "direct models" and "sky emissivity models". Direct models employ equations to calculate the sky temperature using mainly ambient temperature data, while indirect models, on the other hand, are based on the evaluation of sky emissivity. Currently, building energy simulations have adapted to this by using these models that can estimate the sky temperature. However, it is noteworthy that the majority of these models are approximations due to the lack of accurate measured data [21]. These sky models are intrinsically related to local weather conditions and are site-specific in their application.

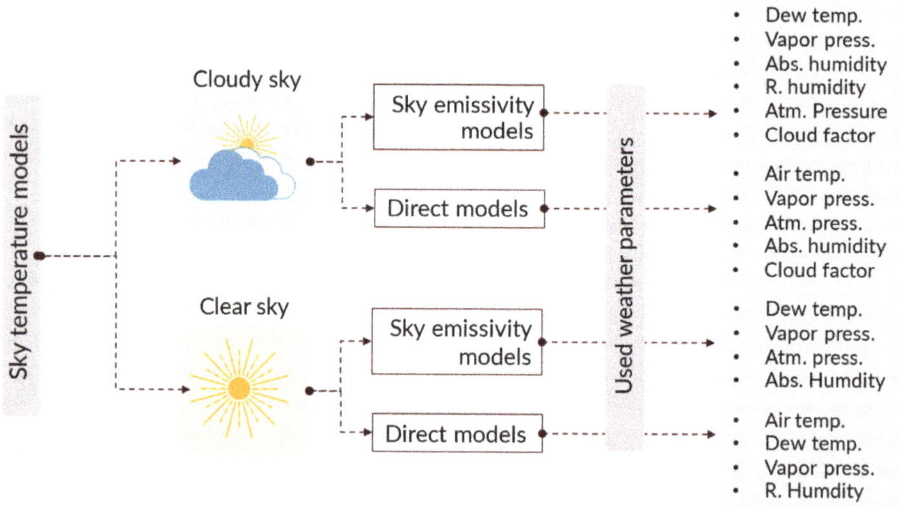


**Figure 1.** Schemes follow the same formatting.

In a previous paper, the author explored the available cooling potential from the sky vault in hot climates, and buildings' ability to exploit this cooling [22]. In this one, the sky temperature was estimated using an equation proposed by Centeno [23] and Clark and Allen [24], due to a lack of measured data in the study site, Djibouti. To address this limitation, a significant advancement has been made with the installation, for the first time in Djibouti, of a pyrgeometer, a tool specifically designed to measure longwave radiation for the direct measurement of sky temperature.

In addition to the empirical models detailed earlier, nature-inspired models have been introduced in the scientific field in recent years. These models are centered on the mathematical modeling of their natural ability for decision-making while facing a complex situation. Nature has inspired many algorithms for solving complex problems characterized as linear and non-linear through a learning procedure on data. One of the well-known models is ANNs. The ability to achieve superior performance and the reliability of adequate ANN predictions have been highlighted in various scientific fields. In regard to building energy consumption, their strength has been stressed in [23–28] for active cooling. But, considering passive cooling along with sky temperature prediction using ANN, a study carried out by Adelard et al. [29] highlighted both the inefficiency of empirical model prediction that resulted in the overestimation of sky temperatures and the satisfaction of the results of neural networks.

The artificial neural network (ANN) is a computing algorithm that maps the human nervous system. ANN is inspired by the biological neural network (BNN). The latter one is a collection of neurons cells that reacts to internal or external changes that disturb biological system equilibrium and provides adequate responses. Just as in BNN, ANNs are composed of multiple processing units that work together to learn, recognize patterns, and predict data [30].

In the literature, Al-Shawwa et al. developed an ANN model to predict environmental temperature based on various influencing factors, including location characteristics, proximity to water, vegetation, and elevation. Testing revealed that the model achieved 100% accuracy in temperature predictions [31]. By incorporating historical data and geographical variables, the training yielded predictive maps with minimal error. Dombaycı and Gölcü presented an artificial neural network (ANN) model to predict daily mean ambient temperatures in Denizli, southwestern Turkey, using data from the Turkish State Meteorological Service [32]. The results indicate that the ANN model is reliable for ambient temperature prediction, achieving optimal performance with six hidden neurons. Ozturk et al. studied the development and testing of cost-effective feedforward artificial neural network (ANN) models for estimating soil temperature at various depths using standard geographical and meteorological data in Turkey [33]. The results show strong correlation coefficients (up to 99.03%) between ANN-estimated and measured soil temperatures. Ruano et al. focused on the design of predictive neural network models for indoor air temperature, using climate and environmental data collected from a secondary school in southern Portugal [34]. The results indicate that predictive models enhance temperature regulation and achieve significant energy savings throughout the year. Veronez et al. presents an artificial neural network (ANN) method for extrapolating land surface temperature (ST) in the Rio dos Sinos Hydrographic Basin, Brazil, using positional variables, temperature, and humidity. Trained with NOAA satellite thermal images, the model achieved an average error of 2.2 °C in ST [35]. Routh et al. investigated the application of an artificial neural network (ANN) to estimate the daily mean, maximum, and minimum temperatures in Dhaka, Bangladesh, for photovoltaic (PV) efficiency with ten years of temperature data from the Bangladesh Meteorological Department [36]. The results indicate that the ANN model effectively predicts temperature trends, showcasing its reliability for forecasting. Artificial neural networks (ANNs) have been consistently demonstrated to provide accurate temperature predictions, highlighting their reliability in forecasting across a wide range.

This study is centered in sky temperature prediction and existing empirical models for estimating sky temperature are often unreliable in hot and humid regions, frequently overestimating sky temperature and leading to inaccurate predictions of heat exchange. To improve accuracy, on-site measurements were conducted using a pyrgeometer to directly measure longwave radiation emitted by the sky, offering a more precise evaluation of the cooling potential in Djibouti's unique climate.

The primary aim of this study is to eliminate the deficiencies in current empirical models by introducing a more reliable, data-driven approach using artificial neural networks (ANNs) based on on-site climate measurements. The ANN model addresses the inadequacies of existing models by accurately capturing the complex, non-linear interactions in local climate data. By enhancing the precision of sky temperature predictions, this study contributes to optimizing the radiative cooling potential of buildings, ultimately improving energy efficiency and supporting sustainable design solutions. Thus, the primary problem addressed in this research is twofold: (1) conducting on-site measurements to develop a reliable data methodology, coupled with the development of an artificial neural network (ANN) modeling approach, and (2) an analysis of the results addressing the inadequacy of existing empirical models to reliably estimate sky temperature in climates like Djibouti. The training of the ANN model and the analysis of the results in order to identify the corresponding architecture for adequately predicting the sky temperature.

## 2. Materials and Methods

### *2.1. Site Description*

Djibouti, the city in which the study site is located, is the capital city of Djibouti, situated in the eastern part of the Horn of Africa, along the Gulf of Tadjourah and the Bab-el-Mandeb strait (Figure 2). The city's climate is classified as BWh (hot deserts climate) in the Köppen–Geiger classification [37], which is typically described as a maritime desert climate.

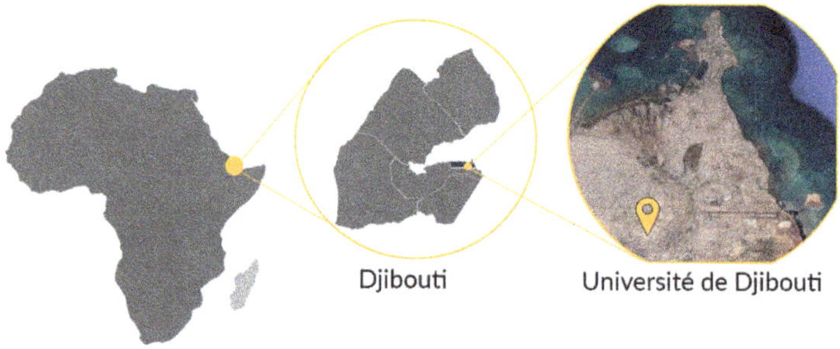


**Figure 2.** Location of the measurement site: faculty of Engineers at the University of Djibouti—lat. 11°32′21.75″ N and long. 43°7′10.5024″ E.

This climate is characterized by consistently high average temperatures, relatively low diurnal and seasonal temperature ranges, high humidity levels, and oceanic winds, except for the hot season and low annual precipitation. Djibouti experiences two primary seasons rather than the traditional four: a hot season and a relatively cool season. The hot season, which lasts from mid-April to October, is marked by high temperatures that can reach up to 34 °C ± 2 in July. The cooler season spans from November to mid-April, with average temperatures dropping to around 20 °C ± 1 in January. The temperature variations between day and night typically range from 5 °C to 10 °C, while the seasonal temperature difference is approximately 10 °C.

High relative humidity levels are present throughout the year, averaging around 62 ± 10% RH. The wind speed typically averages around 3.8 ± 0.7 $m \cdot s^{-1}$ with some variation depending on the season and prevailing weather patterns. Oceanic winds from the east are generally prevailing outside of the hot season, offering some relief from the high temperatures and humidity. High solar radiation and clear weather conditions are present throughout the year with an average monthly solar radiation of approximately 450 ± 100 $W \cdot m^{-2}$.

From an energy perspective, these climatic parameters result in high levels of cooling demand in buildings. Furthermore, the relative effectiveness of potential bioclimatic solutions depends on these climatic conditions. This hot climate therefore poses a significant challenge to finding cooling solutions for buildings. Coch [38] notes that, in general, the more extreme the climatic conditions, the more limited the solutions to passively cool the building.

### 2.2. Meteorological Data Collection and Instruments

The meteorological data used in this study were collected from a weather station installed on a building in the Faculty of Engineers at the University of Djibouti (11°32′21.75″ N, 43°7′10.5024″ E.) from January 2022 to December 2022 (Figure 3), capturing a full range of seasonal variations. Data were collected continuously during this period, providing a comprehensive dataset for the analysis of diurnal and seasonal patterns of sky temperature, as well as their relationship with other meteorological parameters. The weather station was equipped with a Vantage Pro 2 Plus system which continuously collected data for various climatic parameters, including outdoor temperature, relative humidity, dew point, rainfall, wind speed, solar irradiation, etc. This comprehensive set of meteorological parameters allowed for a more in-depth analysis of the factors influencing sky temperature and its relationship with other atmospheric variables at the study site. All the sensors were installed at approximately 8 m above the ground.

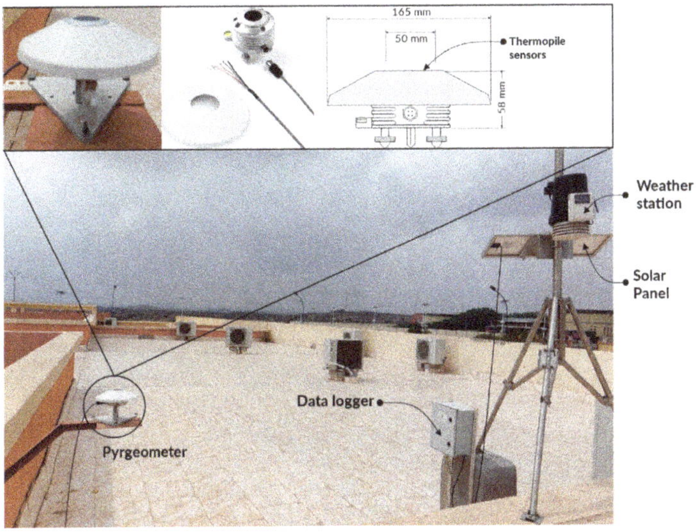


**Figure 3.** Data measurement and collection station: pyrgeometer LP PIRG 01 (AlphaOmega Electronics, Madrid, Spain) with and without protective covering, schematic cross section, and basic elements datalogger CR300 (Campbell Scientific).

Data acquisition was carried out using a CR300 data logger (Campbell Scientific Inc., Stellenbosch, South Africa) housed in a weatherproof enclosure and installed near the sensor to minimize the length of the connecting cables. 10 min averages were recorded on a storage module for subsequent processing. The station was continuously powered by a 12 V battery connected to a 20 W solar panel, as illustrated in Figure 3.

In terms of additional measurements, total sky cover data were gathered at 2 h intervals and ascertained via visual observation. The sky vault was divided into equal quadrants, with cloud cover percentages for each quadrant estimated, averaged, and documented in tenths.

In addition to these measurements, downward longwave radiation data from the sky were obtained using a high precision pyrgeometer LP PIRG 01 (Delta Ohm) with a spectral range of 5 to 50 μm that contained most of the radiation emitted by bodies at ambient temperatures (Figure 3). The radiometer's calibration was provided by the manufacturer, and its sensitivity ranged from 5 to 10 μm/(W·m$^2$).

The pyrgeometer (LP PIRG 01) was built around a thermopile sensor with a matte black paint surface to ensure non-selectivity across various wavelengths. The sensor was protected by a silicon window that shielded the thermopile from weather and defined the instrument's spectral range. This sensor was strategically installed on the rooftop of one of the faculty buildings with the sensing surface facing upward to accurately capture the longwave radiation emitted by the sky, ensuring an unobstructed view of the sky and minimal interference from nearby structures.

A comprehensive list of all the measured climatic parameters is detailed in Appendix A.

### 2.3. Infrared Radiation Measuring Based on Pyrgeometer Sensor

The pyrgeometer operates on the principle of thermopile technology. Radiant energy is absorbed/radiated from the surface of the blackened thermopile, creating a temperature difference between the center of the thermopile (hot junction) and the body of the pyrgeometer (cold junction). The temperature difference between the hot and cold junctions is converted into the potential difference thanks to the Seebeck effect.

If the pyrgeometer temperature is higher than the radiant temperature of the portion of the sky framed by the pyrgeometer, the thermopile will radiate energy and the output signal will be negative (typical situation of clear sky) and if the pyrgeometer temperature is lower than that of the portion of the sky framed, the signal will be positive (typical situation of cloudy sky).

Therefore, to extract the incoming infrared radiation ($\varphi_{r_{in}}$), the output radiance must first be determined, then added to the net input term to extract the desired down-welling component. The actual body temperature of the sensor system, which, for the pyrgeometer, is labeled $T_B$ (K), obtained by the reading of the NTC (10 kΩ) resistance, is thus measured, giving an output intensity of $\varphi_{r_{out}} = \sigma(T_B)^4$ with $\sigma = 5.67 \cdot 10^{-8}$ Wm$^{-2}$K$^{-4}$, the Stefan–Boltzmann constant. It is also necessary to measure the output signal from the thermopile connected to the datalogger. Typically, the output signal from the pyrgeometer is $\left|U_{emf}\right| < 4$ mV. In the following Figure 4, the electrical connections present inside of the pyrgeometer are reported.

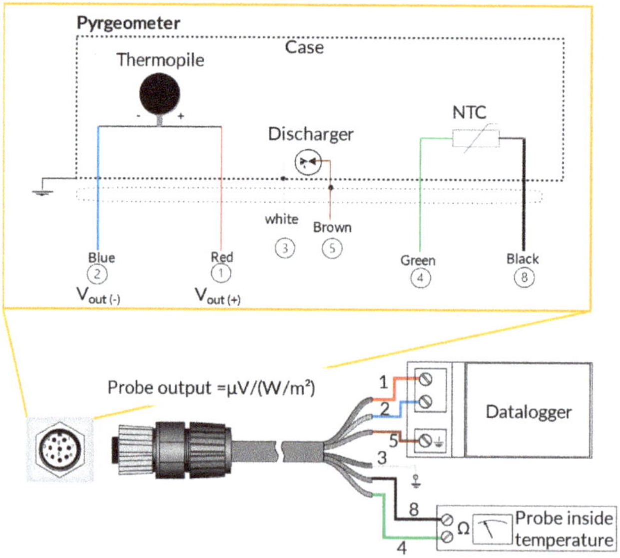


**Figure 4.** Electrical connections present inside the pyrgeometer (AlphaOmega Electronics, Madrid, Spain).

Once the pyrgeometer temperature and the thermopile output signal are known, irradiation $\varphi_{r_{in}}$ (W/m$^2$) is obtained by Equation (1):

$$\varphi_{r_{in}} = \frac{U_{emf}}{C} + \sigma(T_B)^4 \tag{1}$$

where $C$ is the pyrgeometer calibration factor reported on the datasheet in μV/(W/m$^2$). The ratio $\frac{U_{emf}}{C}$ corresponds to the net radiation $\varphi_{r_{net}}$ (positive or negative) measure via the thermopile (W/m$^2$). Finally, the down-welling thermal radiation given by Equation (1) for the whole sky is then used to calculate the sky temperature using Equation (2), expressed below.

$$\varphi_{r_{in}} = \sigma\left(T_{sky}\right)^4 \tag{2}$$

## 2.4. Artificial Neural Network

The ANN's architecture is structured into layers where the connections are ensured by weights that are referred to as the strength from one neuron to another. Starting with the initial layer underlined as the input layer, it contains raw data that has not been processed yet. The number of neurons in the input layer is directly proportional to the input variables inserted. Following the latter, there is the hidden layer, intervening between the external input and the network output, which is depicted as the processing layer. The presence of one or more hidden layers enables the network to extract higher-order statistics, because the latter presents the core of the computing algorithm with all the calculations performed through it. Additionally, highlighting the last level, the output layer portrays the final result of the ANN algorithm, hence the network output. The combination of these three layers forms the ANN architecture as presented in Figure 5.

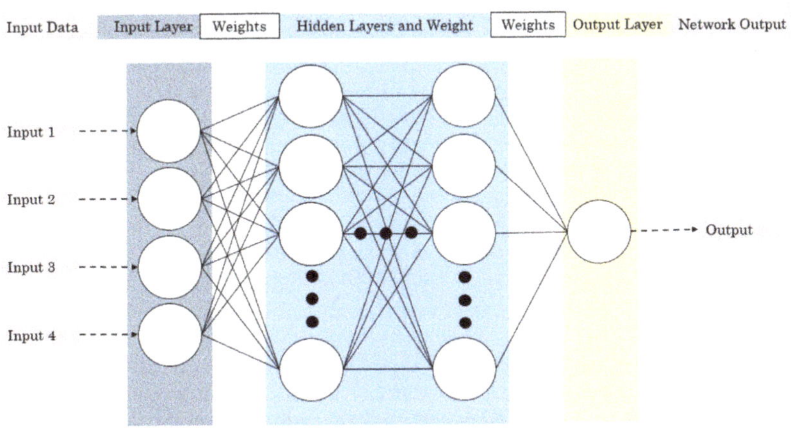


**Figure 5.** ANN architecture.

The neurons represent a fundamental block of the artificial neural network. Hence, knowing the number of neurons in the input and output layer, the challenge is determining the number of neurons in the hidden layer, as well as their corresponding size.

In order to obtain well-defined ANN architecture for sky temperature prediction, this paper provides a comprehensive analysis combining pyrgeometer-based sky temperature measurements and artificial neural networks (ANNs) to improve sky temperature forecasting in hot climates, offering a promising avenue for enhanced energy efficiency in building design and operations. Moreover, this paper outlines detailed steps to determine well-defined ANN architecture for sky temperature prediction in the methodology.

There are two principal computations in the ANN model, forward and back computation. Forward computation processes input data through layers to the output. This study uses a feedforward network (FNN) with multi-layer perceptrons (MLPs), where input flows from the input to the output layer. MLPs, derived from perceptron neural networks, are chosen for their problem-solving abilities. The training algorithm is important for an MLP-based ANN. The chosen optimizer minimizes output layer error by adjusting weights until a predefined error tolerance is met. Common optimizers in engineering are the Levenberg–Marquardt (LM) and resilient back-propagation algorithms. This study opts for the LM algorithm due to its efficiency. The LM algorithm combines gradient descent and Gauss–Newton methods, offering faster convergence, robustness, and the ability to find good local minima, thereby reducing training time significantly.

### *2.5. Training of Artificial Neural Network*

The training of artificial neural networks (ANNs) involves adjusting their weights and biases to minimize prediction errors by passing input data through layers, calculating loss, and applying backpropagation. The process is iterative with a learning algorithm. In the literature for ANNs there are supervised, unsupervised, and reinforcement learning algorithms [39–42]. Each type of learning algorithm serves different purposes: supervised learning uses labeled data, unsupervised learning finds patterns in unlabeled data, and reinforcement learning learns through interactions with an environment. This study uses supervised learning to predict sky temperature, as it provides inputs and corresponding outputs to the network.

MATLAB® and its neural network toolbox were utilized for training the ANN, considering the following parameters:

- Partition division: data are divided into training (largest portion), validation, and test sets to ensure model accuracy and performance;
- Epoch: One epoch is a complete training cycle through the dataset. The number of epochs impacts the network's learning ability; too many can lead to overfitting, while too few can weaken recognition. The optimal number is determined by comparing training errors and testing procedures;
- Momentum value (MU): This controls weight updates, preventing convergence to local minima. Low MU hampers learning, while high MU halts improvements;
- Activation function: This mathematical transformation scales outputs between layers. Non-linear functions like the sigmoid function are preferred for their ability to process complex data mapping, providing high regression (R) values. In this study, the sigmoid activation function will be employed for the hidden neurons, while a linear activation function will be utilized for the output neurons.

Table 1 below lists the values of the previously described ANN parameters. Additionally, there are three other parameters, such as the performance goal, the performance gradient, and the learning rate. The former refers to the output layer error, which is suitably defined as 0. In other words, during network learning, the ANN trains to attain a difference of zero between the desired output and the predicted output. The performance gradient is the rate of change in the performance goal. The learning rate is a parameter that dictates the magnitude of adjustments applied to the model's weights during each iteration of training. Moreover, in the context of the Levenberg–Marquardt (LM) algorithm, the full dataset is considered the batch size. The batch size is defined as the specific number of training examples processed in a single iteration during the training phase of a machine learning model.

**Table 1.** Parameter variables of ANN.

| Training Parameters | Values |
|---|---|
| Transfer function | Sigmoid function |
| Network | Feed forward |
| Training algorithm | Levenberg–Marquardt (LM) |
| Epoch | 150 |
| Performance goal | 0 |
| Data partition (Train, Validation, Test) | 70%-15%-15% |
| Minimum performance gradient | $1 \times 10^{-30}$ |
| Maximum MU | $1 \times 10^{100}$ |
| Learning rate | 0.01 |

## 2.6. Learning and Optimization

The ANN model used in this study was trained through supervised learning. Therefore, after the implementation of the data into the MATLAB®, the variables were distinguished by their type characteristics (input or output). Following the latter, data partition is made. This process is made randomly, using the random generation function in MATLAB. This action provided fairness. With the selection of the number of neurons and the size of the hidden as well as network-training parameters, the training of the ANN started, through which the weight and bias were randomly initialized at first. The latter was updated using the Levenberg–Marquart algorithm through three stages following a condition. After the training with each train, validation, and test sample, the attainment of the error goal was checked by the algorithm. If the latter was not met, the training would have restarted and, in the opposed case the training continues. Hence, when the condition was achieved, the network was stored and proceeded to the next step, and new training started with new ANN architecture, while increasing the number of neurons one by one with 25 neurons in the first hidden layer and 4 neurons in the second hidden layer. The training was done with 100 different ANN architectures. A complete description of the ANN algorithm in the form of a flowchart is shown in Figure 6.

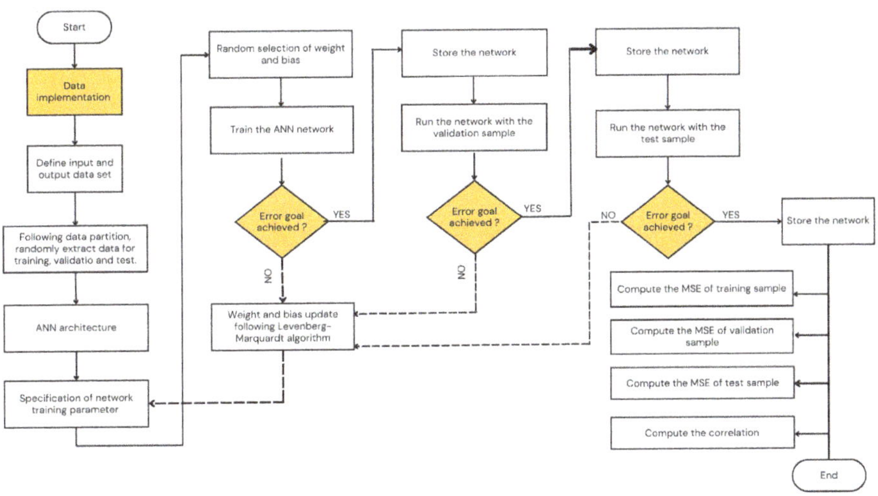


**Figure 6.** ANN training model flowchart.

Hence, to obtain the best predictive results, determining the optimum ANN architecture is required, as mentioned before. There are several methods to pre-determine the latter. Rules of thumb are available, such as the geometric pyramid rule which states that if a network has n input and m output neurons, the hidden layer would have $\sqrt{n \times m}$ neurons [43]. Another way of determining the number or neurons is by using a constructive algorithm [44]. In the ANN model of this research, the above methods will not be used, but rather a 'statistical approach' for computing the optimal number of neurons in the hidden layer along with the size of the hidden size will be taken into account. In fact, such an optimum number will be selected based on the results for the mean square error (MSE) and the correlation coefficient.

If the training was designed to run only one time for each case (ANN architecture), the comparison between the neurons would be unfair due to the fact that the weight and bias are first initialized based on randomness. This means that there might be some training that starts with better random weights and biases, leading to better performance, in contrast to the other simulations, starting with inappropriate random weights and biases. So, to avoid such unfairness, the training was repeated 1000 times for each case (ANN architecture). In this study, the average of the MSE (train, validation, and test sample) with the correlation will be used to determine the optimum ANN architecture for sky temperature prediction. A combination of minimum MSE and maximum correlation, is considered as an indicator of the optimum neuron number. The equation of the mean square error and the correlation is expressed below.

$$MSE = \frac{1}{n}\sum_{i=1}^{n}\left(Y_d - Y_p\right)^2 \quad (3)$$

$$r_{Y_d,Y_p}\frac{\Sigma\left(Y_d Y_p\right) - \Sigma\left(Y_d\right)\Sigma\left(Y_p\right))}{\sqrt{\left(n\Sigma\left(Y_d^2\right) - \Sigma\left(Y_p^2\right)\right)\left(n\Sigma\left(Y_p^2\right) - \Sigma\left(Y_d^2\right)\right)}} \quad (4)$$

$n$ = corresponds to the length of the data and it is equal to 365
$Y_d$ = actual output value
$Y_p$= predicted output value
$r_{Y_d,Y_p}$ = correlation coefficient between $Y_d$ and $Y_p$

## 3. Results and Discussion

### 3.1. Input and Output Variable Description

For selecting appropriate input parameters for buildings, creating a well-defined and working ANN model is a crucial task. These variables should be independent (input–input) with a high correlation property (input–output) in order to build a mathematical relationship. But, more importantly, the selection of input variables depends on the type and characteristic of the desired element to be predicted (output). This means that the selected inputs must have a direct influence on the output. Considering the latter perspective, the Pearson correlation among the climatic data were assessed. Figure 7 shows the correlation with the sky temperature parameter.

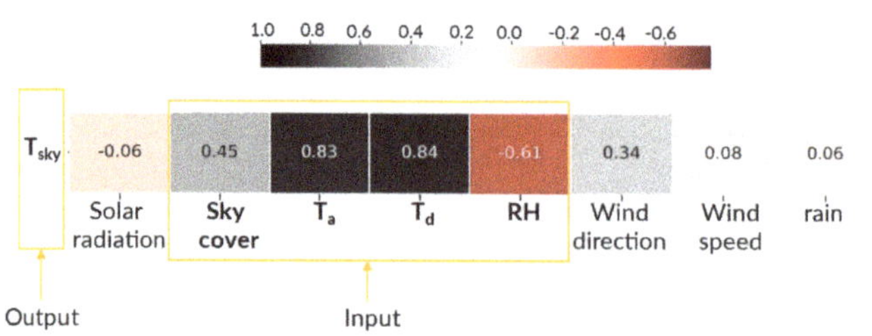


**Figure 7.** Correlation matrix between variables.

Air temperature (Ta) and dew point temperature ($T_d$) showed strong positive correlations of 0.83 and 0.84 respectively, suggesting significant influence on sky temperature ($T_{sky}$), while the sky cover displayed moderate positive correlations. In contrast, a negative correlation of −0.61 between the relative humidity (RH) and the sky temperature was found, whereas the solar radiation, wind speed, wind direction and rain, indicated a weak correlation. Based on this analysis, the selected variables and their nature in the ANN is listed in Table 2.

**Table 2.** Variables of ANN mode.

| Indicator Name | ANN Variable Type |
|---|---|
| Mean $T_a$ | Input |
| Mean $T_d$ | Input |
| Mean RH | Input |
| Mean Cover sky | Input |
| Mean $T_{sky}$ | Output |

The key statistical measures of these selected inputs and outputs are gathered in Figure 8. The air temperature and relative humidity display high variability, evidenced by their standard deviations. Meanwhile, the sky temperature shows important variation and a mean value lower than the air temperature, which affirms its potential independent influence on the building energy performance. The total sky cover ranges between 0 (clear sky) and 1 (completely overcast), with an average of around 0.38, suggesting diverse sky conditions. The boxplots would further illustrate the spread and skewness of the data. The sky temperature's boxplot, for example, reflects a broad range of values from −3.8 °C to 38.3 °C, underlining the high variability of sky temperatures in Djibouti. These boxplots, along with the earlier correlation analysis, not only reaffirm the choice of input parameters for the ANN model but also highlight the complexity of climate and the non-linear dependencies that require the use of sophisticated models that can learn and reproduce these relationships to make accurate predictions.

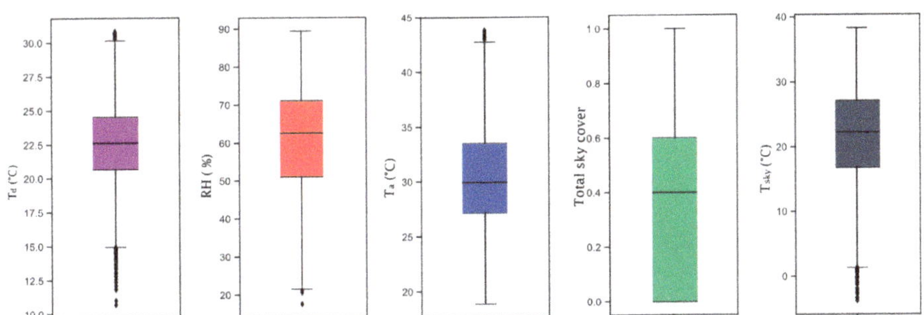


| | Dew Temperature | Relative Humidity (%) | Air Temperature (°C) | Sky Cover | Sky Temperature (°C) |
|---|---|---|---|---|---|
| Mean | 22.5 | 60.2 | 30.4 | 0.4 | 21.5 |
| Median | 22.6 | 62.6 | 29.9 | 0.4 | 22.2 |
| Std | 2.9 | 13.7 | 4.3 | 0.3 | 7.3 |
| Min | 10.7 | 17.5 | 18.9 | 0 | -3.8 |
| Max | 30.8 | 89.4 | 43.8 | 1 | 38.3 |

**Figure 8.** Boxplot representation of the ANN data.

Transitioning from the comprehensive climatic dataset, further investigation is made towards the key element; Figure 9 displays the measured data in Djibouti over the year to analyze the temporal patterns in sky temperature.

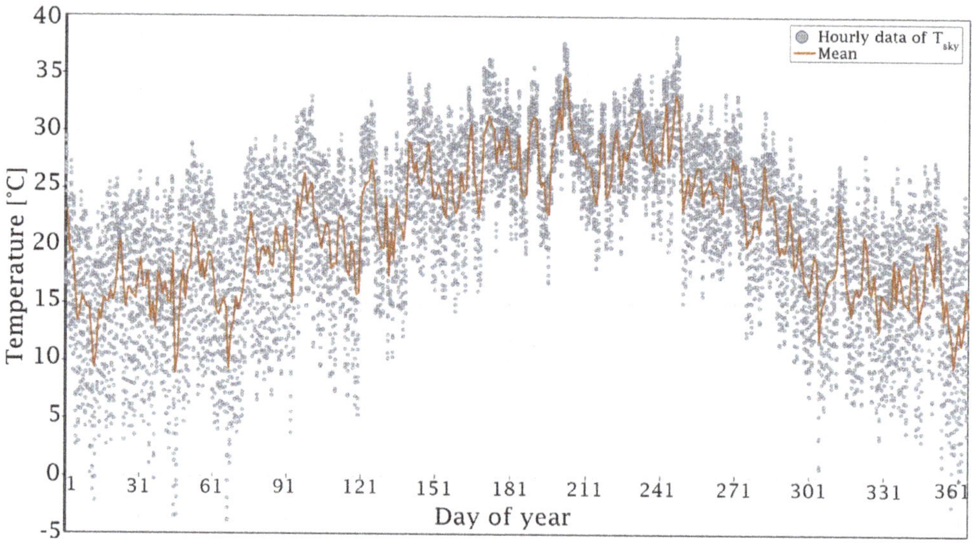


**Figure 9.** Hourly averages values of the sky temperature.

The observed diurnal and seasonal variations, along with any discernible long-term trends, are important for understanding the unique thermal dynamics of the Djiboutian atmosphere.

In fact, the data reveal notable seasonal variations, with cooler conditions during the so-called cool season (from November to April), with the average temperatures ranging between 10 °C and 25 °C. In contrast, the hot summer season (from May to October) features higher average temperatures, varying between 18 °C and 37 °C. Interestingly, these measured sky temperatures are consistently lower than the air temperatures, suggesting that an exterior surface at air temperature will emit more energy towards the sky than it receives. This observation underscores the potential of the sky as a source of freshness, which can be exploited to reduce a building's cooling needs.

Moreover, diurnal patterns, dependent on the season, often display peaks during the afternoon hours when solar radiation is most intense, and lows during the early morning, illustrating their radiative cooling effects. The within-day variability, alongside day-to-day changes, may reflect the influence of evolving weather conditions on sky temperature. This measurement of sky temperature provides a resource for informing and refining the development of predictive models such as ANNs, and ultimately, for enhancing our understanding of the thermal interaction between buildings and the environment in Djibouti's unique climatic context.

### 3.2. Evaluation of Existing Empirical Correlations

Starting from the available empirical correlations, four of them were implemented and compared to the sky temperatures measured in the studied site (via pyrgeometer sensor in Djibouti), considering clear and cloudy sky conditions. For the clear-sky direct model, the correlation developed by Swinbank [12] to estimate the sky temperature considering only the ambient air temperature was selected. In the case of the clear sky emissivity model (indirect), we were applying the correlation proposed by Tang et al. [18]. The use of this particular correlation for predicting sky emissivity is based on a measurement conducted during a short period of time in Israel. It is well-known that clear-sky conditions are not continuous along time and cloudy conditions increase infrared radiation. For this reason, the clear sky correlations have to be adapted to cloudy conditions, although the impact of cloudiness on sky temperature is difficult to evaluate. Considering the cloudy-sky direct models, we have selected the one developed by Aubinet [19]. The latter evaluates the sky temperature using water vapor pressure and ambient temperature. Finally, for cloudy-sky emissivity models, the correlation of Clark and Allen [24] introducing a correction factor ($C_a$) for estimating the influence of the clouds is also selected. This factor can be calculated as the ratio between the measured cloud sky atmospheric radiation and the estimated clear sky atmospheric radiation. Moreover, $C_a$ is a function of the sky cover (N). For clear sky conditions, N is equal to 0 and for overcast sky, it is equal to 10. This model is currently employed in the dynamic code EnergyPlus. These correlations are listed in Table 3.

**Table 3.** Variables of ANN mode.

| Reference | Date | Correlation | Measurement Site |
|---|---|---|---|
| Swinbank [12] | 1963 | $T_{sky} = 0.0552T_a^{1.5}$ | Australia |
| Tang et al. [18] | 2004 | $\varepsilon_{sky} = 0.754 + 0.0044T_d$ | Israel |
| Aubinet [19] | 1994 | $T_{sky} = 94 + 12.6.\ln(P_v) - 13K_0 + 0.341T_a$ | - |
| Clark and Allen [24] | 1978 | $\varepsilon_{sky} = 0.787 + 0.764\left(\frac{T_d}{273}\right)C_a$ <br> $C_a = 1 + 0.0224N - 0.0035N^2 + 0.00028N^3$ | Texas |

The results of the comparison are shown in Figure 10, and several conclusions can be drawn from the corresponding observation. The evaluated correlations are not fully representative. They account roughly between 68% and 76% of the variance in the data (worse agreement with measured data). Moreover, the mean square error (MSE) values, indicative of prediction errors, range from approximately 0.95 to 1.96. These figures suggest that the models experience error margins in predicting sky temperatures. This comparison underscores the limitation of existing empirical correlations in the accurate prediction of measured sky temperature under hot climate conditions, especially in Djibouti. We are therefore far from unanimous on the subject, and progress is needed in this field if applications based on the exploitation of the sky applications are to be developed.

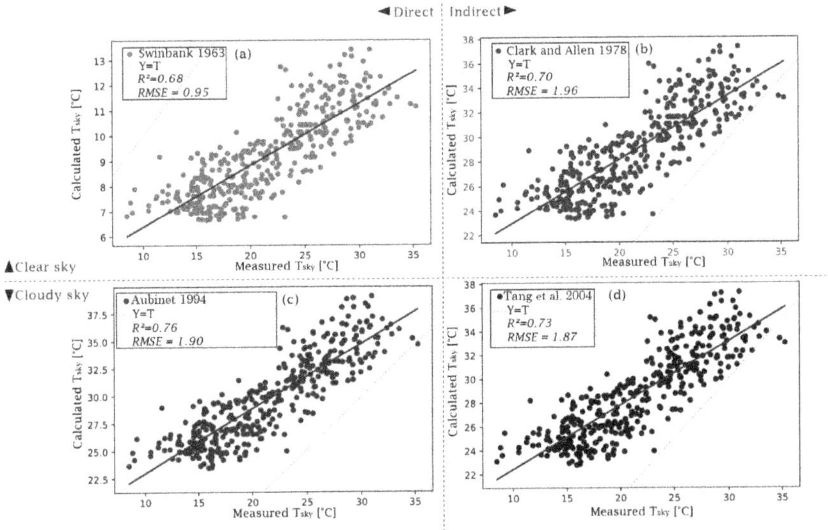


**Figure 10.** Comparison with measured data results for (**a**) Swinbank [12], (**b**) Tang et al. [18], (**c**) Aubinet [19], and (**d**) Clark and Allen [24].

These limitations possibly originate from these models' inability to fully capture the complex, non-linear relationships between various meteorological variables and sky temperature. Consequently, this suggests a more sophisticated modeling approach, such as ANNs, to address these shortcomings.

### 3.3. Performance Analysis of the Trained ANN Architectures and Optimal Neuron Selection

Through the increase in the hidden layer size and its corresponding neurons, 100 different ANN architectures were trained and tested. With the above-mentioned framework, the comparison was made alongside the mean square error (MSE) during training and testing phases and the correlation. These criteria highlight the performance of the trained models, hence, determining the optimal ANN for sky temperature prediction.

The preferred model is that which yields the minimum MSE and the maximum correlation. It is important to note that such an indication indicates that the estimates generated by the model converge to the desired values.

To provide a holistic illustration of the trained networks, Principal Component Analysis (PCA) and Hierarchical Clustering were utilized. These statistical techniques permit the identification of dominant quality parameters that account for the vast majority of the observed variations. Notably, the first two criteria express an impressive 98.96% of the total inertia, indicating that they embody 98.96% of the total variability within the dataset. This value considerably surpasses the reference value of 75.82%, signifying that the explained variability via this plan is significantly meaningful.

Figure 11 illustrates the individual contributions of each ANN architecture to the dimensions. In the first dimension, artificial neural network (ANN) models with high correlation (strong positive coordinates on the horizontal axis) stand in opposition to models demonstrating high MSE during testing (strong negative coordinates on the horizontal axis). Conversely, the second-dimension pits ANNs with high MSE during training against models with strong negative coordinates on the axis. These findings suggest that MSE during testing and correlation are inversely related and contribute significantly to the first dimension, while MSE during training exhibits positive correlation with the second dimension.

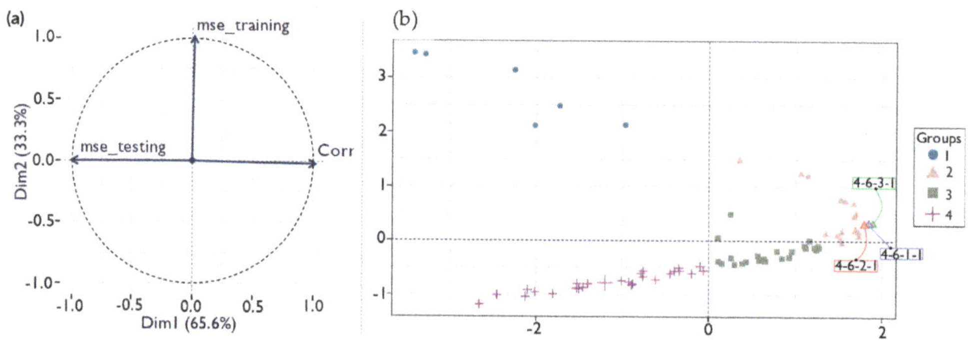


**Figure 11.** Principal Component Analysis (PCA) representation of ANN architecture: (**a**) Correlation plot of variables showing contributions of training and testing datasets to the first two dimensions. (**b**) Score plot indicating the grouping of samples into different clusters (Groups 1, 2, 3 and 4) based on PCA scores, with each group represented by distinct markers and colors.

Supplementary to the PCA, a Hierarchical Clustering on Principal Components (HCPC) was executed. This algorithm clustered ANN models into distinct groups, based on similarity in variable patterns, resulting in four distinct clusters (Figure 11b). Each cluster's characteristics were defined by the location of their original variables in the score plot:

- Cluster 1: characterized by high values of MSE during training and testing, coupled with low correlation values;
- Cluster 2: features high values for correlation and MSE during training but low MSE during testing;
- Cluster 3: composed of models marked by low MSE values during training;
- Cluster 4: displays models with high MSE during testing and low values for both correlation and MSE during training.

Upon the result of the clustering, the three most optimal ANN architectures were, respectively, 4-6-1-1, 4-6-2-1, and 4-6-3-1. In these configurations, '4' symbolized the input, '6' corresponded to the neurons in the first hidden layer, and '1', '2', or '3' denoted neurons in the second hidden layer, while the final digit represented the output. Through these optimum ANN architectures, a distinctive one denoted as the best ANN architecture with the highest optimal degree was to be selected.

Hence, with the latter objective, to effectively compare these architectures, the ANN training was set again with these ANN architectures. Furthermore, to provide a fair comparison criterion, the simulation repetition was increased to 2000 trials for each one. The results are shown in Figure 12.

Figure 12 displays both the architectural presentation for 4-6-1-1, 4-6-2-1, and 4-6-3-1, and the attached regression between the desired and predicted sky temperature. The lowest mean square error for testing is defined for 4-6-3-1 ANN with 0.276 in contrast to 4-6-1-1, and 4-6-2-1 with 0.590 and 0.618, respectively. Likewise, considering the correlation, the highest is attributed to 4-6-3-1 ANN with a 0.920 and the lowest is assigned to 4-6-1-1, and 4-6-2-1 with 0.914 and 0.915, respectively. Considering the previous points, the 4-6-3-1 ANN architecture exhibits an advantage through the evaluated criteria in contrast to the 4-6-1-1 and 4-6-2-1. An adequate and precise sky temperature ANN prediction can relatively be made through the 4-6-3-1 ANN architecture, which suggests that the ANN model is an appropriate model to determine sky temperature.

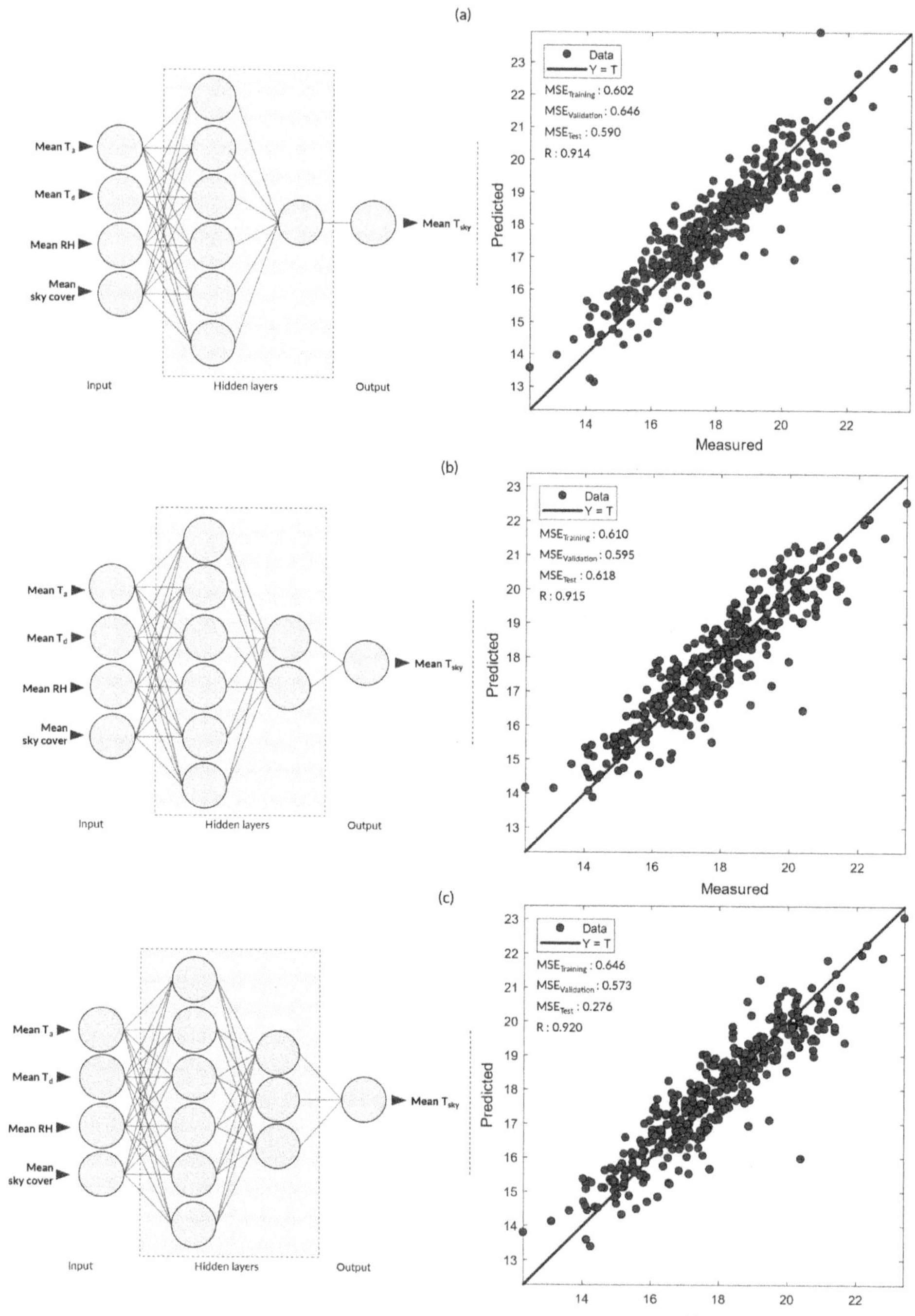


**Figure 12.** Regression between target vs predicted sky temperature with (**a**) 4-6-1-1, (**b**) 4-6-2-1, (**c**) 4-6-3-1.

## 4. Conclusions

The sky's temperature is of the utmost significance in influencing the energy performance of buildings. Considering the latter, the effectiveness of both empirical and numerical approaches have been compared. Debuting from the empirical method, four models defining clear and cloudy skies have been integrated along with the database extracted from the pyranometer. The results have shown that the MSE varies between 0.95 and 1.96 and that the correlation is around 68% to 76%. In contrast to the empirical method and the numerical method, the natural inspire algorithms have shown remarkable results. Modeled and simulated using MATLAB® 2017, the database collected from the pyranometer was trained based on multi-layer feedforward neural networks with a supervised learning approach and Levenberg–Marquardt algorithm for error minimization. Considering the mean air temperature, dew temperature, relative humidity, and sky cover as input data and mean sky temperature as an output data, the ANN was trained. With a training partition of 70% and 15% and pre-selected stopping criterion, the best ANN architecture was selected. Judged according to the lowest average of the mean square errors (MSEs) of the test data, and the correlation, the architecture 4-6-3-1 was regarded as the best structure. Based upon the results from the training, it was noticed that there is a strong correlation of 0.920 between the observed output (target) and the predicted output and a lower MSE of 0.276. Therefore, it can be concluded that in contrast to the empirical method, the artificial neural network exhibits an outperformance behavior in predicting the sky temperature. At this stage, it seems interesting to explore the consequences of considering other ANN models and training algorithms in predicting sky temperature. As a result, this study will continue, to more precisely evaluate the cooling energy available from the sky in Djibouti and to explore the exploited cooling energy in buildings.

**Author Contributions:** Conceptualization, H.A. and A.I.; methodology, H.A.; software, H.A. and A.I.; validation, H.A., A.I. and A.D.T.L.; formal analysis, H.A.; investigation, H.A.; resources, H.A.; data curation, H.A.; writing—original draft preparation, H.A.; writing—review and editing, A.I. and A.D.T.L.; visualization, A.I.; supervision, A.I.; project administration, A.I.; funding acquisition, A.I. All authors have read and agreed to the published version of the manuscript.

**Funding:** This research was funded by the African Center of Excellence for Logistics and Transport, College of Engineering, edition 2021 project titled "Batifrai: Enveloppe du Bâtiment: la valorisation des ressources de rafraîchissement bioclimatique–Bâtifraî".

**Data Availability Statement:** The original contributions presented in the study are included in the article, further inquiries can be directed to the corresponding author.

**Acknowledgments:** The authors acknowledge the African Center of Excellence for Logistics and Transport for their funding of the project titled "Batifrai: Enveloppe du Bâtiment: la valorisation des ressources de rafraîchissement bioclimatique–Bâtifraî".

**Conflicts of Interest:** The authors declare no conflicts of interest. The funders had no role in the design of this study; in the collection, analyses, or interpretation of data; in the writing of the manuscript; or in the decision to publish the results.

## Appendix A

**Table A1.** Meteorological parameters specifications.

| | |
|---|---|
| Hardware | Davis vantage Pro 2 Plus/Delta Ohm pyrgeometer LP PIRG 01/Datalogger CR300 |
| Software | Davis Vantage Pro2 Plus (Wireless)/LoggerNet |
| Coordinates | 11°32′21.75″ N; 43° 7′10.5024″ E |
| Elevation | 8 m from the ground |

**Table A1.** *Cont.*

| Solar radiation | |
|---|---|
| Device | Solar radiation sensor: Apogee Pyranometer SP-110 |
| Calibration | 5000 W/m$^2$ per V_manufacturer calibration |
| Resolution | 1 W/m$^2$ |
| Range | 0–1800 W/m$^2$ |
| Accuracy | 5% |
| Cosine response | ±3% for angles 0–75° |
| Uncertainties | ±1% |
| **Air temperature** | |
| Device | Integrated Sensor Suite (ISS) in Davis Vantage Pro2 Plus |
| Calibration | Manufacturer calibration |
| Resolution | 0.1 °C |
| Range | −40 °C–65 °C |
| Accuracy | 0.5 °C |
| Uncertainties | ±0.5 °C |
| **Humidity** | |
| Device | Integrated Sensor Suite (ISS) in Davis Vantage Pro2 Plus |
| Calibration | Manufacturer calibration |
| Resolution | 1% |
| Range | 1–100% |
| Accuracy | 3–4% above 90% |
| Uncertainties | 3–4% |
| **Dew point temperature** | |
| Device | Integrated Sensor Suite (ISS)/Hum Station in Davis Vantage Pro2 Plus |
| Calibration | Manufacturer calibration |
| Resolution | 0.1 °C |
| Range | −76 °C to +54 °C |
| Accuracy | 1.5 °C |
| Uncertainties | ±1.5 °C |
| **Sky temperature** | |
| Device | Delta Ohm Lp pirg 01 |
| Calibration | Manufacturer calibration |
| Resolution | 5–10 µV/(W/m$^2$) |
| Measuring range | −300–+300 W/m$^2$ |
| Spectral range | 5.5 µm–45 µm |
| Impedance | 33 Ω–45 Ω |
| Field of view | 160° |

## References


1. Algarni, S. Modeling Radiation Heat Transfer for Building's Cooling and Heating Loads: Considering the Role of Clear, Cloudy, and Dusty Conditions in Hot and Dry Climates. Ph.D. Thesis, University of Arkansas, Fayetteville, AR, USA, 2015. Available online: https://scholarworks.uark.edu/etd/1231 (accessed on 8 October 2024).
2. Global Status Report for Buildings and Construction. 2022. Available online: www.globalabc.org (accessed on 23 February 2023).
3. Abdou, I.O. Études de L'amélioration de la Performance Energétique et de Stratégies de Conception Architecturale Bioclimatique en Climat Chaud et Humide. 2020. Available online: https://theses.hal.science/tel-03346755 (accessed on 9 October 2024).
4. Albatayneh, A.; Alterman, D.; Page, A.; Moghtaderi, B. The Significance of Sky Temperature in the Assessment of the Thermal Performance of Buildings. *Appl. Sci.* **2020**, *10*, 8057. [CrossRef]
5. Evangelisti, L.; Guattari, C.; Asdrubali, F. On the sky temperature models and their influence on buildings energy performance: A critical review. *Energy Build.* **2019**, *183*, 607–625. [CrossRef]
6. Idso, S.B.; Jackson, R.D. Thermal radiation from the atmosphere. *J. Geophys. Res.* **1969**, *74*, 5397–5403. [CrossRef]
7. Raman, P.K. Heat radiation from the clear atmosphere at night. *Proc. Indian Acad. Sci.-Sect. A* **1935**, *1*, 815–821. [CrossRef]
8. Satterlund, D.R. An improved equation for estimating long-wave radiation from the atmosphere. *Water Resour. Res.* **1979**, *15*, 1649–1650. [CrossRef]
9. Idso, S.B. A set of equations for full spectrum and 8- to 14-µm and 10.5- to 12.5-µm thermal radiation from cloudless skies. *Water Resour. Res.* **1981**, *17*, 295–304. [CrossRef]

10. Berdahl, P.; Fromberg, R. The thermal radiance of clear skies. *Sol. Energy* **1982**, *29*, 299–314. [CrossRef]
11. Hatfield, J.L.; Reginato, R.J.; Idso, S.B. Comparison of long-wave radiation calculation methods over the United States. *Water Resour. Res.* **1983**, *19*, 285–288. [CrossRef]
12. Swinbank, W.C. Long-wave radiation from clear skies. *Q. J. R. Meteorol. Soc.* **1963**, *89*, 339–348. [CrossRef]
13. Berdahl, P.; Martin, M. Emissivity of clear skies. *Sol. Energy* **1984**, *32*, 663–664. [CrossRef]
14. Berger, X.; Buriot, D.; Garnier, F. About the equivalent radiative temperature for clear skies. *Sol. Energy* **1984**, *32*, 725–733. [CrossRef]
15. Chen, B.; Clark, D.; Maloney, J.; Mei, W.N.; Kasher, J. Measurement of night sky emissivity in determining radiant cooling from cool storage roofs and roof ponds. In Proceedings of the 20th National Passive Solar Conference, Minneapolis, MN, USA, 15–20 July 1995; Volume 80301, pp. 143–148.
16. Maloney, J.; Chen, D.B.; Kasher, J.; Girgis, G.A.; Clark, D. Determination of the Clear Sky Emissivity for Use in Cool Storage Roof and Roof Pond Applications. 1991. Available online: https://www.researchgate.net/publication/237705238 (accessed on 23 February 2023).
17. Niemelä, S.; Räisänen, P.; Savijärvi, H. Comparison of surface radiative flux parameterizations: Part I: Longwave radiation. *Atmos. Res.* **2001**, *58*, 1–18. [CrossRef]
18. Tang, R.; Etzion, Y.; Meir, I.A. Estimates of clear night sky emissivity in the Negev Highlands, Israel. *Energy Convers. Manag.* **2004**, *45*, 1831–1843. [CrossRef]
19. Aubinet, M. Longwave sky radiation parametrizations. *Sol. Energy* **1994**, *53*, 147–154. [CrossRef]
20. Crawford, T.M.; Duchon, C.E. An Improved Parameterization for Estimating Effective Atmospheric Emissivity for Use in Calculating Daytime Downwelling Longwave Radiation. *J. Appl. Meteorol. Climatol.* **1999**, *38*, 474–480. [CrossRef]
21. Wong, R.Y.M.; Tso, C.Y.; Jeong, S.Y.; Fu, S.C.; Chao, C.Y.H. Critical sky temperatures for passive radiative cooling. *Renew. Energy* **2023**, *211*, 214–226. [CrossRef]
22. Omar, A.I.; David, D.; Vergnault, E.; Virgone, J.; Idriss, I. A New Set of Indicators to Evaluate the 1 Bioclimatic Performance of Air Conditioned 2 Buildings in a Hot and Humid Climate 3. 2020. Available online: https://www.elsevier.com/open-access/userlicense/1.0/ (accessed on 23 February 2023).
23. Centeno, V.M. New formulae for the equivalent night sky emissivity. *Sol. Energy* **1982**, *28*, 489–498. [CrossRef]
24. Clark, G.; Allen, C. The Estimation of Atmospheric Radiation for Clear and Cloudy Skies. In Proceedings of the 2nd National Passive Solar Conference (AS/ISES), Philadelphia, PA, USA, 16–18 March 1978; pp. 675–678.
25. Admon, M.R.; Senu, N.; Ahmadian, A.; Majid, Z.A.; Salahshour, S. A new efficient algorithm based on feedforward neural network for solving differential equations of fractional order. *Commun. Nonlinear Sci. Numer. Simul.* **2023**, *117*, 106968. [CrossRef]
26. Sholahudin; Giannetti, N.; Miyaoka, Y.; Saito, K.; Tanaka, K.; Ogami, H. Experimental verification of artificial neural network scalability for performance monitoring of multi-split type air conditioners. *J. Build. Eng.* **2023**, *76*, 107427. [CrossRef]
27. Li, N. Comparison of the characteristics of the control strategies based on artificial neural network and genetic algorithm for air conditioning systems. *J. Build. Eng.* **2023**, *66*, 105830. [CrossRef]
28. Kim, M.H.; Park, H.J. Application of artificial neural networks using sequential prediction approach in indoor airflow prediction. *J. Build. Eng.* **2023**, *69*, 106319. [CrossRef]
29. Adelard, L.; Pignolet-Tardan, F.; Mara, T.; Lauret, P.; Garde, F.; Boyer, H. Sky temperature modelisation and applications in building simulation. *Renew. Energy* **1998**, *15*, 418–430. [CrossRef]
30. Kareem, S.S.; Pathak, Y. Clinical Applications of Artificial Neural Networks in Pharmacokinetic Modeling. In *Artificial Neural Network for Drug Design, Delivery and Disposition*; Elsevier: Amsterdam, The Netherlands, 2016; pp. 393–405. [CrossRef]
31. Al-Shawwa, M.; Al-Absi, A.A.-R.; Hassanein, S.A.; Baraka, K.A. ANN for Predicting Temperature and Humidity in the Surrounding Environment. *Int. J. Acad. Pedagog. Res.* **2018**, *9*, 1–5. Available online: www.ijeais.org/ijapr (accessed on 23 February 2023).
32. Dombayci, Ö.A.; Gölcü, M. Daily means ambient temperature prediction using artificial neural network method: A case study of Turkey. *Renew. Energy* **2009**, *34*, 1158–1161. [CrossRef]
33. Ozturk, M.; Salman, O.; Koc, M. Artificial neural network model for estimating the soil temperature. *Can. J. Soil Sci.* **2011**, *91*, 551–562. [CrossRef]
34. Ruano, A.E.; Crispim, E.M.; Conceição, E.Z.E.; Lúcio, M.M.J.R. Prediction of building's temperature using neural networks models. *Energy Build.* **2006**, *38*, 682–694. [CrossRef]
35. Veronez, M.R.; Wittmann, G.; Reinhardt, A.O.; Da Silva, R.M. Surface Temperature Estimation Using Artificial Neural Network. *Int. Soc. Photogramm. Remote Sens.* **2010**, *38*, 612–617.
36. Routh, T.K.; Yousuf, A.H.B.; Hossain, N.; Asasduzzaman, M.M.; Hossain, I.; Husnaeen, U.; Mubarak, M. Artificial neural network based temperature prediction and its impact on solar cell. In Proceedings of the 2012 International Conference on Informatics, Electronics and Vision, ICIEV 2012, Dhaka, Bangladesh, 18–19 May 2012; pp. 897–902. [CrossRef]
37. Peel, M.C.; Finlayson, B.L.; McMahon, T.A. Updated world map of the Köppen-Geiger climate classification. *Hydrol. Earth Syst. Sci.* **2007**, *11*, 1633–1644. [CrossRef]
38. Coch, H. Chapter 4—Bioclimatism in vernacular architecture. *Renew. Sustain. Energy Rev.* **1998**, *2*, 67–87. [CrossRef]
39. Sanger, T.D. Optimal Unsupervised Learning in a Single-Layer Linear Feedforward Neural Network. *Neural Netw.* **1989**, *2*, 459–473. [CrossRef]

40. Fabiyi, S.D. A Review of Unsupervised Artificial Neural Networks with Applications. *Int. J. Comput. Appl.* **2019**, *181*, 22–25.
41. Hilaiwah, A.I.S.; Allah, H.A.A.A.; Abbas, B.A.; Sutikno, T. Live to learn: Learning rules-based artificial neural network. *Indones. J. Electr. Eng. Comput. Sci.* **2021**, *21*, 558–565. [CrossRef]
42. Sathya, R.; Abraham, A. Comparison of Supervised and Unsupervised Learning Algorithms for Pattern Classification. *Int. J. Adv. Res. Artif. Intell.* **2013**, *2*, 020206. [CrossRef]
43. Talaee, P.H. Multilayer perceptron with different training algorithms for streamflow forecasting. *Neural Comput. Appl.* **2014**, *24*, 695–703. [CrossRef]
44. Mayerich, D.; Sun, R.; Guo, J. Deep Learning. In *Microscope Image Processing*, 2nd ed.; Elsevier: Amsterdam, The Netherlands, 2023; pp. 431–456. [CrossRef]

energies

MDPI

Article

# Multi-Objective Optimization of Insulation Thickness with Respect to On-Site RES Generation in Residential Buildings

Agis M. Papadopoulos [1,*], Konstantinos Polychronakis [2], Elli Kyriaki [1] and Effrosyni Giama [1]

1 Process Equipment Design Laboratory, Department of Mechanical Engineering, Aristotle University, 54124 Thessaloniki, Greece; kelli@auth.gr (E.K.); fgiama@auth.gr (E.G.)
2 Department of Mechanical Engineering, National Technical University of Athens, 15772 Athens, Greece; polichronakis_konstantinos@mail.ntua.gr
* Correspondence: agis@auth.gr

**Abstract:** This paper investigates the optimization of insulation thickness with respect to the integration of renewable energy systems in residential buildings in order to improve energy efficiency, maximize the contribution of renewables and reduce life cycle costs. Using the DesignBuilder and EnergyPlus software, this study models a representative two-story residential building located in Athens, Greece. The building envelope features extruded polystyrene thermal insulation and windows with unplasticized polyvinyl chloride frames and low-e glazing. Six scenarios with hybrid renewable energy systems are analyzed, including air- and ground-source heat pumps, solar thermal systems and a biomass fired boiler, so as to assess energy consumption, economic feasibility and internal air temperature conditions. A Pareto-fronts-based optimization algorithm is applied to determine the optimal combination of insulation thicknesses for the walls, the roof and the floor, focusing on minimizing the life cycle cost and maximizing the percentage of renewable energy utilized. The results demonstrate that scenarios involving biomass boilers and solar thermal systems, both for heating and cooling, when combined with reasonable thermal protection, can effectively meet the recent European Union's directive's goal, with renewable energy systems contributing more than 50% of the total energy requirements, whilst maintaining acceptable internal air temperature conditions and having a life cycle cost lower than contemporary conventional buildings.

**Keywords:** thermal insulation; renewable energy; energy efficiency; optimization

## 1. Introduction

In the European Union (EU), for the year 2022, residential buildings accounted for approximately 26% of the total final energy consumption and services, which, to a large extent, is also taking place in other buildings, accounting for another 14% [1]. The first Energy Performance of Buildings Directive was introduced in 2001, aiming at using less and cleaner energy whilst improving indoor environmental conditions [2]. Still, the increasing environmental burden, the rise in fuel prices, especially since the geopolitical instability due to the war in the Ukraine, and the dependency on supplier countries have led to the necessity of reaching the levels of nearly zero-energy buildings, both in new constructions and, by means of refurbishment, in the building stock, given the fact that 75% of European buildings still have low energy efficiency [3]. Towards this direction, three important directives were adopted by the European Parliament in the last two years: (a) the revised Energy Performance of Buildings Directive (EU/2024/1275), aiming to achieve a fully decarbonized building stock by 2050, (b) the revised Energy Efficiency Directive (EU/2023/1791), introducing the 'energy efficiency first' as a fundamental principle of EU energy policy, giving it legal standing for the first time and (c) the revised Renewable Energy Directive (EU/2023/2413), establishing an overall renewable energy target of at least 42.5% for by 2030 but aiming for 45%. These directives are for Europe, and, hence, also for Greece, and they are significant milestones on the path to carbon neutrality by 2050 [4]. Furthermore, when studying the directives in detail, one cannot fail to notice that EU policy focuses not only on the upgrading building efficiency but also, in a more holistic approach, raising the quality of living for people in the buildings, as well as improving the urban environment [5]. According to the latest statistics mentioned in the directive, two-thirds of the energy used to heat and cool buildings still comes from fossil fuels. Natural gas plays the largest role in heating buildings, accounting for around 39% of the energy used for space heating in the residential sector. Oil is the second most important fossil-based heating fuel, accounting for 11%, while coal accounts for about 3%. It is therefore understood that the gradual abolition of fossil fuels in the heating and cooling sector is of particular importance for the decarbonization of the building sector. Member States should phase out conventional boilers, and, as a first step, no financial incentives should be provided for the installation of these boilers, while it should still be possible to provide financial incentives for the installation of hybrid heating systems with a significant share of renewable energy, such as combinations of a boiler with solar thermal energy or a heat pump [6]. So, in order to accomplish the required reduction in carbon emissions by up to or more than 80% and simultaneously increase energy efficiency, the share of renewable energy systems (RES) implemented in the energy sector and, more specifically, in the building sector should be increased [7]. More specifically, the large-scale application of hybrid renewable energy systems is a key parameter in achieving zero carbon emissions, so renewable energy sources must satisfy building energy demands [8]. Moreover, a prerequisite for this strategy is also the limitation of the use of fossil fuels in electricity generation. In Greece, significant progress has been made over the past decade; in 2023, renewables accounted for 45% of the total electricity generation, and when counting big hydroelectric plants, this figure reached 57% [9], compared to less than 15% in 2010. However, it is clear that there are more challenges associated with the electrification of heat, especially in terms of the transition from conventional heating technologies to renewable energy systems, which, in some cases, may require an expanded electricity infrastructure, which could result in high initial investment costs [10]. Also, control systems will play a significant role in reducing energy consumption, so attention should be given to finding optimal control systems, especially when hybrid renewable energy systems are considered, so as to ensure that the efficiency of the RES system is maximized and that the maximum possible RES energy is used. Moreover, the integration of control systems in RES systems would ensure that there would be more efficient collaboration between energy storage, the system itself and the demand, which would again lead to a more efficient system overall [11]. However, control systems on their own are not enough. In order to ensure power system stability, more sophisticated and complex control techniques, like the demand response, are required, in order to provide flexible solutions to the inelastic demand side. The incorporation of RES in power systems enhances the existing issues associated with grid operation due to their unpredictability and intermittency, and this is why the demand response technique is investigated as a solution to fixing or eliminating these problems [12].

Hybrid renewable energy systems with or without thermal energy storage, as well as hybrid combined heat and power renewable energy systems, are a vivid research field. Daneshazarian and Berardi studied a hybrid renewable energy system consisting of photovoltaic collectors for covering electricity demand, vacuum tube collectors and a ground-source heat pump. This system was combined with an underground thermal energy storage system with nanomaterials so as to increase its energy efficiency. The results showed that the incorporation of the nano-enhanced thermal energy storage system increased the overall performance of the hybrid system by 27.1%, and the energy consumption was reduced by 36.7% [13]. In another research work, Wang et al. investigated a combined system where solar thermal collectors, photovoltaic panels, a hot water storage tank, a battery and a diesel engine were installed in a residential building. The solar thermal system was used to produce domestic hot water, space heating and space cooling by using an absorption refrigeration system. It was concluded that after the optimal design of the proposed system, the internal temperature of the studied residential building was controlled, and thermal comfort was achieved [14]. A hybrid system with photovoltaic panels, a wind turbine, a ground-source heat pump, a fuel cell, a diesel generator and a battery was examined by Cheraghi and Jahangir. More specifically, they implemented a multi-objective optimization analysis of the aforementioned system, which was installed in a residential building. It was seen that the power system that used only the diesel engine as a backup source had a lower levelized cost of energy but higher carbon dioxide emissions. Additionally, they stated that fuel cells are an expensive technology but have many environmental advantages [15]. Moreover, Jaffarian et al. performed a multi-objective optimization analysis of a hybrid energy system for residential buildings. More specifically, different energy storage systems (a latent thermal energy storage system with phase-change materials, hydrogen storage with fuel cells and battery storage with Li-On cells) combined with renewable energy sources (photovoltaic/thermal collectors, heat pumps and a desiccant wheel) were studied. It was calculated that, in warm climates, a renewable energy fraction of 85.35% can be achieved, while in cold climates, a renewable energy fraction of 59.23% can be achieved, leading to 1088.24 kWh and 731.37 kWh annual electricity savings, respectively [16]. Last but not least, a novel solar–biomass micro combined heat and power system for residential applications was studied by Skiadopoulos et al. The proposed system covered 97.10% and 25.85% of the annual demand for space heating and electricity, respectively. In addition, the annual carbon footprint of the studied building was reduced by 31.22 kg equivalent carbon dioxide ($CO_{2eq}$), while the annual primary energy saving was 46.58 kWh/$m^2$ after the implementation of the proposed system [17].

Within the framework of this paper, strategies for optimizing insulation thickness and integrating renewable energy sources to improve energy efficiency and reduce life cycle costs (LCCs) in residential buildings are studied. Utilizing state-of-the-art simulation tools such as DesignBuilder V7.0.2.006 which uses EnergyPlus V9.4.0.002 sourced by Gloucestershire, United Kingdom, this study models a two-story residential building located in climate zone B. This study evaluates various combinations of insulation and RES to determine solutions that not only achieve optimal thermal comfort but also align with the EU's directive requiring a minimum of 49% renewable energy integration [18]. By systematically analyzing and optimizing different scenarios, this study aims to provide actionable insights for enhancing residential energy systems and meeting stringent energy performance standards, thereby supporting the broader goals of sustainable building practices and regulatory compliance.

With respect to the scenarios examined, following reasons led to choosing the specific systems: There are a great variety of RES systems that one can examine, and within the limited space of a paper, we focused on combinations that are currently usually used in Greece. Solar thermal systems are the most popular RES system in urban areas and the most widespread RES system in the Greek building sector due to the country's favorable climate conditions and the presence of a strong industry since the 1970s. They have the highest acceptance rate by the public and are therefore the primary option for hot water production and supporting space heating systems. Heat pumps are becoming gradually more popular, despite their high initial costs. Biomass-fired boilers are not used in urban areas due to their emissions and due to practical problems, like the lack of storage space for pellets. They are, however, used in suburban and rural areas. Furthermore, natural gas will act as a bridge fuel towards the climate neutrality target and will remain the conventional mainstay in the coming two decades for residential heating, both alone and as a back-up system.

The use of building applied photovoltaics undoubtedly has benefits and is an appealing option. In this paper, however, we focused on the use of thermal RES (solar thermal, biomass, geothermal energy) as part of the effort to support heating and cooling decarbonization, which accounts for a large proportion of the energy consumption in European households. The 'electrification' of buildings' heating systems is a parallel track to meeting this goal, and there is strong interest in it, which is expressed in many research projects and the resulting literature.

Furthermore, this study's results can be used for the calculation and improvement of the smart readiness indicator of residential buildings, an indicator which will be mandatory in the next years. The smart readiness indicator of buildings assesses the technological readiness of a building by evaluating the functionality level of various smart devices, aiming at energy savings, the ability of the building to respond to users' needs and energy flexibility [19]. Last but not least, it is worth noting that there are many papers in the literature regarding the calculation of the optimal insulation thickness for various building elements [20], but investigations into the optimal insulation thickness in relation to the achieved renewable energy percentage are missing.

## 2. Materials and Methods

The computational analysis was conducted using the DesignBuilder V7.0.2.006 software which uses EnergyPlus V9.4.0.002, sourced by Gloucestershire, United Kingdom, for the energy simulation This study focused on a two-story residential building with a ground floor and two similar upper levels, as illustrated in Figure 1. The building was located in Athens, which belongs to climate zone B according to the Greek National Legislation [21]. The total conditioned floor area was 876.9 square meters, with a conditioned volume of 2178.3 cubic meters. The building was oriented to the north, and the window-to-wall ratio was approximately 30%.

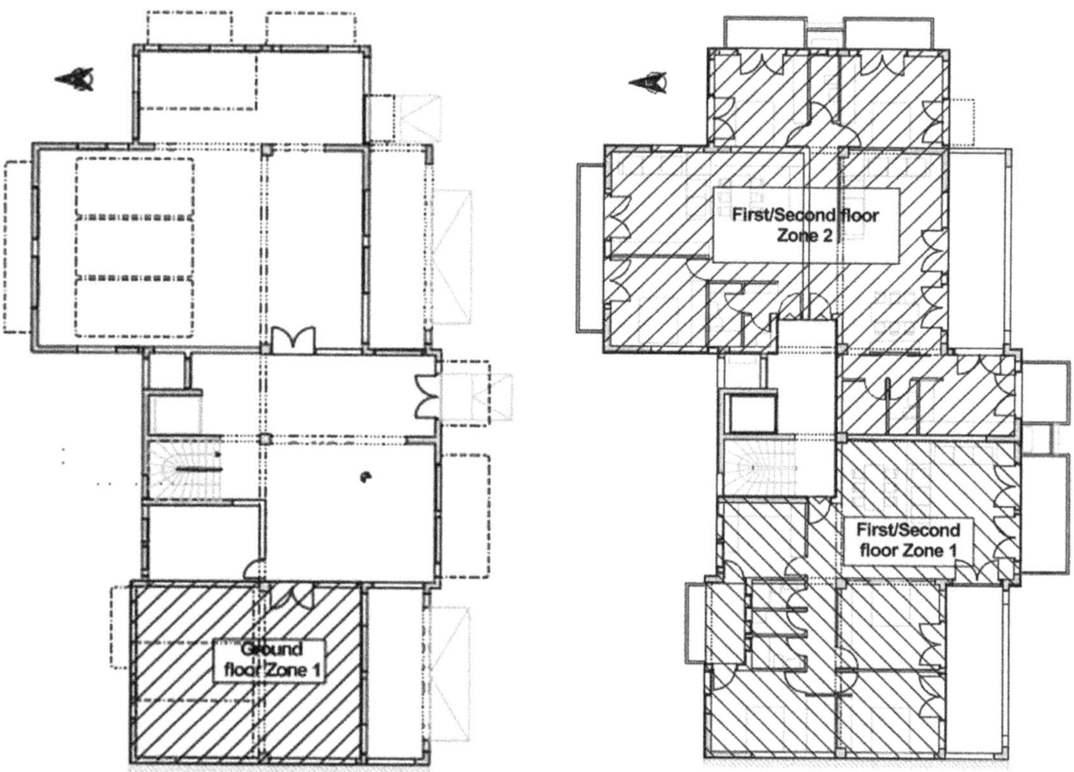


(a) (b)

**Figure 1.** (**a**) Ground floor, (**b**) first/second floor.

The building was initially modeled based on provided architectural designs, and then the structural elements were defined, and the appropriate materials were selected to minimize thermal losses. So, extruded polystyrene was used to insulate the external walls, roof and floor, with a thermal conductivity coefficient $\lambda = 0.033[\mathrm{W/mK}]$ [22]. Also, triple-glazed windows were used with $U_{glazing} = 0.8\left[\mathrm{W/m^2K}\right]$. Regarding the window frame, unplastic polyvinyl chloride (UPVC) material was selected, with a U-value of $U_{window\ frame} = 1.6\left[\mathrm{W/m^2K}\right]$. Furthermore, regarding the demand for lighting, it was assumed that most rooms needed between 200–300 lux illuminance, which was covered by light-emitting diode (LED) lamps of 2.5 $\mathrm{W/m^2}$ each. After the modeling of the building, the next step was to select the heating, ventilation and air conditioning (HVAC) systems. In this study, the calculations regarding the internal conditions were based exclusively on the internal air temperature within the building's thermal zones. The indoor temperature was maintained between 22 °C and 24 °C during summer and between 19 °C and 21 °C during winter. External factors, such as radiant heat gains or losses, were excluded from the assessment to ensure consistency across all scenarios. This approach focused on stabilizing the internal air temperature as the primary metric, which allowed for a controlled comparison of the different energy supply systems, independent of the outdoor environmental conditions. In every scenario, a steady-state heating design analysis was performed to calculate the maximum heating load of the building, and a steady-state cooling design analysis was performed to calculate the maximum cooling load. To ensure consistency in comparing the impacts of the various energy supply methods on the building's cooling and heating loads, an initial sizing of each HVAC system was conducted using the uninsulated building model as a baseline. This baseline allowed for a standardized calculation of the heating and cooling load requirements, facilitating a uniform starting point for assessing the different energy systems. For each subsequent scenario that included various insulation thicknesses, the *autosize* function was employed. This function dynamically adjusted the capacity of each HVAC system to match the specific heating and cooling demands created by the different insulation levels, ensuring that each system's output power was aligned with the corresponding load. By maintaining the initial sizing based on an uninsulated structure and employing the *autosize* adjustment for each insulation variant, the analysis was conducted on a comparable scale. This methodology allowed for consistent evaluation across the energy supply methods while accounting for variations in the building envelope performance and its influence on operational costs.

To calculate renewable energy integration, six different scenarios of hybrid systems were studied with different combinations of boilers, heat pumps, solar thermal systems and solar absorption chillers. In this analysis, the price of electricity was assumed to be 0.12 EUR/kWh, and the prices for biomass (pellet) were set at 0.11 EUR/kWh and 0.085 EUR/kWh, depending on the source. In Table 1, the technical characteristics of each scenario are presented.

### 2.1. Detailed HVAC Modeling

In every scenario, a detailed HVAC model was designed to appropriately size each part of the system, with the main goal of achieving acceptable internal conditions within the building. The controlled parameter was the air temperature in each thermal zone, which was divided into five distinct areas, with the first on the ground floor and the others on the first and second floors, corresponding to the four apartments. The heating setpoint was established at 20 °C and the cooling setpoint at 24 °C based on values from the Greek legislation of the Technical Chamber of Greece for residential buildings in climate zone B [23]. To optimize the system performance and maintain stable conditions, a hysteresis of ±2 °C was implemented. Consequently, the heating system was set to activate when the temperature falls below 18 °C and deactivate when it exceeds 22 °C. Similarly, the cooling system was set to begin operation when the temperature rises above 26 °C and stop when it drops below 22 °C.

**Table 1.** Technical specifications of the HVAC systems for each scenario.

| Case | Heating | Cooling | Domestic Hot Water | Technical Specifications |
|---|---|---|---|---|
| 1 | Biomass boiler, 20 $m^2$ solar collector | Local air conditioning units | 20 $m^2$ solar thermal collector, biomass boiler as auxiliary source | 40 kW Boiler 89% efficiency EERlocal units = 3 |
| 2 | Water-to-water geothermal heat pump, condensing natural gas boiler as auxiliary source | Water-to-water geothermal heat pump | 20 $m^2$ solar thermal collector thermal, condensing natural gas boiler as auxiliary source | COP = 4, EER = 5.5, Pheating = 10 kW Pcooling = 9 kW |
| 3 | Air-to-air heat pump, condensing natural gas boiler as auxiliary source | Air-to-air heat pump | 20 $m^2$ solar thermal collector, condensing natural gas boiler as auxiliary source | COP = 3.2, EER = 4, Pheating = 16 kW Pcooling = 13 kW |
| 4 | Air-to-water heat pump, condensing natural gas boiler as auxiliary source | Local air conditioning units | 20 $m^2$ solar thermal collector, condensing natural gas boiler as auxiliary source | COP = 4.5, Pheating = 11 kW EERlocal units = 3 Pcooling = 16 kW |
| 5 | 160 $m^2$ solar collector, biomass boiler as auxiliary source | Solar absorption chiller | 20 $m^2$ solar thermal collector, biomass boiler as auxiliary source | 30 kWbBoiler 89% efficiency 55 kW solar absorption chiller COP = 0.85 |
| 6 | Water-to-air geothermal heat pump | Water-to-air geothermal heat pump | 20 $m^2$ solar thermal collector | COP = 3, EER = 4.5, Pheating = 15 kW Pcooling = 12 kW |

### 2.2. Optimization Algorithm

The optimization process utilized the JEA optimization engine, which offers a range of configurable parameters to fine-tune the performance and effectiveness of the search for optimal solutions. The generation population size determines the number of solutions evaluated per iteration, with larger populations generally required for more design variables and influenced by the number of available processor cores. As the Pareto archive builds, the max population size limits the number of solutions per generation. The crossover rate, set high at 1.0, combines genetic information from parent solutions to create new offspring, mimicking biological crossover. The mutation rate, with a default of 0.4, ensures genetic diversity and helps prevent premature convergence, balancing the need for fast convergence and thorough exploration of parameter space. The tournament size, typically set to 2, governs the selection process, where larger sizes reduce the chance of weak individuals being selected, thus pushing towards desired objectives more effectively. The objective bias setting controls the trade-off between the Pareto ranking and constraint feasibility. Additionally, the optimization can be constrained by the max evaluations and max wall time, ensuring the process halts after a specified number of iterations or hours, respectively, preventing excessive computational time and resources from being expended unnecessarily. In this study, a Pareto-front-based optimization algorithm was employed to determine the optimal combination of the insulation thickness, RES integration and life cycle cost (LCC). The objective function used was the minimization of the LCC and the maximization of the RES, considering a 20-year life cycle and the percentage of RES integration [24]. Two different optimization algorithms were used in this study, one for the scenarios that included a heat pump and a second one for the scenarios that included a boiler. However, in all scenarios, the equations used regarding domestic hot water, lighting, equipment fans and pumps were the same.

#### 2.2.1. Equations for the Heat Pump Scenarios

The energy performance of the heat pump scenarios was calculated by Equations (1)–(8).

$$E_{HP,C} = \frac{Q_C}{EER} \text{ [kWh]} \quad (1)$$

$$E_{HP,H} = \frac{Q_H}{COP}\ [\text{kWh}] \quad (2)$$

$$E_{HP,Total} = E_{HP,C} + E_{HP,H}\ [\text{kWh}] \quad (3)$$

$$RES_{HP,C} = Q_C - E_{HP,C}\ [\text{kWh}] \quad (4)$$

$$RES_{HP,H} = Q_H - E_{HP,H}\ [\text{kWh}] \quad (5)$$

$$RES_{Total,HP} = RES_{HP,H} + RES_{HP,C}\ [\text{kWh}] \quad (6)$$

$$RES_{Total} = RES_{HP,H} + RES_{HP,C} + RES_{DHW}\ [\text{kWh}] \quad (7)$$

$$Q_{Total} = E_{HP,Total} + Q_{DHW} + Q_L + Q_E + E_{pumps} + E_{fans}\ [\text{kWh}] \quad (8)$$

### 2.2.2. Equations for the Boiler Scenarios

Moreover, Equations (9)–(12) were used to calculate the energy performance of the boiler scenarios.

$$Q_B = \frac{Q_H}{n_B}\ [\text{kWh}] \quad (9)$$

$$RES_{Biomass} = Q_H\ [\text{kWh}] \quad (10)$$

$$Q_{Total} = Q_B + Q_C + Q_{DHW} + Q_L + Q_E + E_{pumps} + E_{fans}\ [\text{kWh}] \quad (11)$$

$$RES_{Total} = RES_{Biomass} + RES_{DHW}\ [\text{kWh}] \quad (12)$$

### 2.2.3. Domestic Hot Water Equation

The percentage of RES regarding DHW was calculated by Equation (13).

$$RES_{DHW} = Q_{DHW} - E_{DHW}\ [\text{kWh}] \quad (13)$$

### 2.2.4. Total Renewable Percentage Equation

The overall RES percentage was estimated by Equation (14).

$$RES_{\%} = \frac{RES_{Total}}{Q_{Total}} \times 100\% \quad (14)$$

### 2.2.5. Life Cycle Cost Equation

Regarding the economic assessment, LCC was calculated by Equation (15).

$$LCC = C_{initial} + \sum_{t=1}^{n} \frac{\left(C_{O,t} + C_{M,t} + C_{replacement,t}\right)}{(1+r)^t} - \frac{R_{end}}{(1+r)^n}\ [€] \quad (15)$$

### 2.2.6. Initial Cost Equation

The initial Cost was estimated by Equation (16).

$$C_{initial} = C_{external\ walls} + C_{internal\ walls} + C_{windows\ frame} + C_{windows\ glazing} + C_{lighting} + C_{insulation} + C_{HVAC}\ [€] \quad (16)$$

### 2.2.7. HVAC Cost Equation (Ground-Source Heat Pumps Cases)

The HVAC cost for the ground source heat pump scenarios was calculated by Equation (17).

$$C_{HVAC} = C_{purchase} + C_{drilling}\ [€] \quad (17)$$

### 2.2.8. Operational Cost Equation

The operational cost was calculated by Equation (18)

$$C_{O,t} = C_{HVAC,t} + C_{lighting,t} + C_{devices,t}\ [€] \quad (18)$$

#### 2.2.9. HVAC Operational Cost Equation (Only Electricity)

The HVAC operation cost, only for electricity, was estimated by Equation (19).

$$C_{HVAC,t} = E_{HVAC} \times C_{electricity,t}\ [€] \quad (19)$$

#### 2.2.10. HVAC Operational Cost Equation (Electricity and Biomass)

The HVAC operational cost both for electricity and biomass was estimated by Equation (20).

$$C_{HVAC,t} = \left(E_{cooling} + E_{pumps} + E_{fans} + E_{DHW}\right) \times C_{electricity,t} + (E_{heating}) \times C_{biomass,t}\ [€] \quad (20)$$

#### 2.2.11. Objective Function and Constraints

The objective function and the constraints are presented in Equation (21).

$$\begin{gathered} Min_{s_{walls}, s_{roof}, s_{floor}}\ ( LCC - RES_{\%}) \\ subject\ to: \\ C_{initial} \leq C_{initial,max} \\ C_{O,yearly} \leq C_{O,yearly,max} \\ Q_{total} \leq Q_{total,max} \\ 0 \leq s_{walls} \leq 20\text{ cm} \\ 0 \leq s_{roof} \leq 20\text{ cm} \\ 0 \leq s_{floor} \leq 20\text{ cm} \end{gathered} \quad (21)$$

#### 2.2.12. Python Code for Data Extraction

A Python code was developed to extract data from an SQL database created after each EnergyPlus simulation. The code uses the simulation output file to consider every energy demand of the building. The workflow diagram of the code is presented in Figure 2.

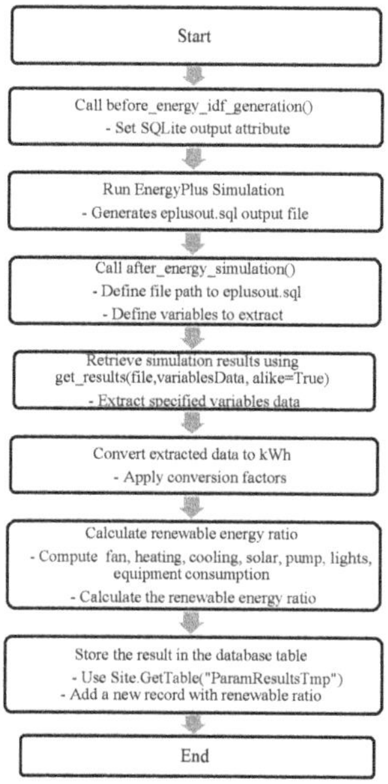


**Figure 2.** Workflow diagram of the Python script.

The script automates the process of extracting, processing and storing the energy consumption data from EnergyPlus simulations. By calculating the renewable energy ratio, it provides valuable insights into the sustainability and energy efficiency of the simulated building. This information can be crucial for designing energy-efficient buildings and optimizing their energy systems.

## 3. Results

The results of every optimization process indicate the optimal combination of the insulation thickness for the external walls, the ground floor and the roof of the building.

In Figures 3–7, the results for every scenario are shown.

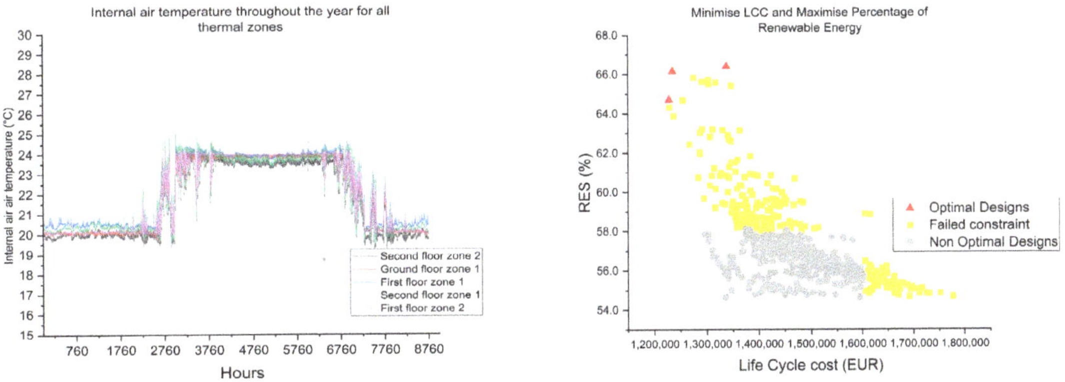


(a) (b)

**Figure 3.** (**a**) Internal air temperature throughout the year, (**b**) optimization results for Scenario 1.

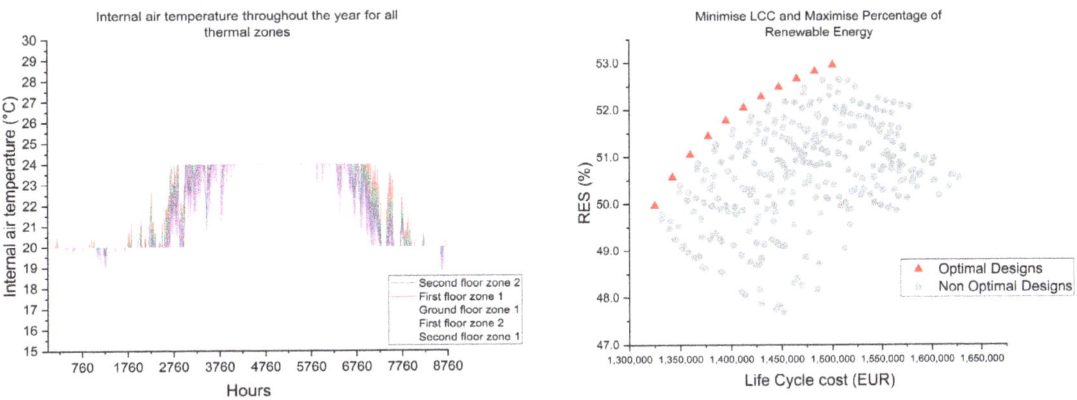


(a) (b)

**Figure 4.** (**a**) Internal air temperature throughout the year, (**b**) optimization results for Scenario 2.

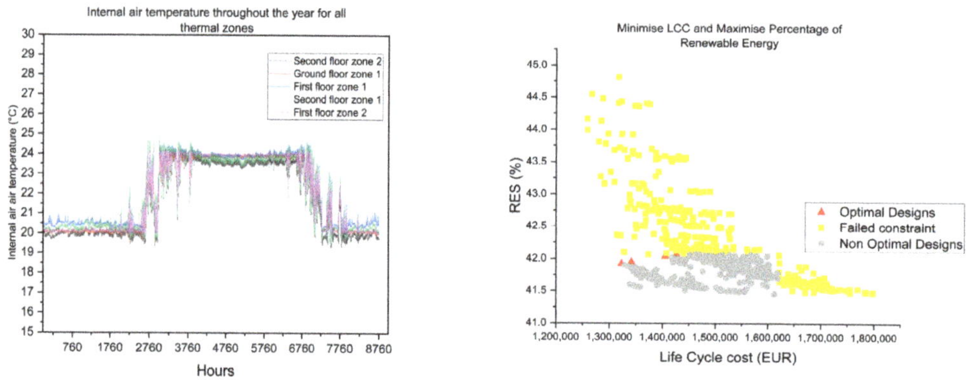


(a) (b)

**Figure 5.** (**a**) Internal air temperature throughout the year, (**b**) optimization results for Scenario 3.

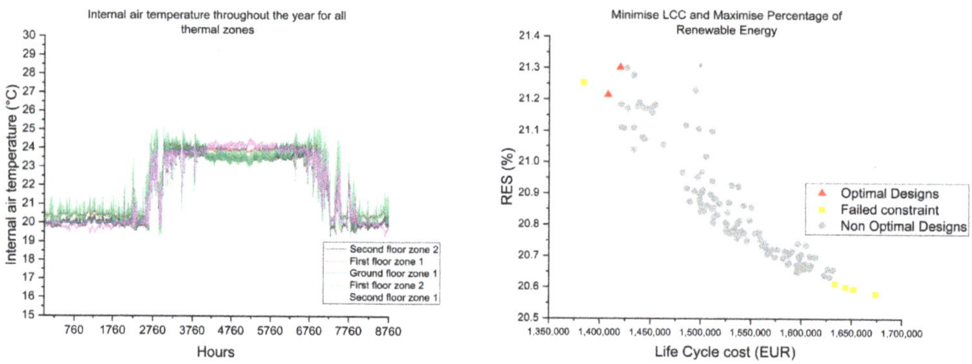


(a) (b)

**Figure 6.** (**a**) Internal air temperature throughout the year, (**b**) optimization results for Scenario 4.

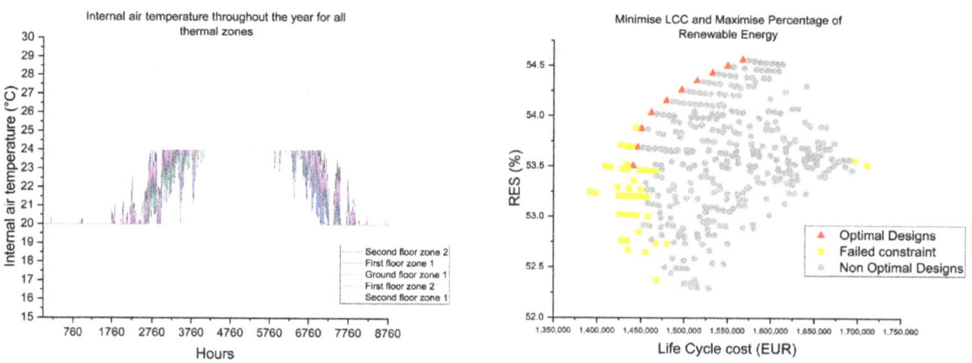


(a) (b)

**Figure 7.** (**a**) Internal air temperature throughout the year, (**b**) optimization results for Scenario 5.

### 3.1. Scenario 1: Biomass Boiler, Solar Thermal Collector and Packaged Terminal Air Conditioning Unit

In this case, the heating demand was effectively met by using a biomass boiler combined with floor heating as the terminal unit. The system maintained stable thermal comfort levels throughout the year. The cooling demand was covered by a packaged terminal air conditioning unit, which operated efficiently in all thermal zones. Domestic hot water requirements were fulfilled by a solar thermal collector equipped with an internal heating element. The integration of renewable energy reached approximately 65%, primarily due to the contribution of the biomass boiler. However, the optimization process revealed that several configurations exceeded the operational cost constraints and the construction cost constraints for insulation thicknesses over 16 cm.

### 3.2. Scenario 2: Ground-Source Water-to-Water Heat Pump and Solar Thermal Collector

In this scenario, the ground-source heat pump (GSHP) demonstrated stable performance in terms of both heating and cooling, while a low-temperature natural gas boiler was installed as a backup. Fan coil units were chosen as terminal units, allowing for the efficient management of heating and cooling. The optimization did not encounter any violations of constraints, mainly due to the reduced electricity consumption and operational costs achieved by the GSHP. This case effectively minimized operational costs while maintaining high energy efficiency.

### 3.3. Scenario 3: Air-to-Air Heat Pump and Solar Thermal Collector

This scenario, relatively unstable thermal comfort levels were obtained, though they were still within acceptable limits. The instability was attributed to the selected air system. Optimization attempts frequently failed to meet the operational cost constraints, primarily due to the significant electricity consumption caused by the heat pump's mechanical components, including the supply fan and the heating coil.

### 3.4. Scenario 4: Air-to-Water Heat Pump and Solar Thermal Collector

The air-to-water heat pump seemed to deliver very stable thermal comfort levels due to the usage of water convectors for heating and the packaged terminal air conditioning unit for the cooling demand. On the other hand, the renewable energy percentage was low because of the internal heating element in the heat pump that was necessary in order to cover the heating demand when the outside air temperature was low. So, it is obvious that in this case, the electricity consumption was increased, while the renewable energy percentage was reduced.

### 3.5. Scenario 5: Solar Absorption Chiller, Biomass Boiler and Solar Thermal Collector

The scenario with the solar absorption chiller had the most complex and detailed HVAC model. The solar thermal collectors' area (160 $m^2$) had multiple purposes; at first, it was used to cover the domestic hot water demand, and second it was used as a heating source for the solar absorption chiller, and it also participated in the coverage of the heating demand along with the biomass boiler. The optimization indicates a high renewable energy integration percentage with a small number of combinations failing the constraint of the operational costs. Additionally, the insulation thickness of over 13 cm also failed the constraint of the construction costs.

### 3.6. Scenario 6: Ground-Source Water-to-Air Heat Pump and Solar Thermal Corrector

Last but not least, the scenario with the water-to-air heat pump presented a relatively unstable performance (but between the acceptable margins) in terms of thermal comfort levels due to the air system being used as the terminal unit. As it can be seen in Figure 8, the renewable energy percentage was lower than the scenario with the water-to-water heat pump due to the required higher electricity consumption. The optimization indicates that the combinations with an insulation thickness of over 15 cm failed the constraint of the construction costs. The overall results regarding the optimal insulation thickness for each scenario are summarized in Table 2.

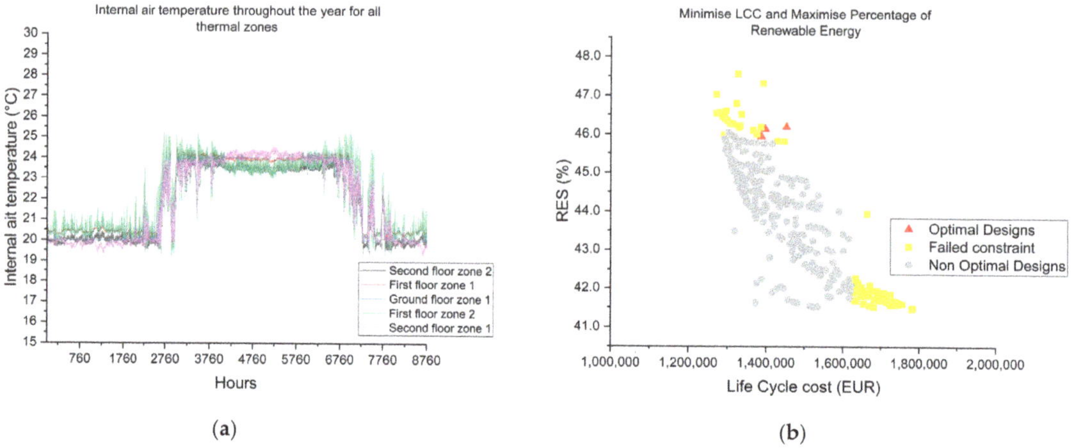


**Figure 8.** (**a**) Internal air temperature throughout the year, (**b**) optimization results for Scenario 6.

**Table 2.** Optimal insulation thicknesses for external walls, roof and floor in every case.

| Optimal Insulation Thickness | | | |
|---|---|---|---|
| **Scenario** | **External Walls [cm]** | **Roof [cm]** | **Floor [cm]** |
| Biomass boiler, solar collector and local air conditioning units | Varies from 0 to 3 cm | Varies from 0 to 8 cm | Varies from 5 to 15 cm |
| Water-to-water heat pump and solar thermal collector | Varies from 5 to 6 cm | Varies from 5 to 6 cm | Varies from 2 to 6 cm |
| Air-to-air heat pump and solar collector | Varies from 2 to 9 cm | Varies from 0 to 9 cm | Varies from 0 to 4 cm |
| Air-to-water heat pump and solar collector | Varies from 5 to 13 cm | Varies from 5 to 6 cm | Varies from 0 to 4 cm |
| Water-to-air heat pump, solar thermal collector and local air conditioning units | Varies from 6 to 12 cm | Varies from 4 to 5 cm | Varies from 6 to 9 cm |
| Solar absorption chiller, biomass boiler and solar thermal collector | Varies from 5 to 8 cm | Varies from 7 to 15 cm | Varies from 5 to 6 cm |

## 4. Discussion

In Figure 9, it is clearly shown that the scenarios (a) with the biomass boiler, (b) with the solar absorption chiller combined with a biomass boiler and (c) with the solar thermal collectors and the water-to-water heat pump could meet the EU directive's goal of 45% renewable energy integration. The water-to-air heat pump achieved a slightly lower percentage of renewable energy integration due to the required additional fan systems powered by electricity, but it still stands as an excellent option because of its low operational costs and stable performance. The air-to-air heat pump and the air-to-water heat pump seemed to have the lowest integration of renewable energy due to the significantly higher electricity consumption.

Furthermore, in Figure 10, the economic feasibility of each scenario is presented based on the life cycle analysis. It is obvious that the scenario with the biomass boiler had the lowest life cycle costs. On the other hand, the most cost-intensive solution was the solar absorption chiller due to its significantly lower COP and its high initial investment costs. Regarding the heat pumps, it was observed that for all four scenarios, the differences were small. It was seen that the ground-source heat pumps had high initial investment costs, mainly due to drilling costs, whereas air-source heat pumps had higher operational costs due to the required electricity consumption when the outside temperature was low. Although the indoor and outdoor temperatures remained constant across all scenarios, the building envelope's performance had to be adjusted based on the characteristics of each energy supply system. Systems with higher operational efficiency, such as ground-source heat pumps, benefit from better insulation, as it helps to reduce operational costs by minimizing heat losses. Conversely, less efficient systems, like air-to-air heat pumps, result in higher electricity consumption, requiring adjustments to the envelope to balance the life cycle costs. Therefore, optimizing the building envelope directly contributes to lowering operational costs and ensuring that renewable energy systems can meet the desired energy targets while maintaining cost-effectiveness.

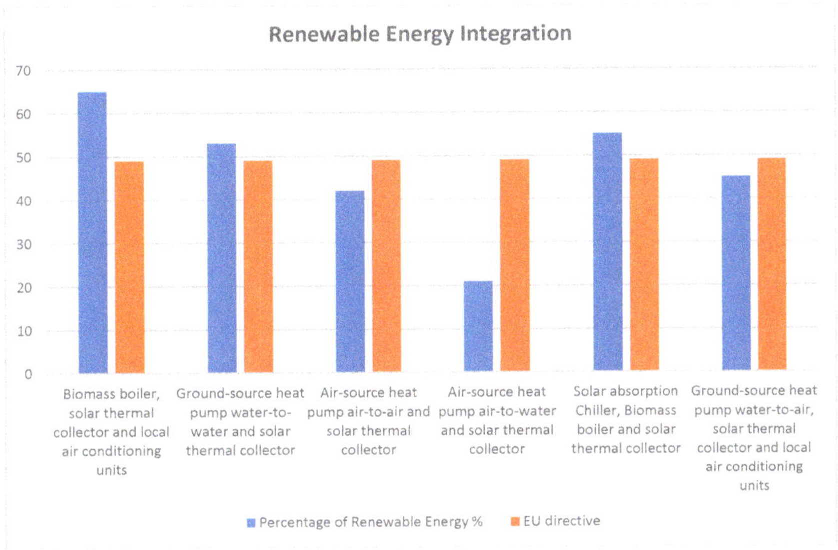


**Figure 9.** Comparison of each case with the EU directive.

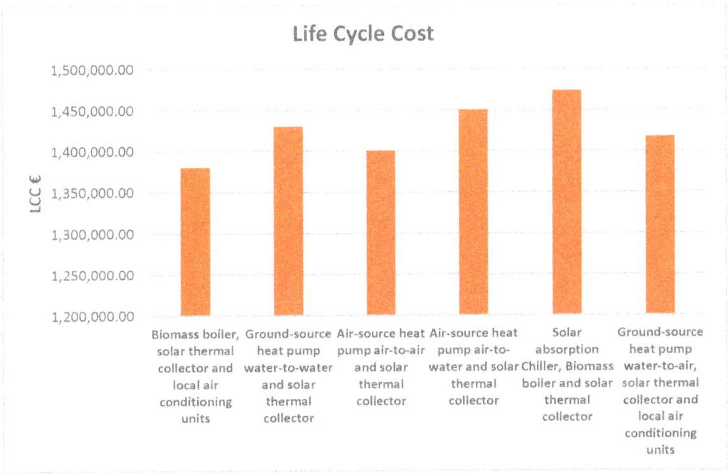


**Figure 10.** Comparison of life cycle costs for each scenario.

## 5. Conclusions

This research underscores the importance of optimizing insulation and incorporating renewable energy sources in residential buildings so as to maximize energy efficiency and cost savings. The use of advanced computational tools and optimization algorithms offers a robust framework for evaluating and implementing sustainable building solutions. The results highlight the importance of selecting appropriate materials and systems to enhance energy efficiency and economic feasibility. More specifically, regarding thermal insulation, it was observed that the optimal thermal insulation thickness in combination with the maximization of the RES integration varied from 0 cm up to 13 cm for external walls, 0–15 cm for roofs and 0–15 cm for floors, depending on the heating and cooling system. In terms of the energy simulations, it was found that thermal insulation was the variable with the largest effect on the energy consumption of the building. The increase in thickness limited the energy requirements for heating to a great extent, but at the same time, there was a slight increase in the cooling requirements, as heat transfer was minimized between the interior and exterior of the building. Based on the energy results regarding RES integration, it was concluded that the optimal solution was the biomass boiler. The ground-source heat pumps also had an equally high rate of RES integration along with a steady operation. On the other hand, the air-source heat pumps presented an unstable operation when the outside air temperature was low, and as a result, the RES integration percentage was reduced. Furthermore, regarding the calculated life cycle costs, the optimal scenario was once again the one with the biomass boiler, while the absorption chiller scenario was the worst. Additionally, the four heat pump scenarios presented similar results in the life cycle costs analysis with small differences. More specifically, the ground-source heat pumps had higher initial investment costs, whereas the air-source heat pumps had higher operational costs. Lastly, future research should investigate the integration of more renewable energy technologies like photovoltaics and/or micro wind turbines and, more importantly, hybrid solutions, including, for example, photovoltaics, heat pumps and storage, alongside their economic feasibility. Moreover, an environmental assessment should take place throughout the whole life cycle of the studied systems so as to include the environmental aspect in the analysis. Additionally, future studies should consider the variability in HVAC system performance ratios, specifically the efficiency, coefficient of performance and energy efficiency ratio, as these metrics respond to changes in a building's thermal envelope and different weather conditions across climate zones. This approach would provide a more nuanced understanding of energy supply optimization across diverse climates and building structures. Finally, incorporating thermal comfort metrics, such as the Predicted Mean Vote (PMV) and Predicted Percentage of Dissatisfied (PPD), would enable a comprehensive evaluation of occupant thermal comfort.

**Author Contributions:** Conceptualization, A.M.P., K.P., E.K. and E.G.; methodology, A.M.P., K.P. and E.K.; software, K.P. and E.K.; validation, A.M.P. and K.P. and E.G.; formal analysis, A.M.P., K.P. and E.K.; investigation, A.M.P. and K.P.; resources, A.M.P., K.P., E.K. and E.G.; data curation A.M.P. and E.K.; writing—original draft preparation, A.M.P., K.P., E.K. and E.G. writing—review and editing, A.M.P., K.P. and E.K.; visualization, A.M.P., K.P.; supervision, A.M.P.; project administration, A.M.P.; funding acquisition, A.M.P. All authors have read and agreed to the published version of the manuscript.

**Funding:** This research received no external funding.

**Data Availability Statement:** Data will be made available on request.

**Conflicts of Interest:** The authors declare no conflicts of interest.

## Abbreviations

The following abbreviations are used in this manuscript:

| | |
|---|---|
| EU | European Union |
| RES | Renewable energy systems |
| $CO_{2eq}$ | Equivalent carbon dioxide [kg] |
| $LCC$ | *Life cycle cost* [EUR] |
| UPVC | Unplastic polyvinyl chloride |
| LED | Lighting-emitting diode |
| HVAC | Heating, ventilation and air conditioning |
| $Q_H$ | *Annual heating load* [kWh] |
| $Q_C$ | *Annual coaling load* [kWh] |
| $Q_{DHW}$ | *Annual domestic hot water load* [kWh] |
| $Q_L$ | *Annual lighting energy consumption* [kWh] |
| $Q_E$ | *Annual equipment energy consumption* [kWh] |
| $EER$ | *Energy effiency ratio* [−] |
| $COP$ | *Coefficient of perfomance for heating* [−] |
| $n_B$ | *Efficiency of the boiler* |
| $Q_B$ | *Boiler fuel consumption* [kWh] |
| $E_{pumps}$ | *Annual electricity consumption of pumps* [kWh] |
| $E_{fans}$ | *Annual electricity consumption of fans* [kWh] |
| $E_{heating}$ | *Annual electricity consumption for heating* [kWh] |
| $E_{cooling}$ | *Annual electricity consumption for cooling* [kWh] |
| $E_{DHW}$ | *Annual electricity consumption for domestic hot water* $[kWh]$ |
| $Q_{Total}$ | *Total energy demand* [kWh] |
| $RES_{Total,HP}$ | *Renewable energy heat pump case* [kWh] |
| $RES_{Total,Biomass}$ | *Renewable energy biomass boiler case* [kWh] |
| $C_{initial}$ | *Initial investment cost* $[EUR]$ |
| $C_{O,t}$ | *Operational cost for the year t* $[EUR]$ |
| $C_{M,t}$ | *Maintenance cost for the year t* $[EUR]$ |
| $C_{replacement,t}$ | *Replacement cost for the year t* $[EUR]$ |
| $R_{end}$ | *Residual value an the end of the cycle* $[EUR]$ |
| $r$ | *Discount rate (reflecting the time value of money)* |
| $n$ | *Total number of year in the life cycle* |
| $PMV$ | *Predicted Mean Vote* [−] |
| $PPD$ | *Predicted Percentage of Dissatisfied* [−] |

## References


1. Eurostat. Available online: https://ec.europa.eu/eurostat/web/interactive-publications/energy-2023 (accessed on 15 September 2024).
2. Papadopoulos, A.M. Fourty years of regulations on the thermal performance of the building envelope in Europe: Achievements, perspectives and challenges. *Energy Build.* **2016**, *127*, 942–952. [CrossRef]
3. Eurostat. Available online: https://ec.europa.eu/eurostat/ (accessed on 10 September 2024).
4. European Union. Directive (EU) 2023/2413 of the European Parliament and of the Council. *Official Journal of the European Union*, 18 October 2023.
5. Giama, E.; Kyriaki, E.; Fokaides, P.; Papadopoulos, A.M. Energy policy towards nZEB: The Hellenic and Cypriot case. *Energy Sources Part A Recovery Util. Environ. Eff.* **2021**, 1–14. [CrossRef]
6. European Union. Directive (EU) 2024/1275 of the European Parliament and of the Council on the energy performance of buildings (Recast). *Official Journal of the European Union*, 24 April 2024.
7. Giama, E.; Kyriaki, E.; Papaevaggelou, A.; Papadopoulos, A.M. Energy and Environmental Analysis of Renewable Energy Systems Focused on Biomass Technologies for Residential Applications: The Life Cycle Energy Analysis Approach. *Energies* **2023**, *16*, 4433. [CrossRef]
8. Lü, X.; Lu, T.; Karirinne, S.; Mäkiranta, A.; Clements-Croome, D. Renewable energy resources and multi-energy hybrid systems for urban buildings in Nordic climate. *Energy Build.* **2023**, *282*, 112789. [CrossRef]
9. Saffari, M.; Keogh, D.; De Rosa, M.; Finn, D.P. Technical and economic assessment of a hybrid heat pump system as an energy retrofit measure in a residential building. *Energy Build.* **2023**, *295*, 113256. [CrossRef]
10. Indepent Power Transmission System Operator (IPTO). Annual Press Release for 2023, Athens, 2024. (In Greek, Englisch Abstract). Available online: https://www.admie.gr/en/nea/deltia-typoy/2023-record-year-clean-energy-greece (accessed on 12 September 2024).

11. Jung, Y.G.; Lee, K.H.; Park, B.R.; Kim, T.W.; Moon, J.W. Development and performance evaluation of intelligent algorithm for optimal control of a hybrid heat pump system during the cooling season. *Energy Build.* **2024**, *306*, 113934. [CrossRef]
12. Chantzis, G.; Giama, E.; Nižetić, S.; Papadopoulos, A.M. The potential of demand response as a tool for decarbonization in the energy transition. *Energy Build.* **2023**, *296*, 113255. [CrossRef]
13. Daneshazarian, R.; Berardi, U. Nano-enhanced thermal energy storage coupled to a hybrid renewable system for a high-rise zero emission building. *Energy Convers. Manag.* **2023**, *291*, 117301. [CrossRef]
14. Wang, X.; Mi, Z.; Li, K.; Huang, X.; Bao, W.; Song, J.; Wang, C.; Chen, G.; Cao, P. Design and transient analysis of renewable energy-based residential net-zero energy buildings with energy storage. *Renew. Energy* **2024**, *220*, 119512. [CrossRef]
15. Cheraghi, R.; Jahangir, M.H. Multi-objective optimization of a hybrid renewable energy system supplying a residential building using NSGA-II and MOPSO algorithms. *Energy Convers. Manag.* **2023**, *294*, 117515. [CrossRef]
16. Jafarian, M.; Assareh, E.; Ershadi, A.; Wang, X. Optimal integration of efficient energy storage and renewable sources in hybrid energy systems: A novel optimization and dynamic evaluation strategy. *J. Energy Storage* **2024**, *101*, 113880. [CrossRef]
17. Skiadopoulos, A.; Kosmadakis, G.; van Heule, X.; Lecompte, S.; De Paepe, M.; Manolakos, D. Hybrid solar-biomass residential micro–Combined heat and power systems driven by the Partially Evaporating Organic Rankine Cycle—A case study in Southern Europe. *Therm. Sci. Eng. Prog.* **2024**, *55*, 102906. [CrossRef]
18. Official Journal of the European Uion. Available online: https://eur-lex.europa.eu (accessed on 10 September 2024).
19. Chatzikonstantinidis, K.; Giama, E.; Fokaides, P.A.; Papadopoulos, A.M. Smart Readiness Indicator (SRI) as a Decision-Making Tool for Low Carbon Buildings. *Energies* **2024**, *17*, 1406. [CrossRef]
20. Fokaides, P.; Papadopoulos, A.M. Cost-optimal insulation thickness in dry and mesothermal climates: Existing models and their improvement. *Energy Build.* **2014**, *68*, 203–212. [CrossRef]
21. *TOTEE 20701-3/2010*; Technical Instruction of the Technical Chamber of Greece. Ministry of Environment, Energy and Climate Change: Athens, Greece, July 2010. (In Greek)
22. Kaemmerlen, A.; Vo, C.; Asllanaj, F.; Jeandel, G.; Baillis, D. Radiative properties of extruded polystyrene foams: Predictive model and experimental results. *J. Quant. Spectrosc. Radiat. Transf.* **2010**, *111*, 865–877. [CrossRef]
23. *TOTEE 20701-1/2017*; Technical Instruction of the Technical Chamber of Greece. Ministry of Environment and Energy: Athens, Greece, September 2017. (In Greek)
24. Purdue University. Available online: https://engineering.purdue.edu (accessed on 25 August 2024).

*energies*

MDPI

*Article*

# Development and Application of a Platform for Optimising Heating System Operation Based on the Building User's Temperature Perception

**Andre Beblek [1,*], Florian Felix Sehr [1], Viktor Grinewitschus [1], Carolin Baedeker [2] and Aaron Immanuel Wolber [3]**

[1] EBZ Business School, University of Applied Sciences, 44795 Bochum, Germany; f.sehr@ebz-bs.de (F.F.S.); v.grinewitschus@ebz-bs.de (V.G.)

[2] Research Unit Innovation Labs, Division Sustainable Production and Consumption, Wuppertal Institute for Climate, Environment, and Energy, 42103 Wuppertal, Germany; carolin.baedeker@wupperinst.org

[3] iBG Inbestergesellschaft GmbH, 50672 Köln, Germany; aaron@inbestergesellschaft.de

* Correspondence: a.beblek@ebz-bs.de; Tel.: +49-(0)-234-94-47-837

**Abstract:** The energy challenges of overcoming climate change and economic and global political crises require not only the increased integration of renewable energies but also an optimisation of energy generation and use and, as a result, a reduction in energy consumption in various sectors. Thermal energy consumption in buildings in particular accounts for a significant proportion of final energy consumption. With respect to commercial buildings, a central problem in optimising the system settings is the lack of or only limited information about the actual room temperatures as well as the comfort requirements and temperature perception of the users in the rooms on the one hand and the operational management and settings specified by the facility management, for example, the heating curves of the heating circuits, on the other. The aim here is to create a bidirectional flow of data and information so that the compromise between the necessary room temperatures and the comfort of the users and the most energy-efficient operational management possible can be achieved. In this context, the paper presents a developed and tested web platform that makes it possible to optimise the operation of the system technology, particularly the heating system in the building, from an energy point of view and to involve the building user (e.g., office employees) and to pass on information to the facility management, thus pursuing a holistic approach. In the associated Living Lab project (called ComfortLab), it was possible to obtain over 6500 votes on temperature perception and combine this with building operation and the parameters relevant to facility management. This made it possible to bridge the gap between user requirements and room temperatures on the one hand and energy consumption and the inlet temperature of the heating system and supply circuits on the other. The use of the platform makes it possible to optimise the setpoint specification, specifically the inlet temperature of individual heating circuits, considering both regular building operation at times of presence and the setting of weekend and night setback times. The results show a diversified picture regarding temperature perception and possible room temperature reductions of several degrees Celsius and energy savings in the double-digit percentage range.

**Keywords:** occupant-centric controls; building control; heating systems; building automation; energy efficiency; feedback systems; heating energy; facility management

## 1. Introduction

The current draft of the Energy Performance of Buildings Directive (EPBD) [1] provides for more buildings to be equipped with building automation systems from 2025 onwards, enabling more comprehensive monitoring of new or extensively renovated buildings, as in many cases buildings do not achieve the specific consumption levels that would be expected based on their building envelope and building technology. This is particularly important because buildings, and especially heating energy consumption, account for a significant proportion of energy consumption. In Germany, for example, this currently amounts to around 28% [2]. Overall, the EU-wide legislation aims to decarbonise the building stock and thus the associated energy supply by 2050. A mix of different measures will be necessary to achieve this ambitious goal. These include the integration of renewable energies, both in the electricity and heating sectors and the flexible and intelligent management of energy generation and consumption. Energy storage also plays a key role in decoupling the timing and can ensure an optimised balance between the supply and consumption of energy flows.

Studies on real buildings show that there is an interaction between user behaviour and the operational management of the building technology and the associated energy consumption [3,4]. Users not only have a direct influence on consumption, e.g., through inappropriate room temperature settings or incorrect ventilation behaviour. In commercially operated buildings, complaints from users about a real or perceived undersupply of heat also lead to adjustments to the system technology by the facility management, which lead to additional consumption and encourage energy wastage through user behaviour. For example, a ventilation heat loss caused by permanent tilt ventilation can be compensated with correspondingly high inlet temperatures at the radiator if sufficient heat is provided. The comfort gain (room temperature and fresh air supply) maintained by the user in this way leads to disproportionately high heat energy consumption.

The plant automation (e.g., control of the heat generator, speed specification of circulation pumps, etc.) and the room automation (e.g., control of the lighting, control of electronic thermostatic valves) are organised as two trades in terms of structure and planning [5]. Although information can be exchanged via interfaces, this is only done in a few cases and mostly in informative form (for example, as the integration of warning messages from room automation in the building management system) and not to optimise operational management (for example, to adapt setpoint specifications for the system technology from the room automation data determined). Energy management is traditionally another system to which information is also transferred via corresponding interfaces and used for evaluation. A central challenge in the operation and optimisation of the operational management of the system technology is therefore the great uncertainty about the real perception of the employees and the situation in the offices on the one hand and the decentralised information situation in the various systems on the other. In traditional building automation systems and platforms, there is no structured way of obtaining and processing feedback apart from unrepresentative individual opinions and problems. This applies not only to settings during working hours, but also to information about the effect and the utilised and still existing potential of setback operations outside working hours. In practice, this often leads to a generalised oversupply of heating energy, especially when individual complaints reach facility management.

This is where the ComfortLab platform presented in this paper comes in. The associated solution is to obtain standardised feedback on the temperature perception of the largest possible number of employees with as little individual effort as possible and to take this into account as a specification for the operational management of the system technology. The platform was developed in a BBSR-funded (Federal Institute for Research on Building, Urban Affairs and Spatial Development) research project by the EBZ Business School and the Wuppertal Institute with the involvement of the design studio 'inbestergesellschaft' on the basis of the Living Lab approach [6,7]. The project was carried out in the heating periods 2021/2022 (pre-test) and 2022/2023 (field test) with over 200 participants in the BMUV properties (Federal Ministry for the Environment, Nature Conservation, Nuclear Safety and Consumer Protection) in Berlin and Bonn, which cover a total gross floor area of around 100,000 $m^2$. In the predecessor project (details can be found in the corresponding publication [8]) the main focus was on the user's individual interaction with the building in terms of heating and ventilation behaviour. To this end, the effect of feedback systems and surveys was analysed and successfully evaluated. In the follow-up project, this approach was continued, but with the focus on reaching a significantly larger number of employees to provide better findings for the operational management of the system technology as a whole and as support for instructions for facility management.

The current status of other research and publication activities was analysed on the basis of a literature search. Several sources were found that take partial aspects of the ComfortLab approach into account. The integral overall approach between querying the temperature perception, analysing the building technology and providing feedback to the facility management, considering an appealing user interface that is understandable for non-technical users, has not yet been tested and represents a novelty with regard to the available publications. The following is a brief description of current research activities and results and how they relate to this publication and the ComfortLab approach.

In [9], a literature search of studies was carried out in which user-centred operational management is described. Various aspects and technologies were considered and summarised here. An existing challenge is seen in the necessary amount of user-related data; the number of studies that take building operators and facility management into account is also summarised as very low and requires further investigation. Di Giuda et al. [10] have developed a system to support adjustments to the system technology, in which data from IoT sensors (e.g., cameras for presence detection) and post-occupancy evaluations were used to support facility management in optimising operations and building use based on defined KPIs. This did not include any kind of direct feedback from employees. The authors in [11] summarise the structure of IoT systems in the context of building management systems and propose an 'adaptation level' at the management level that will, for example, incorporate user activities into operational management in the future. In [12], an interface was developed that visualises information on energy consumption for students. Surveys conducted at the same time in the co-design process were met with broad interest; 48 students were surveyed, and the feedback was used to revise the visualisation. Feedback into the operational management or the derivation of recommendations for action for facility management was not investigated. Garg et al. [13] have developed a survey-based interface for households, which is intended to encourage energy savings through feedback. Here too, optimisation or adaptation of the system technology was not investigated.

In [14], the authors examined building automation systems in a review paper with regard to the possibilities of adapting building operation while taking real-time feedback into account. Only a few studies were found that show mathematical relationships between physical factors of the indoor climate and productivity, most of which are still based on static models (e.g., the PMV), and there is potential for an individual approach. Jayathissa et al. [15] have developed an app that can be installed on Fitbit watches. In a four-week test phase, almost 1500 votes were collected from 15 people. Hierarchical K-Means Clustering was used to form four clusters for comfort perception. The use requires configuration and availability of the smart watches and does not offer any direct recommendations for actions for the facility management regarding system technology settings. In [16], a system was developed that provides building users with recommendations for reducing their energy consumption on the basis of operating data; the premise here was a high degree of scalability to other buildings and systems. Feedback from the user or the derivation of system settings was not considered.

In [17], Jin et al. simulatively tested the possibility of reducing the consumption of ventilation systems, plug-in loads and lighting by means of presence detection through WLAN activity; on a daily average, this could lead to energy savings of 10% in the study. There was no transfer to real buildings or a field test. In [18], the acceptance of a building automation system by facility managers was evaluated, with the system providing machine-learning-based information about faults in the system technology. Compliance with the user's comfort requirements was defined as a goal alongside energy- and cost-efficient operational management but was not specifically taken into account during development. Desogus et al. [19] are actively developing a platform that enables facility management to optimise building operation based on measured comfort indicators. The evaluation of comfort is based on measured parameters and not on feedback. Based on a partial dataset from the ASHRAE Global Thermal Comfort Database II [20], the authors of [21] investigated how the load control of HVAC systems can be realised under the predicted thermal satisfaction based on the dataset. No original data samples were collected, and no tests were carried out on real systems.

The summary of the current research situation shows that individual aspects, such as recommended actions for system settings and feedback from users via end devices (e.g., smart watches), already exist and have been implemented in the research landscape. The ComfortLab platform, on the other hand, offers a device- and user-independent way of intuitively assessing the temperature well-being of the largest possible number of people, combining these data with technical parameters from building automation and making them available to facility management. The platform is both flexible in terms of building technology and cost- and resource-efficient in terms of the necessary data collection for the largest possible group of people. The following research questions were defined as guiding principles and were to be answered:

- What are the requirements in terms of thermal comfort in the object under investigation and what parameters need to be collected for the envisaged solution?
- What is the status quo of conventional operational management, considering different operating periods?
- Is it possible to develop a simple and intuitive, web-based solution that addresses as many participants as possible and obtains their temperature perception over a sufficiently long period of time on one site and considers the requirements of the facility management on the other site?
- What reduction in the flow temperature in heating circuits as a central control variable is possible, and what effects does this have on heat energy consumption?

The rest of the paper is structured as follows: Section 2 presents the background to the measurements and platform development as well as the functionality and technical parameters. Section 3 presents and interprets the results of the comfort measurements and the use and acceptance of the platform. The resulting energy consumption and room temperatures as well as the results from the final survey, which was carried out with the participants after the test period, are included. Section 4 concludes the article with a summary and critical evaluation of the results, as well as open research questions and planned future developments.

## 2. Materials and Methods

This section describes the platform and the associated development. The platform was developed and tested in several iteration loops in four heating periods. Initially, measurements were taken to carry out a basic analysis of the comfort, ventilation and presence situation (Section 2.1). These results were used as the basis for the initial assessment of the operational management of the system technology and for evaluating the necessary technical parameters for the development. This is followed by a description of the platform used to collect the participants' data and provide it together with the system parameters for facility management (Section 2.2).

### 2.1. Comfort and Room Climate Measurements

To record the current comfort situation and the indoor climate, measurements were carried out in selected rooms using measuring devices to record the relevant parameters for determining the PMV/PPD in accordance with DIN EN ISO 7730 [22]. The measured values include relative humidity, air velocity, air temperature and globe temperature. The clothing index and metabolic rate were assumed to correspond to standard values for office work (metabolic rate: 1.3 MET, clothing insulation: 0.9).

The aim of these measurements was to obtain an assessment of whether there were any fundamental problems in the building, such as draughts, that needed to be considered. Furthermore, as in the previous project, the PIAF ventilation assistant was used in selected offices [23]. This records measurement data on the room climate (carbon dioxide concentration, room air temperature, relative humidity) and uses acoustic and visual signals to alert the user to necessary and sensible ventilation processes. This enabled the occupancy of the offices and the room temperature level to be recorded in selected offices over the entire test period. Technical details on the PIAF ventilation assistant can be found in the corresponding publication [8]. In addition to the data from the system technology, this forms the database that was used in the development of the ComfortLab platform. As part of the comfort monitoring, 36 rooms in Bonn with different types of use (meeting rooms, offices) were measured; details of the measurement technology used can be found in [24]. The values were collected in the period from February to May 2021, so that although only part of the heating period was recorded, there are still sufficient periods with low temperatures below the heating limit. The data storage interval was set to 5 min; measurements were taken for each room for a period of two weeks, after which the measuring devices were installed in the next room. Figure 1 shows an example setup including the data logger and the sensors for measuring indoor air quality, radiant temperature and air velocity.

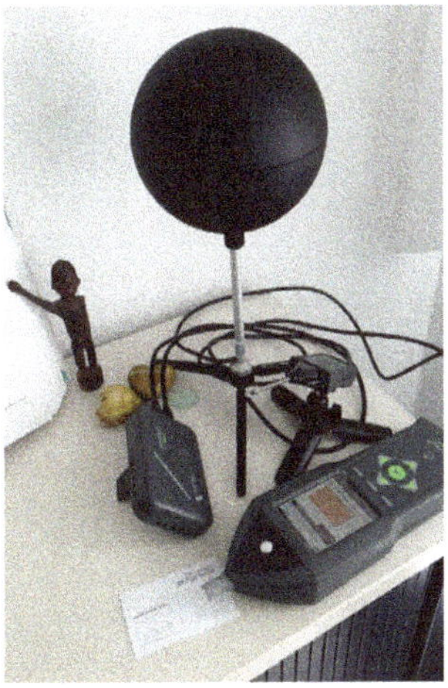

**Figure 1.** Example of the comfort measurement setup.

The devices were positioned in such a way that the effects of sunlight, draughts or other disturbances were avoided.

### 2.2. Technical Implementation

The ComfortLab platform consists of a decentralised structure with different subsystems as shown in Figure 2. The data interface was set up in such a way that the use of common protocols (e.g., BACnet, OPC, Modbus) of the various building automation systems in individual buildings was possible. The transfer to the ComfortLab platform from the individual properties (left in Figure 2) takes place via IoT and web technologies (e.g., MQTT, REST). Edge devices serve as intermediaries between the property and the platform, recording local information from the building management system and forwarding it (Wi-Fi, LTE). This allows different buildings with different automation solutions to be connected. The second part is the system consisting of a database server, MQTT broker and REST interface (middle part of Figure 2). This is where the data are stored, normalised and made available in various aggregation levels (e.g., daily consumption of individual heating circuits, hourly average inlet temperatures of heating circuits). Different units, intervals, etc., from the individual properties are thus harmonised, and data can be compared. Weather data were obtained from the German Weather Service via a REST interface (upper part of Figure 2). For the Bonn property, data from the weather station "Köln/Bonn (Flugh.)" at a distance of about 17 km were used; for Berlin, data from the station "Berlin-Tempelhof" at a distance of about 1 km were used. The final part is the ComfortLab platform, operated on a separate system in the field test (right part in Figure 2). This combines the information and enables the temperature perception and visual overview to be provided by and for the employees (ComfortStripes) as well as the technical representation for facility management (ComfortHub).

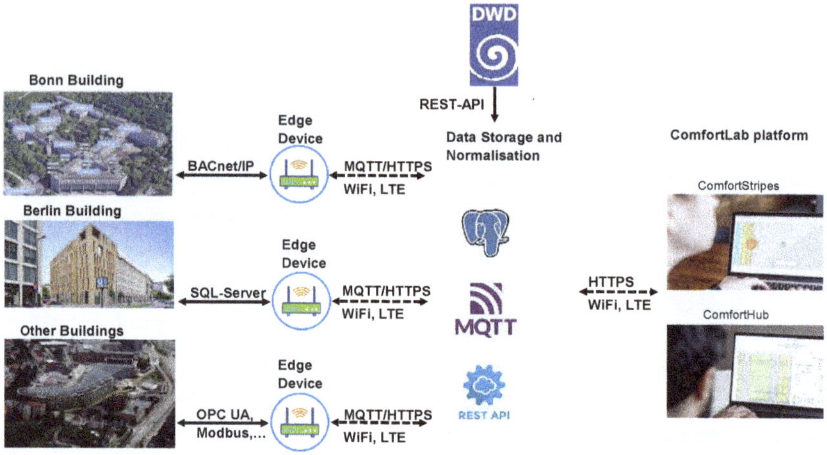


**Figure 2.** Technical structure of the ComfortLab platform.

As office staff and facility management have different objectives and a different technical understanding of the indoor climate and system parameters, the presentation of the data collected in the ComfortLab platform was adapted for both groups. The representation for both groups is based on the idea of Ed Hawkins' "Warming Stripes", which are used to visualise the long-term rise in global warming temperatures [25]. The "Warming Stripes" show the average annual temperatures of a location or region. On the multi-coloured, sequential colour scale, deep blue stripes represent years with low average temperatures, while deep red stripes show years with high average temperatures. The absolute values differ from location to location, but over the course of many years, there is a general progression from blue to red bars and a corresponding rise in temperature.

This idea was adapted for the ComfortLab platform: Blue colours stand for cold room temperatures and a cold temperature sensation; red colours, for high room temperatures and a warm temperature sensation. An example of the representation of employees and facility management is shown in Figure 3.

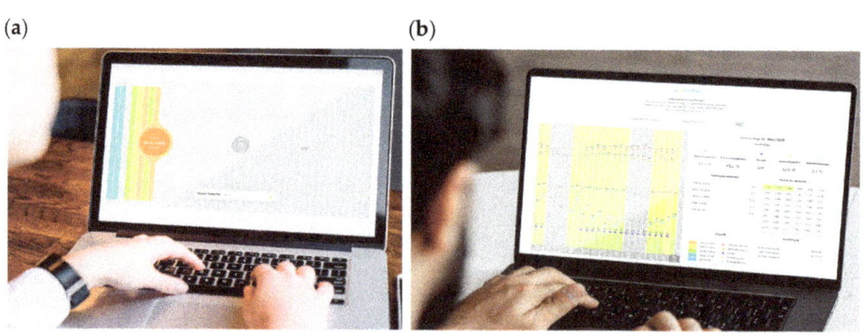


**Figure 3.** (**a**) ComfortStripes view for employees: The bars (blue: cold, orange: warm) show the temperature perception and the measured room temperatures over time; (**b**) ComfortHub view for facility management: the values are aggregated at heating circuit level.

In order to reach as many office employees as possible and encourage them to participate, the simplest possible form of interaction was developed—the "ComfortStripes", which is shown in the left part of Figure 3. The vote on the temperature sensation and the recorded room temperature is visualised over time using a coloured bar. Blue stands for cool, and red for warm temperatures. To collect the data, employees were notified twice a day by e-mail and asked to vote. As wireless temperature sensors could not be installed due to data protection concerns, the room temperature readings from a desktop thermometer had to be entered in addition to the temperature perception data. This step is prone to error and only represents a snapshot, so the aim is to automate this recording in the future. This was successfully tested as a prototype in a selection of rooms. Each participant could view their previous votes and entered room temperatures with access protection using a web browser.

As many office buildings in Germany are heated via radiators and the air is exchanged via manual window openings, targeted humidification and dehumidification or active cooling of the room air, for example, as is possible with air conditioning systems, is not possible or only possible to a very limited extent. This reduces the possibilities for local technology in the office to influence the indoor climate parameters, which are included in the calculation according to the Fanger PMV comfort model, for example. As this limits the possibilities for influencing thermal comfort, a simplified specification of five instead of seven possible levels (warmer, warm, neutral, cold, colder) was chosen for the ComfortLab platform. This should further simplify everyday responses and further increase acceptance and motivation for long-term participation among employees. In addition to stating their temperature perception and the measured temperature, participants were able to enter their days and times of attendance. This information can be used to estimate the occupancy of the building and individual parts of the building and to incorporate the control of the heating circuits. In this way, the data made it possible to optimise operational management during working/operating times as well as during non-attendance times and provide assistance with night and weekend heat supply reduction.

The display for the facility management, shown as an example in the right part of Figure 3, was expanded to include an aggregation level at the heating circuit level as well as the evaluation of inlet and return temperatures and energy consumption depending on the outside temperature. This enabled the facility management team to view the reconciliations for individual heating circuits, together with the entered room temperatures, and to make the necessary adjustments to the system technology on this basis. The aggregation of the individual offices into a heating circuit/line level enables an overall view of the system parameters and recorded temperatures and comfort settings. For example, the temperature level of all offices connected to a heating circuit can be recorded and evaluated. This enables a targeted assessment of whether, for example, the overall temperature level in the corresponding heating circuit is too low or whether there are any individual, local problems in certain rooms. This represents a contrast to the classic approach, in which individual complaints lead to the temperature level of the heat supply being raised at an overall level (heating circuit or entire building) to eliminate individual complaints due to a lack of alternative information.

## 3. Results

Following the presentation of the infrastructure and the use of the ComfortLab platform, this section presents the key results from the comfort and indoor climate measurements (Section 3.1) and the use of the ComfortLab platform as well as the temperature perception (Section 3.2). Section 3.3 explains and classifies the effects on system operation and the resulting energy consumption. The results are then compared with the findings from the final survey, in which the ComfortLab participants were asked about their experiences following the test period (Section 3.4).

### *3.1. Results of Comfort and Room Climate Measurements*

As the parameters for the ComfortLab platform should be recorded with as little effort as possible in order to enable as many employees as possible to participate, comfort measurements were initially taken in a representative selection of offices, as described in Section 2.1. Should fundamental problems arise, for example with draughts, this would have to be considered in the research design and voting behaviour. Otherwise, the indoor air temperature in the offices should form the basis for the survey as the dominant and relatively easy-to-determine measured value in a large number of offices. As the offices were largely unoccupied in 2021 and partly also in 2022 due to the ongoing COVID-19 pandemic and the resulting predominant home office work of most employees, the measurements represent the indoor climate situation without the effect of the presence of people (settings on thermostats, ventilation processes, etc.) and the necessary temperature control of the offices. In principle, a general lowering of the temperature level would have been possible here, but definitely during non-working hours. The corresponding time periods are defined as follows for all evaluations:

- Working days' daytime operation: Monday to Friday from 7 a.m. to 6 p.m.
- Working days' night setback: Monday to Friday from 6 p.m. to 7 a.m.
- Weekend operation: Saturday and Sunday all day.

Figure 4 shows the frequency distribution of the calculated operative temperature as the mean value of the measured room temperature and the mean radiation temperature from the comfort measurements of all rooms as hourly average values for the three time periods. Statistical parameters from the operative temperature distribution are listed in Table 1. Only values from days with average outdoor temperatures below 15 °C corresponding to the heating limit of the building are considered. It can be seen that in daytime operation (left diagram, olive green bars), in night-time operation (middle diagram, slate grey bars) and on weekends (right diagram, dark magenta bars), the average operating temperature on heating days is above 23 °C and also rises to up to 27 °C at times during the night and on weekends.

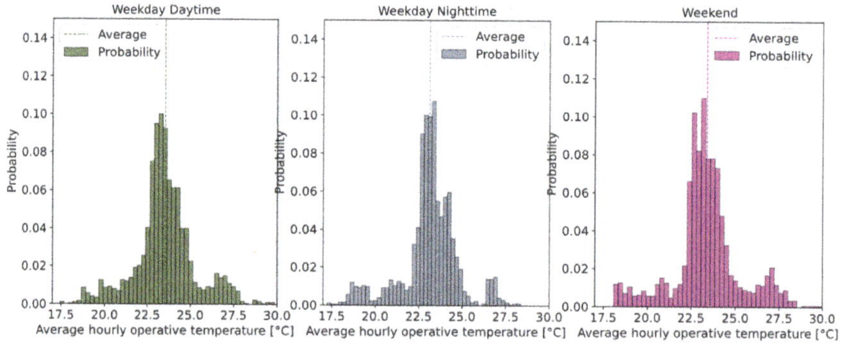


**Figure 4.** Probability distributions of the operative temperature (average of the measured air and radiant temperature) during the different time periods.

**Table 1.** Statistical parameters of the operative temperature distribution.

| Parameter | Weekday Daytime | Weekday Nighttime | Weekend |
|---|---|---|---|
| Minimum [°C] | 17.44 | 15.45 | 18.10 |
| Maximum [°C] | 32.35 | 28.37 | 33.75 |
| Mean [°C] | 23.55 | 23.11 | 23.39 |
| Median [°C] | 23.44 | 23.20 | 23.26 |
| Standard Deviation [°C] | 1.82 | 1.70 | 2.00 |
| Skewness | 0.29 | −0.55 | 0.21 |

Accordingly, there should be no problems with the possible heating power provided by the heating system. Even without real presence, the PMV value calculated from the measured values and parameters (metabolic rate: 1.3 MET, clothing insulation: 0.9) allows conclusions to be drawn about the comfort situation for different periods of time. Above all, during the assumed non-working hours and the associated non-occupancy of the building, the temperatures can be lowered considering a minimum building physics temperature. Figure 5 visualises the PMV calculated from the measured values as well as the metabolic rate and the clothing index for the three time periods, analogous to Figure 4. Statistical parameters from the PMV distribution are listed in Table 2.

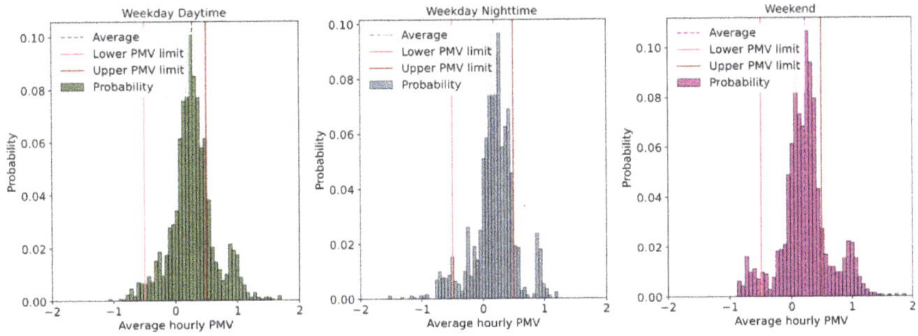


**Figure 5.** Probability distributions of the calculated predicted mean vote during the different time periods.

**Table 2.** Statistical parameters of the PMV distribution.

| Parameter | Weekday Daytime | Weekday Nighttime | Weekend |
|---|---|---|---|
| Minimum | −1.08 | −1.56 | −0.88 |
| Maximum | 1.71 | 1.22 | 1.89 |
| Mean | 0.28 | 0.18 | 0.23 |
| Median | 0.26 | 0.20 | 0.23 |
| Standard Deviation | 0.36 | 0.35 | 0.40 |
| Skewness | 0.16 | −0.61 | 0.10 |

The red lines (−0.5 to 0.5) represent the range in which the PMV should be when people are present. The comfort index changes only slightly within the different time periods but is also in the rather high range at night and on weekends, sometimes even in the "too warm" range. The few low PMV values are associated with individual cold radiation temperatures but are the exception. Other comfort problems, such as draughts, were not recorded by the measurements and should not have a significant negative impact on comfort. To keep the survey for the ComfortLab test feasible in terms of cost and effort, the room air temperature was therefore chosen as the decisive factor. As the office users were already provided with desktop thermometers to record the room temperature and sensitise them to heating and ventilation behaviour due to the 19-degree strategy specified by the EnSikuMaV [26], these values could also be used pragmatically for the platform.

In addition to the short-term comfort measurements, the PIAF assistance system was installed in almost 30 selected offices in Bonn and Berlin for the entire duration of the test to record the long-term development of room temperatures and indoor air quality, at least on a random basis, and to be able to assess the occupancy of the offices with a high temporal resolution. PIAF records relevant room air parameters and uses acoustic and visual signals to provide users with recommendations on correct ventilation behaviour. The background and details are described in the publication on the previous project [8]. The recorded data on room air temperature and $CO_2$ content were used to gain further insights into the occupancy density and room temperatures, as well as to estimate reduction and savings potential. Figure 6 shows the average hourly indoor air temperature above the hourly maximum value of the carbon dioxide concentration for all offices in Bonn, divided into the pre-test period (heating period 2021/2022: circle) and the ComfortLab field test (heating period 2022/2023: cross). In the status before the ComfortLab platform was used, relatively high room temperatures dominated during the entire heating period, even when the building was not occupied. The evaluation of the comfort measurements, which could only be collected for short periods of time due to the complex and cost-intensive measurement setup, can thus be further validated with the PIAF measurements accompanying the heating period.

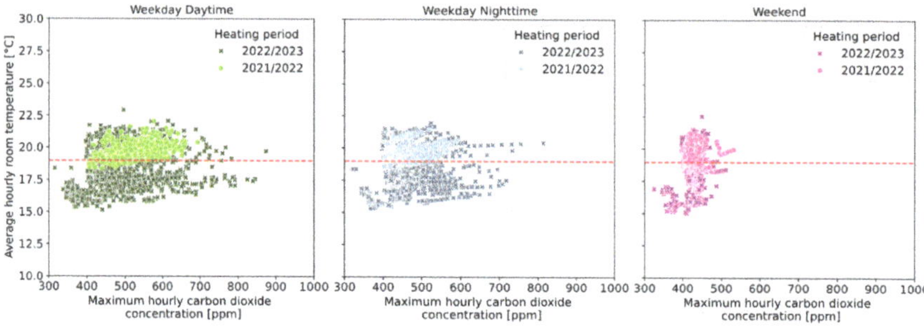


**Figure 6.** Room temperatures and carbon dioxide concentration for different time periods.

The evaluation of the measurements for selected periods and rooms shows that no fundamental comfort problems are to be expected in the rooms. It should be possible to reduce the temperatures to a significant extent, especially a much greater reduction during non-occupancy periods. The possibility of reducing energy consumption by lowering the heating system in residential and non-residential buildings and the practical application have been examined in detail in other studies. In principle, the application in commercial buildings is easier to realise due to the more regular use (often no occupancy on weekends and regular occupancy during the day) compared to residential buildings [27,28].

### 3.2. *ComfortLab Platform Results*

The following section summarises the results of the use of the ComfortLab platform in the pre-test in the 2021/2022 heating period and in the field test in the 2022/2023 heating period. The acceptance of the participants and the evaluation of the activity, the individual adjustments to temperature perception and the effects on room temperatures and energy consumption are explained.

The basic idea of the ComfortLab platform and the possible technical realisation was tested in the pre-test in the 2021/2022 heating period with a small number of selected participants (15 in Bonn and 11 in Berlin). An initial version of the platform was presented to them and made available for use. The findings were discussed in a workshop with the participants, and the feedback was used to adapt the platform and design for the field test in winter 2022/2023. This field test took place from 22 November 2022, to 31 March 2023, in the two BMUV buildings in Bonn and Berlin.

A total of 193 participants (111 in Bonn, 82 in Berlin) tested the platform and submitted at least one vote and the corresponding room temperature measurement, with a total of 6509 votes cast (2823 in Berlin, 3686 in Bonn). As the offices in both properties were rather sparsely occupied in both winters, the number of participants as a group of people present can be considered very high. Of the total number of participants, 44 cast at least 50 votes, 70 cast at least 30 votes and 152 cast at least 10 votes during the survey period; a minimum number of votes was considered in the respective evaluations.

To enable a comparison of the two heating periods, the tunings from the pre-test of the first heating period (winter 2021/2022), in which there were no special regulatory requirements for room heating, were compared with the tunings from the field test in the second heating period. Figure 7 shows the change in the tuning in the two test winters.

In winter 2021/22, the vote in Berlin (86%) and Bonn (70%) was predominantly neutral, with the room temperatures measured by the participants at around 22 °C (the room temperatures measured using existing sensors in new buildings in Berlin are more in the region of 23 °C). The distribution of votes in the field test tended to shift slightly to the left towards the cold range, although the fine vote still represents the most frequent vote in both properties.

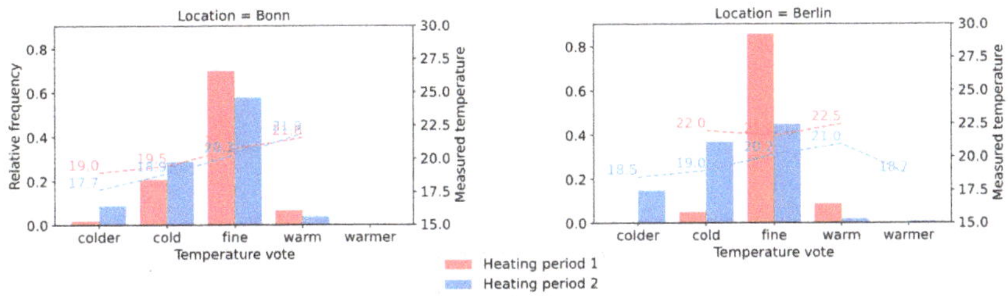


**Figure 7.** Comparison of the frequency of the perceived (bars, left $y$-axis) and measured (dotted line, right $y$-axis) temperature for both heating periods (heating period 1: red, heating period 2: blue) for both locations (Bonn: (**left plot**), Berlin: (**right plot**)).

However, it should be noted that in the status quo, the available heating capacity was sufficient to dissipate heat even at cold outside temperatures, for example, by tilting the windows, thus maintaining comfort in terms of temperature and air quality. The average room temperature associated with the respective vote shows an expected correlation between room temperature and perception but is below the temperatures in the pre-test for the individual voting options in the winter field test and is more pronounced in Berlin than in Bonn. The "warmer" votes in the second winter in Berlin came from one participant who felt it was too warm even at just under 19 °C. This effect was not investigated further as an individual opinion.

The evaluation of the feel-good temperature (corresponding to the measured temperature with a vote of "fine") for all participants in Figure 8 shows that there are large individual differences, with the average measured temperatures ranging between 18 °C and 23 °C. The higher temperatures come from the winter in the pre-test or offices in which, for example, the solar radiation also allowed temperatures above 19 °C in the second winter, corresponding to the limit set by the EnSikuMaV. Only offices in which at least 10 "fine" votes were submitted are listed. While some participants (e.g., office 23, 56, 60) only cast a "fine" vote for one temperature, for other participants (office 15, 86), up to a 3 °C difference in room temperature is acceptable.

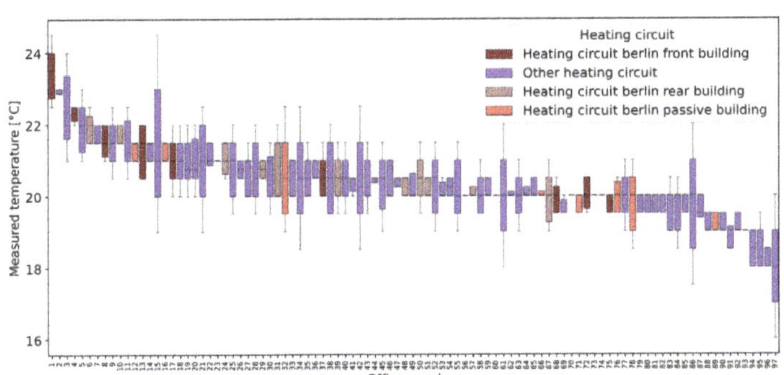


**Figure 8.** Boxplot of the measured room temperatures for times with fine voting of all offices having at least 5 fine votes coloured by different heating circuits.

The comparison of the votes of the participants who took part in both the pre-test in winter 2021/2022 and the pilot test in 2022/2023 and who cast at least five "fine" votes in Figure 9 shows that low temperatures were acceptable for most participants in the second winter and the associated measured room temperature in the test winter was just under 1 °C lower than in the previous winter. There seems to be a corresponding willingness to adapt to lower room temperatures by wearing warmer clothing, for example.

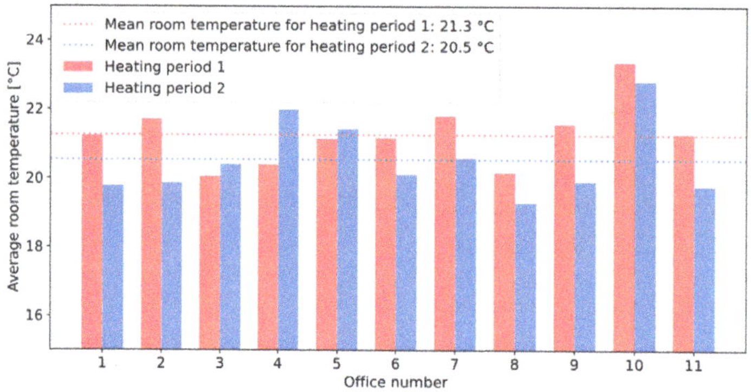


**Figure 9.** Comparison of the measured room temperatures for participants of both heating periods (heating period 1: red, heating period 2: blue) with a vote for 'fine'.

With reference to the time of day or the day of the week and the temperature perceived as fine, no significant differences can be identified. If the fine temperature is considered for mornings or afternoons or for individual days of the week, there is only a very small difference across locations. Within the heating periods, the difference between mornings and afternoons is a maximum of around 0.2 °C at the two locations, and a comparison of Monday mornings, for example (return to the office after the weekend), with other working days also shows no significant differences. Therefore, the temperature perception in this context does not appear to show any noticeable dependence on the time or day of the week. A slight dependency on the outside temperature can be seen; at outside temperatures above 10 °C, the fine temperature is on average around 21 °C, whereas at outside temperatures below 5 °C, it is closer to 20 °C.

### 3.3. *Effects on System Operation and Energy Consumption*

The winter of 2022/2023, during which the ComfortLab platform was field-tested, was overshadowed by the Russian invasion of Ukraine. Due to the resulting tense situation in the energy markets, the German government issued the EnSikuMaV to secure the energy supply. This ordinance limited the heating of public buildings to 19 °C, although higher room temperatures may also have occurred on heating days due to solar radiation, for example. In the properties investigated, the inlet temperature in the heating circuits was noticeably reduced to implement the regulation. The effects and changes in consumption in the field test winter can therefore not be directly transferred to the settings made by the ComfortLab platform. Nevertheless, the recorded data provide a revealing insight into what temperature level is appropriate in the long term and what effects this has on energy consumption. The passive house sector in Berlin is used as a reference for the development of room temperatures, as networked room temperature sensors integrated into the building automation system are installed in around 50 rooms there and high-resolution temperatures from recent years are available.

Figure 10 shows that there were no or only very slight setback times and temperature adjustments in the previous heating periods, regardless of the day of the week, as was already seen from the comfort and PIAF measurements in Bonn. For example, the evaluation of the average temperatures on weekends shows that even in the heating periods (orange-coloured areas) on weekends, hardly any temperature reductions are recognisable, and most room temperatures are above 22 °C, which were achieved by the necessary heating.

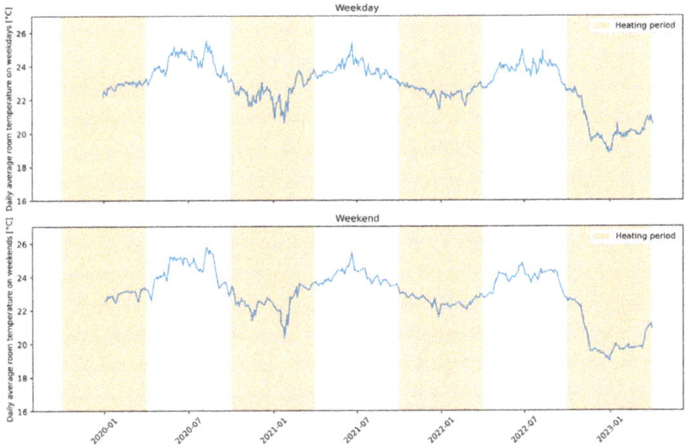


**Figure 10.** Average room temperatures in the passive building in Berlin on weekends during different heating periods.

In the 2022/2023 heating period with the implementation of the 19-degree strategy, a significant reduction in the temperature level can be seen, as shown in Figure 10. This also applies analogously to night-time operation on weekdays and regular daytime operation.

This also results in a significant reduction in heating energy consumption. If the weather-adjusted consumption is considered (Bonn: exclusive heat supply by district heating, Berlin: supply of the property by district heating and additionally by means of heat pumps for the passive house sector), this has fallen by 15% (Bonn) and 30% (Berlin) compared to the weather-adjusted consumption of recent years. The weather-adjusted consumption in previous years was relatively constant with fluctuations in the low single-digit percentage range, so the reduction can be attributed to the reduction strategy implemented with the EnSikuMaV.

To assess how the settings have affected the inlet temperatures, Figure 11 shows an example of the development of the measured inlet temperatures from the heating curve setting of a heating circuit (old building front house) in Berlin.

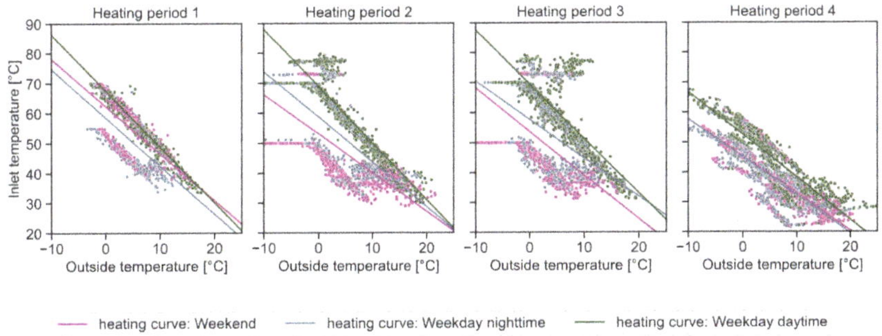


**Figure 11.** Heating curves for different time periods in different heating seasons for one heating circuit in the Berlin building.

If the design temperature is set at −10 °C and the heating limit at 15 °C, the following settings result (Table 3) in the previous heating periods and in winter 2022/2023 (field test).

**Table 3.** Settings of heating curves.

| Outside Temperature | Operating Point | Resulting Inlet Temperature |
|---|---|---|
| −10 °C | Daytime operation before | 85 °C |
| −10 °C | Daytime operation field test | 65 °C |
| −10 °C | Set-back operation before | 65–75 °C |
| −10 °C | Set-back operation field test | 55 °C |
| 15 °C | Daytime operation before | 40 °C |
| 15 °C | Daytime operation field test | 30 °C |
| 15 °C | Set-back operation before | 35 °C |
| 15 °C | Set-back operation field test | 25 °C |

The values determined result in a gradient in daytime operation of around 1.8 (before) or 1.4 (field test); i.e., the inlet temperature of the heating circuit increases by around 1.8 °C or 1.4 °C for every degree Celsius decrease in the outdoor temperature. At the assumed heating limit of 15 °C average daily outdoor temperature, a reduction in the inlet temperature of around 10 °C can be seen in daytime and setback mode. Overall, therefore, both a reduction in the level and the steepness of the heating curve can be seen. In addition to the reduction in useful energy consumption on the heat transfer side, this also has the advantage of a better coefficient of performance and an associated reduction in electricity demand resulting from lower inlet temperatures in the area supplied by heat pumps. Reducing the inlet temperature in heating systems when using heat pumps is therefore particularly advantageous from an energy point of view [29].

Due to the overlap between the implementation of the 19-degree strategy and the ComfortLab field test, the reductions in consumption due to the two possible effects cannot be directly separated. However, the evaluation of the data shows that a slight increase, which would lead to average room temperatures of just under 20 °C, would already bring about a significant improvement in the perception of temperature. Although this would eliminate a small part of the reductions in consumption achieved, it could, on the other hand, support acceptance of long-term temperature reductions to the necessary minimum and achieve further significant reductions in heat consumption. Studies in which room temperature reductions have been investigated estimate possible reductions in consumption in the range of around 6% per degree reduction in room temperature during the reduction period, although this depends on various parameters such as the building physics, the heating system and the type of building use [30–32]. Based on this reference value, a permanent reduction in daytime temperatures in the passive house sector in Berlin from 23 °C to 20 °C would result in potential energy savings of 18% for the period; a further reduction in temperatures on weekends and at night would also be of a similar magnitude in relation to the room temperature values. Due to the parallel 19-degree strategy and the ComfortLab field test, no explicit savings from the use of the platform are shown in this paper. This will be made up for in future applications without the influence of side effects. As the results of the final survey in the following section show, employees are willing to participate in the ComfortLab and adaptation on a permanent basis.

### 3.4. Results from the Final Survey

Following the test phase of the field trial, participants were invited by email to take part in a 10-min online survey to share their experiences and challenges with using the ComfortLab platform. The survey was active from 5 to 17 May 2023, and 110 people (60% Bonn; 40% Berlin) took part and completed the survey. The retrospective survey was intended to re-evaluate the results of the field test and provide the opportunity to find out about open issues that were not or not sufficiently addressed during the trial period.

Interestingly, in retrospect, almost 72% of respondents stated that the room temperature was (much) too cold overall during the winter; see the left plot in Figure 12. This is a significantly higher proportion than the detailed evaluation of the individual votes; see Figure 7. Reasons for this may include the increased participation of non-satisfied respondents in the survey. In retrospect, the overall situation may also have been perceived more negatively than the everyday situation, also in view of the fact that the room temperatures did not change. Nevertheless, almost 57% of participants felt that the 19-degree strategy was a fundamentally sensible measure to secure the energy supply, at least as a short-term and limited measure; see the right plot in Figure 12. Many participants would have liked a slight adjustment of the room temperature to 20 °C, for example, as can be seen from the evaluations of the votes. This was not possible due to the fixed specification of the 19-degree strategy. Around 45% of respondents were still able to adapt to the lower temperatures by taking measures such as wearing warmer clothing.

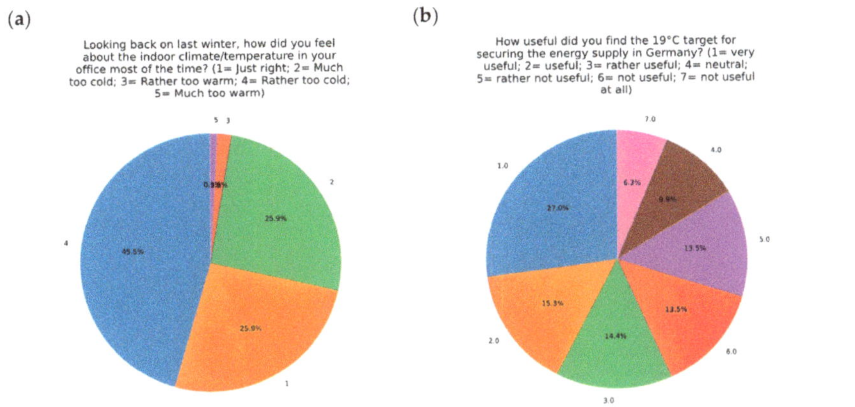


**Figure 12.** Answers to the questions of how the room temperatures were perceived last winter in retrospect (**a**) and how useful the 19-degree strategy was to ensure the energy supply (**b**).

An important point when using the platform is the willingness of participants to take part for a sufficiently long period of time. Almost 64% of participants stated that they did not find participation disruptive in any way, and only just under 14% found participation disruptive to their workflow in any form, as shown in the left plot in Figure 13. Over 72% could imagine participating for a long period of time in general, as shown in the right plot in Figure 13. This underlines the participants' willingness to play an active role in optimising both the temperature settings in their own offices and the associated optimisation of operational management and reduction in energy consumption.

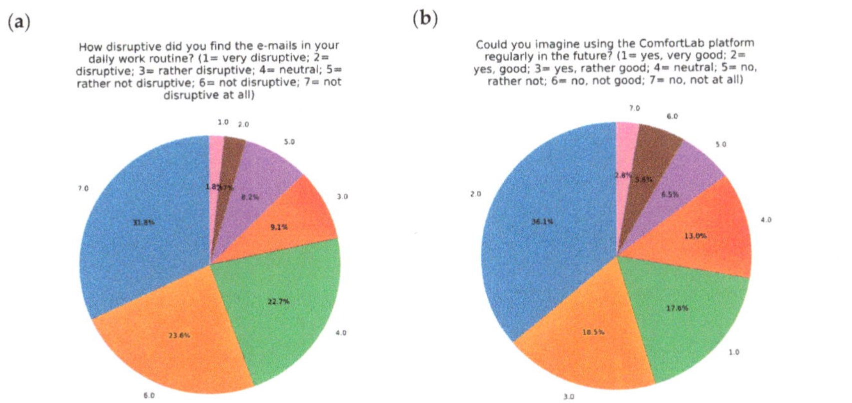


**Figure 13.** Answers to the question of whether the survey was disruptive to the work routine (**a**) and whether the participants can imagine regular and long-term participation in ComfortLab in the future (**b**).

The acceptance of the users is underlined by the fact that almost 80% of those surveyed stated that it is any kind important to them that their feedback on the indoor climate and well-being is considered by the building services and incorporated into the operational management, see the left plot in Figure 14. This goes hand in hand with the assessment of the overwhelming number of participants (57%) that similar temperatures would also be acceptable in the future; see the right plot in Figure 14. Many participants therefore require at least a slight increase in temperatures instead of an inflexible, blanket requirement as specified in the EnSikuMaV.

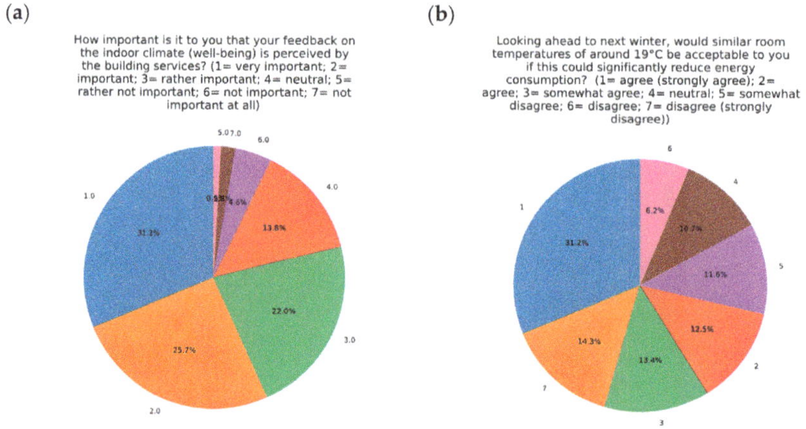


**Figure 14.** (**a**) Answers to the questions about the extent to which participants feel that their feedback on the indoor climate is taken into account in the management of the company (**a**) and the acceptance of lower room temperatures in the long term (**b**).

Overall, it can be concluded from the survey that although the across-the-board reduction has enabled significant savings to be made, this has come at the expense of the satisfaction of a large proportion of the workforce. At the same time, employees are willing to participate, even in the long term. As the 19 °C room temperature had to be implemented as a blanket requirement during the field test winter, acceptance should be even higher if the temperatures are adjusted in a targeted manner. All remaining survey questions and the results are listed in Appendix A.

## 4. Discussion

In this paper, a platform was presented that combines an integration of the temperature perception of office employees, technical parameters from the building automation system and recommendations for settings on the heating system for the operating personnel. This is intended to close the existing gap between the situation in offices (temperatures and individual perceptions) and technical parameters (inlet and return temperatures of heating circuits, energy consumption). The aim is to capture as broad a picture of the employees' opinions as possible so that individual opinions are not processed and the technical operating staff can use the information for the iterative optimisation of operational management.

Fundamental comfort problems (e.g., draughts) were not identified in the reference building after exemplary sampling of the indoor air parameters and calculation of the PMV. Based on this, the room temperature was determined as a representative and comprehensive measured value for the ComfortLab platform. This was the first key question that we were able to answer for the property as an example.

With reference to the indoor climate values recorded from various available sources (PIAF measurements, comfort measurements, historical building management system data from the passive house sector in Berlin), it became apparent that the system operation in the status quo was not operated from an energy efficiency perspective, but that maximum comfort with high room temperatures was the requirement. Reduced operation, for example at night or on weekends, could not be detected to any significant extent. The exemplary measurement of indoor air quality as an indicator of building occupancy showed a relatively low attendance rate due to the COVID-19 pandemic, and reduction potential was also identified here. These results cover our second key question.

In an iterative process, the platform was implemented with displays for employees (ComfortStripes) and facility management (ComfortHub). Based on the Warming Stripes, which show the change in global temperature and were adapted to building measurement values, employees were shown a simplified representation of their own room temperatures and temperature perception over time. For facility management purposes, the individual adjustments were aggregated to provide an overall view at the heating circuit level and combined with system data (inlet temperatures, energy consumption, outside temperature). With reference to our third key question, this goal was achieved in that customised solutions were developed and tested for the two groups (office staff and facility management).

The field test in winter 2022/2023 met with great interest among employees and attracted a three-digit number of employees to participate, with over 6000 votes collected. The application shows that the ComfortLab platform can be used to record the individual temperature perception of a large number of participants with little effort. The results, overlaid by the 19-degree strategy specified in the EnSikuMaV, show that although a general reduction in the temperature level can significantly reduce energy consumption, this is accompanied in some cases by a significant deterioration in the thermal well-being of employees. Although considerable weather-adjusted savings of up to 30% were achieved in the example of the Berlin property, these are more of a short-term success. The 19-degree strategy was a temporary measure in the wake of the energy crisis in 2022. It was no longer implemented in the subsequent heating period 2023/2024, which is why it can be assumed that the settings in many properties were raised to the previous level again. This goes beyond the project duration and could therefore not be considered further. These findings only partially answer our fourth key question. Due to the time overlap between the field test and the mandatory implementation of the 19 °C strategy, the effect of temperature reduction and energy savings cannot be completely separated. Further research is required here, and a further field test without restriction to specific system settings is necessary.

The future development of the platform offers scope for some optimisations and adjustments. In the current field test, the test subjects had to read and transmit the room temperature independently. In addition to a certain susceptibility to error, this also only allows a snapshot to be taken at the time of the assessment, without taking into account and evaluating the development of the room temperature over time. Above all, this resolution is not sufficient for evaluation during non-attendance times (evenings, weekends, vacation periods). For this reason, communicative temperature sensors should be used that can automatically transmit their data to the building automation system, for example, and be integrated into the platform. Some of the rooms in the Berlin property in the passive house sector already have these, and a prototype integration into the ComfortLab platform was tested as part of the research project. In view of the requirements of the EPBD, the automated recording of indoor air parameters will be mandatory in the future for new buildings and refurbishments, so information on temperatures and air quality in rooms will be included in the building for standardised data collection. On the other hand, manual input by employees in the workflow can also increase engagement with and awareness of the platform and heating and ventilation behaviour.

An existing challenge is greater interaction with facility management and evaluation of the implementation of the strategies. Due to the 19-degree strategy, the ability of the operating staff to influence and take into account the individual perceptions of the participants when setting system parameters was greatly reduced. Therefore, the field test had to be limited to the evaluation of theoretically possible and sensible reductions.

In general, the feedback from facility management on the presentation and information content was positive, and this is to be tested in more detail in the future.

The current approach of the ComfortLab platform provides an overview of the recorded room temperatures and the comfort level, aggregated at the heating circuit level. Although this makes it possible to intuitively adjust settings on the heating curve, the plan is to automate this step in the future and to automate adjustments to the heating characteristic curves for day, night and weekend operation by automatically evaluating the settings and parameter changes. By integrating the various building automation protocols with which the measured values are recorded, it is already possible to transfer setpoint specifications to the building automation system.

Another way of using the data from the ComfortLab platform could be to recommend the re-occupancy of offices based on the required comfort temperatures and the utilisation of certain heating circuits. If, for example, offices connected to a heating circuit are only very underutilised for certain days, it may make sense to lower these heating circuits and motivate employees to move to free office space in other heating circuits. Due to the very high home office rate in some cases, this will also be increasingly the case in the future and open opportunities for lowering heating levels in parts of the building. In [33], a reinforcement-learning-based system was tested that provides employees with recommendations for action to reduce energy consumption. Among other things, a move recommendation was tested in the field test, which recommends that employees move to another office to reduce overall energy consumption. This could also be implemented in the ComfortLab platform to divide heating circuits into different temperature levels.

These four possible further developments of the platform are to be pursued and tested in future projects in the Living Lab approach, thus further strengthening the application of the platform.

**Author Contributions:** Methodology, A.B.; validation, A.B., F.F.S.; formal analysis, A.B.; resources, A.B., C.B., A.I.W.; data curation, A.B., F.F.S.; writing—original draft, A.B.; supervision, V.G.; project administration, V.G. All authors have read and agreed to the published version of the manuscript.

**Funding:** This work was funded by the Federal Institute for Research on Building, Urban Affairs and Spatial Development (BBSR) on behalf of the Federal Ministry of Housing, Urban Development and Building (BMWSB) within the research program "Zukunft Bau" (AZ 10.08.17.7-19.38a, AZ 10.08.17.7-19.38b).

**Data Availability Statement:** Not available due to data privacy.

**Conflicts of Interest:** Author Aaron Wolber was employed by the iBG Inbestergesellschaft GmbH. The remaining authors declare that the research was conducted in the absence of any commercial or financial relationships that could be construed as a potential conflict of interest.

## Appendix A

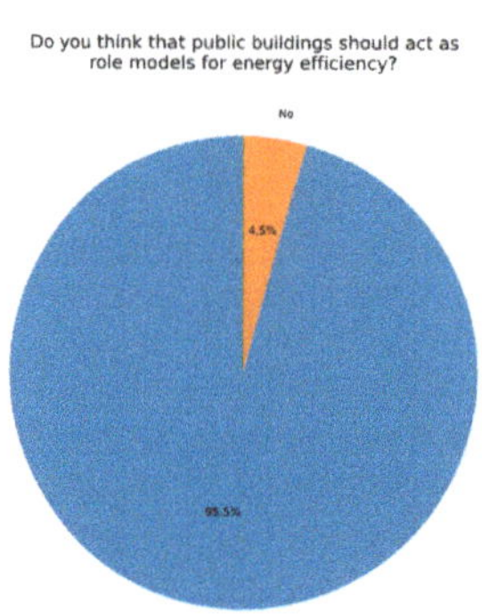


**Figure A1.** Do you think that public buildings should act as role models for energy efficiency?

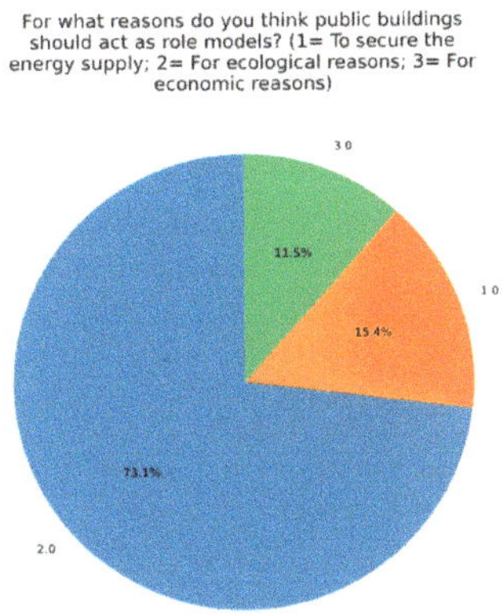


**Figure A2.** For what reasons do you think public buildings should act as role models?

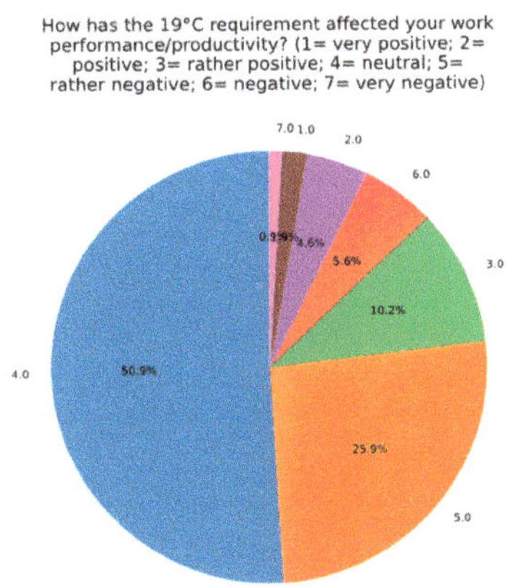


**Figure A3.** How has the 19 °C requirement affected your work performance/productivity?

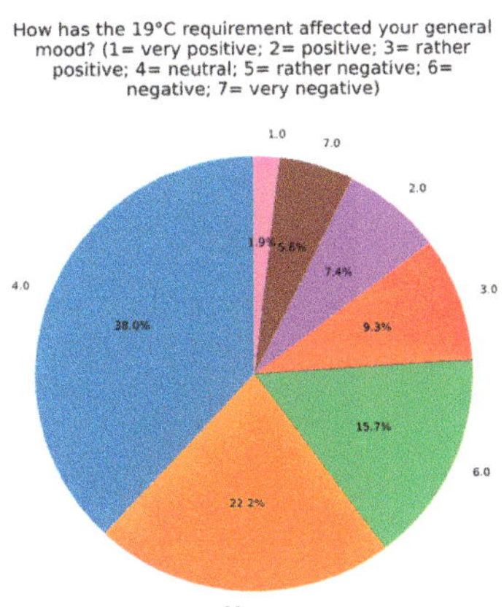


**Figure A4.** How has the 19 °C requirement affected your general mood?

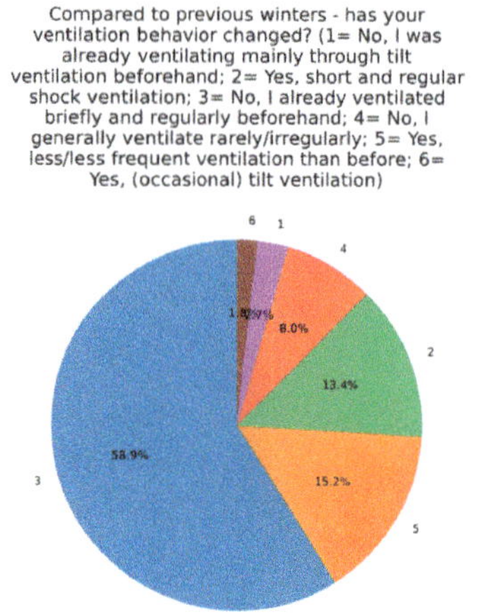


Figure A5. Compared to previous winters—has your ventilation behavior changed?

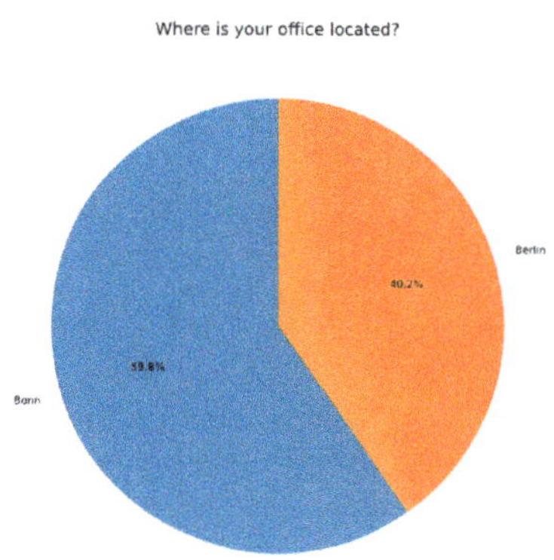


Figure A6. Where is your office located?

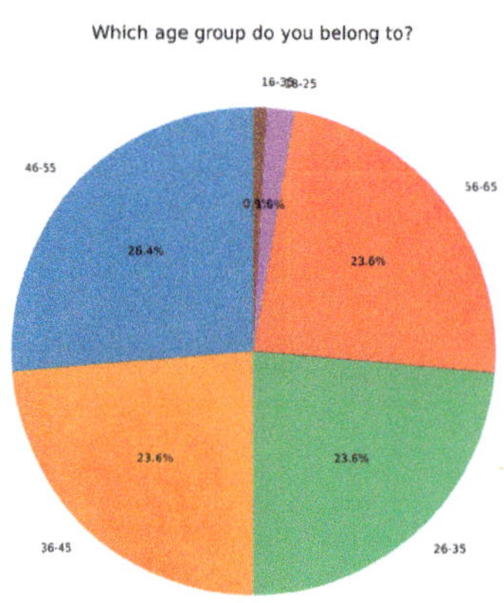


Figure A7. Which age group do you belong to?

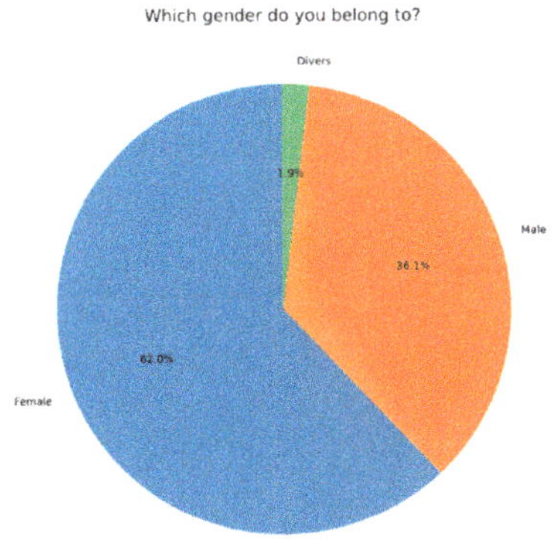


**Figure A8.** Which gender do you belong to?

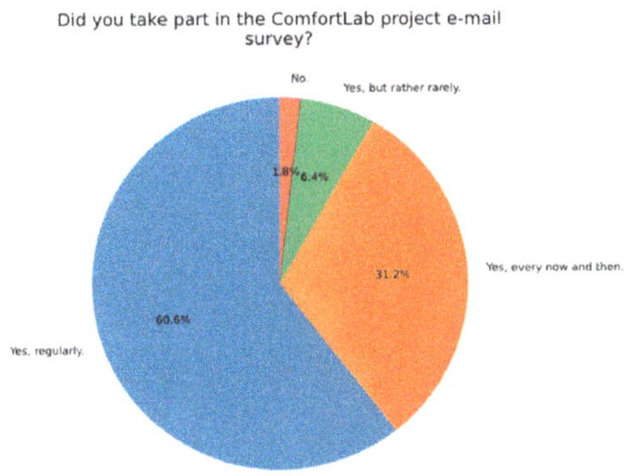


**Figure A9.** Did you take part in the ComfortLab project e-mail survey?

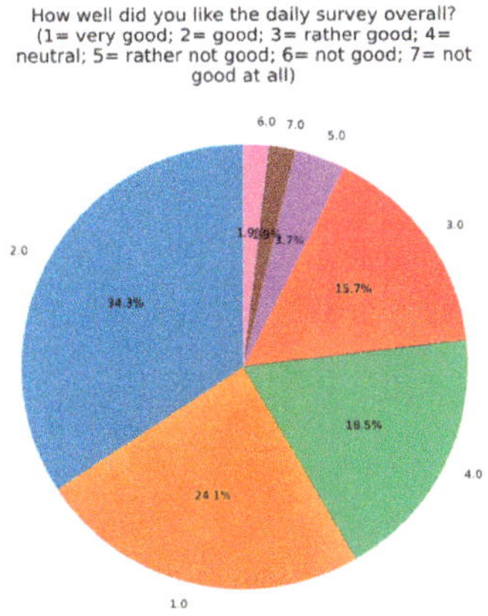


**Figure A10.** How well did you like the daily survey overall?

## References


1. *Energy Performance of Buildings Directive: EPBD*; EU: Brussels, Belgium, 2024.
2. Working Group on Energy Balances. Building-Related Final Energy Consumption for Room Heating, Room Cooling, Hot Water and Lighting. Available online: https://www.umweltbundesamt.de/en/data/environmental-indicators/indicator-energy-consumption-for-buildings#environmental-importance (accessed on 30 June 2024).

3. Du, Y.; Wang, C.; Li, H.; Song, J.; Li, B. Clustering Heat Users Based on Consumption Data. *Energy Procedia* **2019**, *158*, 3196–3201. [CrossRef]
4. Cholewa, T.; Dell'Isola, M.; Canale, L.; Ficco, G.; Michnikowski, P.; Siuta-Olcha, A.; Olszewska, A.; Sadowska, G.; Dudzińska, M.R. On the influence of heat cost allocation on operation of heating system in buildings and possible, additional decrease of supply temperature. *Energy Build.* **2022**, *254*, 111599. [CrossRef]
5. VDI. *VDI 3814 Part 1:2019-01: Building Automation and Control Systems (BACS)—Fundamentals*; Verein Deutscher Ingenieure (VDI): Düsseldorf, Germany, 2019.
6. Liedtke, C.; Baedeker, C.; Hasselkuß, M.; Rohn, H.; Grinewitschus, V. User-integrated innovation in Sustainable LivingLabs: An experimental infrastructure for researching and developing sustainable product service systems. *J. Clean. Prod.* **2015**, *97*, 106–116. [CrossRef]
7. Baedeker, C.; Liedtke, C.; Welfens, M.J. Green Economy as a Framework for Product-Service Systems Development: The Role of Sustainable Living Labs. In *Living Labs 2016*; Keyson, D.V., Guerra-Santin, O., Lockton, D., Eds.; Springer International Publishing AG: Cham, Switzerland, 2016; pp. 35–52. ISBN 978-3-319-33526-1.
8. Baedeker, C.; Piwowar, J.; Themann, P.; Grinewitschus, V.; Krisemendt, B.; Lepper, K.; Zimmer, C.; von Geibler, J. Interactive Design to Encourage Energy Efficiency in Offices: Developing and Testing a User-Centered Building Management System Based on a Living Lab Approach. *Sustainability* **2020**, *12*, 6956. [CrossRef]
9. Lorenz, C.L.; André, M.; Abele, O.; Gunay, B.; Hahn, J.; Hensen, P.; Nagy, Z.; Ouf, M.M.; Park, J.Y.; Yaduvanshi, N.S.; et al. A Repository of Occupant-Centric Control Case Studies: Survey Development and Database Overview. *Energy Build.* **2023**, *300*, 113649. [CrossRef]
10. Di Giuda, G.; Tagliabue, L.; Pellegrini, L.; Meschini, S.; Seghezzi, E.; Schievano, M.; Marenzi, G. Developing a Decision Support System and a Building Management System for building portfolio management. In Proceedings of the EUBIM 2022-BIM International Conference, Valencia, Spain, 18–21 May 2022. [CrossRef]
11. Bovet, G.; Hennebert, J. Will Web Technologies Impact on Building Automation Systems Architecture? *Procedia Comput. Sci.* **2014**, *32*, 985–990. [CrossRef]
12. Ceccarini, C.; Mirri, S.; Prandi, C. Designing Interfaces to Display Sensor Data: A Case Study in the Human-Building Interaction Field Targeting a University Community. *Sensors* **2022**, *22*, 3361. [CrossRef] [PubMed]
13. Garg, M.; Garg, V.; Srivastava, P.; Agarwal, R. Interface design for residential energy feedback, in the Indian context. *Energy Inf.* **2023**, *6*, 12. [CrossRef]
14. Garzia, F.; Verbeke, S.; Pozza, C.; Audenaert, A. Meeting User Needs through Building Automation and Control Systems: A Review of Impacts and Benefits in Office Environments. *Buildings* **2023**, *13*, 2530. [CrossRef]
15. Jayathissa, P.; Quintana, M.; Sood, T.; Nazarian, N.; Miller, C. Is your clock-face cozie? A smartwatch methodology for the in-situ collection of occupant comfort data. *J. Phys. Conf. Ser.* **2019**, *1343*, 12145. [CrossRef]
16. Pruvost, H.; Calleja-Rodríguez, G.; Enge-Rosenblatt, O.; Jimenez-Redondo, N.; Peralta-Escalante, J.-J. A recommendation system for energy saving and user engagement in existing buildings. In Proceedings of the 2021 European Conference on Computing in Construction, Online, 26–28 July 2021; University College Dublin: Dublin, Ireland, 2021; pp. 261–268.
17. Jin, X.; Wang, G.; Song, Y.; Sun, C. Smart building energy management based on network occupancy sensing. *J. Int. Counc. Electr. Eng.* **2018**, *8*, 30–36. [CrossRef]
18. Degen, H.; Budnik, C.J.; Chitre, K.; Lintereur, A. How to Explain It to Facility Managers? A Qualitative, Industrial User Research Study for Explainability. In *HCI International 2021—Late Breaking Papers: Multimodality, eXtended Reality, and Artificial Intelligence*; Stephanidis, C., Kurosu, M., Chen, J.Y.C., Fragomeni, G., Streitz, N., Konomi, S., Degen, H., Ntoa, S., Eds.; Springer International Publishing: Cham, Switzerland, 2021; pp. 401–422. ISBN 978-3-030-90962-8.
19. Desogus, G.; Frau, C.; Quaquero, E.; Rubiu, G. From Building Information Model to Digital Twin: A Framework for Building Thermal Comfort Monitoring, Visualizing, and Assessment. *Buildings* **2023**, *13*, 1971. [CrossRef]
20. Ličina, V.F.; Cheung, T.; Zhang, H.; de Dear, R.; Parkinson, T.; Arens, E.; Chun, C.; Schiavon, S.; Luo, M.; Brager, G.; et al. Development of the ASHRAE Global Thermal Comfort Database II. *Build. Environ.* **2018**, *142*, 502–512. [CrossRef]
21. Lin, G.; Casillas, A.; Sheng, M.; Granderson, J. Performance Evaluation of an Occupancy-Based HVAC Control System in an Office Building. *Energies* **2023**, *16*, 7088. [CrossRef]
22. *DIN EN ISO 7730*; Ergonomics of the Thermal Environment—Analytical Determination and Interpretation of Thermal Comfort Using Calculation of the PMV and PPD Indices and Local Thermal Comfort Criteria. ISO: Geneva, Switzerland, 2005.
23. Zimmer, C. Twistid GmbH & Co. KG. Available online: https://www.twistid.de (accessed on 30 June 2024).
24. ELK Elektronische Systeme und Softwaretechnik GmbH. BAPPUevo. Available online: https://www.bappu.de/index.php/en/products/bappu (accessed on 30 June 2024).
25. Ed Hawkins. Warming Stripes. Available online: https://showyourstripes.info/ (accessed on 30 June 2024).
26. Measures Laid down in the Ordinance on Securing the Energy Supply through Rapid Impact Measures: EnSikuMaV. 2022. Available online: https://www.bmwk.de/Redaktion/DE/Downloads/E/ensikumav.pdf?__blob=publicationFile&v=4 (accessed on 30 June 2024).
27. Dolinay, V.; Vasek, L. Benefits of night temperature setback in the control strategy of the district heating system. *MATEC Web Conf.* **2019**, *292*, 1052. [CrossRef]

28. Ben, H.; Walker, S.; Brown, C.; Kolo, I.; Falcone, G. Decarbonising heating and cooling using temperature setback and geothermal energy. *J. Phys. Conf. Ser.* **2023**, *2600*, 22015. [CrossRef]
29. Jurkschat, S.; Sehr, F.F.; Fransen, K.; Beblek, A.; Grinewitschus, V. Measurement Data-Based Estimation of the Suitability of Existing Properties for the Operation of x to Water Heat Pumps Using a Seed of 100 Multi-Family Houses and Different Power Shifting Approaches. *Energies* **2023**, *16*, 7352. [CrossRef]
30. Rohde, C.; Dütschke, E.; Gigli, M.; Bles, M. Behavioural Climate Change Mitigation Options; Domain Report Housing. Available online: https://climate.ec.europa.eu/document/download/49f2248d-e2cd-40b8-b6dd-1b0982ea94dc_en?filename=housing_report_en.pdf (accessed on 30 June 2024).
31. IEA. *Playing My Part: How to Save Money, Reduce Reliance on Russian Energy, Support Ukraine and Help the Planet*; IEA: Paris, France, 2022.
32. ul Deen, S.; Farooq, S. European Energy Crises Climate Action and Emerging Market of Carbon-Neutral LNG. *J. Eur. Stud. (JES)* **2023**, *39*, 33. [CrossRef]
33. Wei, P.; Xia, S.; Chen, R.; Qian, J.; Li, C.; Jiang, X. A Deep-Reinforcement-Learning-Based Recommender System for Occupant-Driven Energy Optimization in Commercial Buildings. *IEEE Internet Things J.* **2020**, *7*, 6402–6413. [CrossRef]

*energies*

MDPI

*Review*

# Artificial Intelligence and Digital Twins for Bioclimatic Building Design: Innovations in Sustainability and Efficiency

**Ekaterina Filippova [1,*], Sattar Hedayat [1], Tina Ziarati [1] and Matteo Manganelli [1,2,*]**

[1] Faculty of Civil and Industrial Engineering, Sapienza University of Rome, 00184 Rome, Italy; hedayat.1996509@studenti.uniroma1.it (S.H.); ziarati.1966279@studenti.uniroma1.it (T.Z.)
[2] Nuclear Department, National Agency for New Technologies, Energy and Sustainable Economic Development (ENEA), 40121 Bologna, Italy
* Correspondence: filippova.2021500@studenti.uniroma1.it (E.F.); matteo.manganelli@enea.it (M.M.)

**Abstract**

The integration of artificial intelligence (AI) into bioclimatic building design is reshaping the architecture, engineering, and construction (AEC) industry by addressing critical challenges in sustainability and efficiency. By aligning structures with local climates, bioclimatic design addresses global challenges such as energy consumption, urbanization, and climate change. Complementing these principles, AI technologies—including machine learning, digital twins, and generative algorithms—are revolutionizing the sector by optimizing processes across the entire building lifecycle, from design and construction to operation and maintenance. Amid the diverse array of AI-driven innovations, this research highlights digital twin (DT) technologies as a key to AI-driven transformation, enabling real-time monitoring, simulation, and optimization for sustainable design. Applications like façade optimization, energy flow analysis, and predictive maintenance showcase their role in adaptive architecture, while frameworks like Construction 4.0 and 5.0 promote human-centric, data-driven sustainability. By bridging AI with bioclimatic design, the findings contribute to a vision of a built environment that seamlessly aligns environmental sustainability with technological advancement and societal well-being, setting new standards for adaptive and resilient architecture. Despite the immense potential, AI and DTs face challenges like high computational demands, regulatory barriers, interoperability and skill gaps. Overcoming these challenges will be crucial for maximizing the impact on sustainable building, requiring ongoing research to ensure scalability, ethics, and accessibility.

**Keywords:** artificial intelligence; digital twin; bioclimatic building design; building performance; internet of things (IoT); energy efficiency

## 1. Introduction

In recent years, the construction industry has undergone significant transformation, driven by advances in both sustainable design and digital technologies. Bioclimatic design, as a key aspect of sustainable building, optimizes natural resources such as solar radiation, wind, and thermal regulation to enhance climate-responsive structures, minimize energy consumption, and reduce environmental impact. The growing emphasis on sustainability is being reinforced by the adoption of the Sustainable Development Goals (SDGs), with the construction sector increasingly aligning its practices with green building standards, sustainable materials, lifecycle assessments, and collaborative research efforts. This shift towards sustainability is driving a more environmentally responsible and inclusive future, with stakeholders throughout the industry working together to meet global sustainability targets.

At the same time, the integration of artificial intelligence (AI), particularly through machine learning (ML) and deep learning (DL), has revolutionized decision-making in the construction sector. These technologies enable advanced decision support systems that assist construction professionals by providing intelligent capabilities to enhance project outcomes [1]. As industry faces the dual challenges of digitalization and sustainability, AI has emerged as a powerful tool that can address both. The strategic application of AI across the entire lifecycle of building projects—from design and construction to operation and maintenance—presents unique opportunities to reduce inefficiencies and drive significant improvements in performance [2].

AI plays a crucial role in optimizing building designs to improve energy efficiency and reduce environmental impact. Through comprehensive data analysis and scenario simulations, AI enables architects and engineers to make well-informed decisions that align with rigorous green building standards. Beyond enhancing sustainability, AI is also advancing construction materials and technologies, driving innovation and resilience in structural design. The use of AI algorithms and machine learning techniques has led to the development of novel materials that enhance the durability, safety, and efficiency of buildings. These innovations not only promote sustainable construction practices but also address the challenges posed by urbanization, climate change, and the evolving needs of infrastructure development [3].

Despite the growing interest in AI and ML applications in construction engineering, there is a notable research gap that requires attention. While some studies have explored specific aspects of AI and ML within the industry, a comprehensive review that synthesizes the various methodologies and examines their collective impact is still lacking [1].

This work aims to fill that gap by providing an in-depth analysis of the latest developments in AI and ML within the architecture, engineering and construction (AEC) industry, as well as identifying potential areas for future research and improvement. By exploring the intersection of AI and bioclimatic design, this work aims to provide insights into how cutting-edge technologies can address environmental challenges, improve building performance, and shape the future of sustainable architecture.

*Methodology*

To achieve the objectives of this work, a systematic literature review was conducted using the PRISMA (preferred reporting items for systematic reviews and meta-analyses) protocol, focusing on recent advancements in artificial intelligence and machine learning in the context of AEC and bioclimatic building design and their implications for sustainable architecture.

The literature search was conducted primarily through the Sciencedirect platform and, to a lesser extent, via ResearchGate. Articles were retrieved using targeted keywords such as "bioclimatic building design", "digital innovations", "artificial intelligence", "artificial intelligence in AEC", "building performance modelling", and "digital twins in bioclimatic design".

The initial literature search yielded 167 articles. These records were collected using targeted keyword combinations and limited to publications not older than 2019. After removing 96 articles based on publication date and language criteria, 71 records remained for further screening. These were assessed against the primary inclusion and exclusion criteria, which reduced the number of articles to 58. After applying secondary filters including detailed evaluation of titles, abstracts, and keywords an additional 10 studies were excluded. The exclusion criteria encompassed articles that were not peer-reviewed, lacked direct relevance to the built environment, or fell outside the defined temporal or linguistic parameters. The final number of articles included for full review and analysis was 47.

Each record was independently assessed by one reviewer, and in certain cases, AI-generated summaries were utilized to facilitate efficient information extraction. The results were compiled and verified collaboratively by the review team, following PRISMA guidelines to ensure transparency and replicability of the selection process. Figure 1 displays the overview of the process of selecting papers.

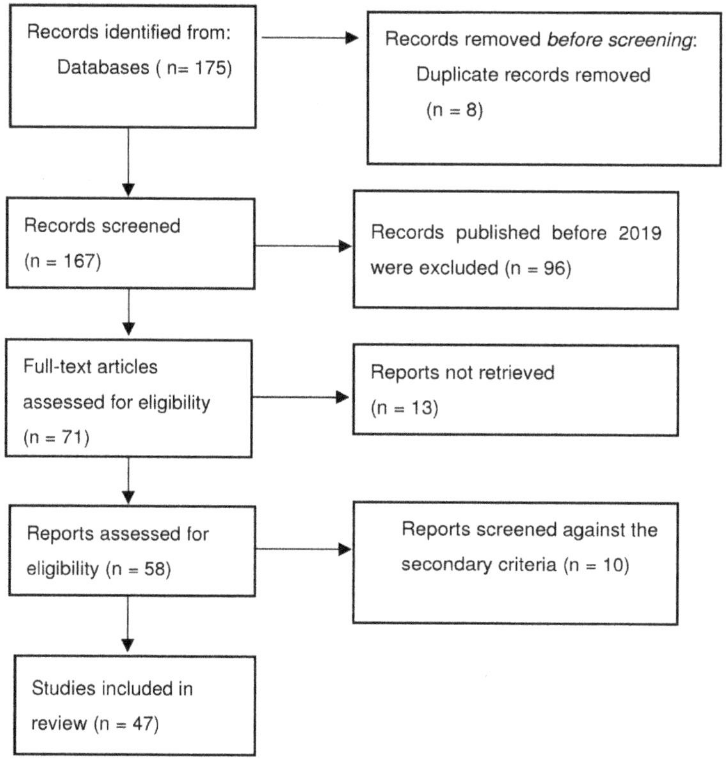


**Figure 1.** The PRISMA selection process of relevant literature.

## 2. Main Principles and Strategies of Bioclimatic Building Design

Bioclimatic design integrates climate-responsive principles into architecture, optimizing solar radiation, temperature, wind, and humidity in order to reduce energy consumption. It aims to minimize dependence on artificial systems while ensuring environmental integration. Precise climate analysis is essential for effective design, with tools like bioclimatic diagrams aiding in visualizing climate-passive design relationships and informing material selection and energy-efficient solutions (Figure 2) [4].

The core principles of bioclimatic building design, including thermal comfort, daylighting, and energy efficiency, align building design with local climate conditions, reduce energy consumption, enhance comfort, and promote environmental and economic sustainability, addressing global challenges like energy use and climate change [5].

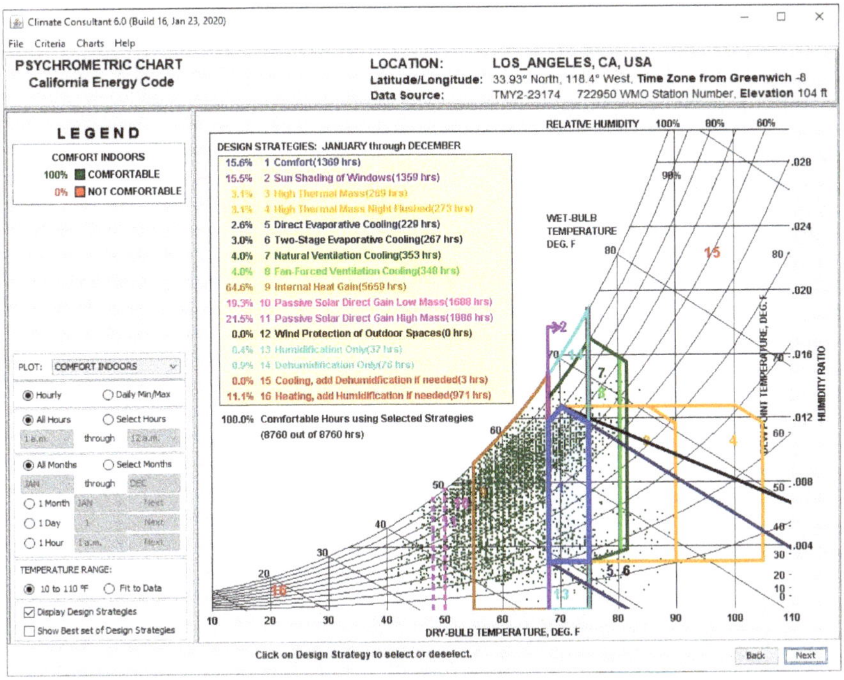


**Figure 2.** Example of psychrometric Building Bio-Climatic chart from Climate Consultant 6.0.

In the past decade, the building sector has drawn significant attention due to its high energy consumption and carbon emissions, accounting for 35% of global energy use and 38% of $CO_2$ emissions in 2019 [6]. Bioclimatic design helps mitigate climate change by reducing reliance on non-renewable energy sources and minimizing emissions [7]. Incorporating locally available renewable resources further decreases dependence on large-scale energy supplies. Case studies demonstrate that the integration of specific bioclimatic strategies—such as Trombe walls (TI), ventilated façades (VB), and green roofs (GR)—into residential buildings can yield substantial reductions in energy demand. For example, simulations conducted with EnergyPlus in six climatically diverse locations worldwide have reported energy savings of up to 60%, particularly in hot and cold climates where Trombe walls effectively shield the building envelope from extreme external conditions [8]. Other studies provide evidence of the performance gains achievable through bioclimatic design and AI-driven optimization in the AEC sector. For instance, Elaouzy & El Fadar (2022) [5] reported that the adoption of key bioclimatic strategies across diverse climates can reduce building energy consumption by as much as 58%, depending on strategies and regional conditions. Adewale et al. (2024) [2] summarized findings on AI Applied across the building lifecycle, showing that AI applications can reduce material waste by as much as 40% and machine learning can reduce energy consumption by as much as 30%. At the design stage, Elwy & Hagishima (2024) [9] demonstrated that surrogate models integrated with BIM and AI maintained comparable accuracy in evaluating energy performance allowing for a dramatic reduction in computation time (by a factor 10 to 20). Shahinmoghadam et al. (2021) [10] presented an example of BIM- and IoT-based virtual reality tool capable of delivering real-time thermal comfort assessments, validating the operational benefits of AI-enhanced digital environments.

This approach should be applied not only to new buildings but also as a key component of renovation strategies, as outlined in the Revised Energy Performance of Buildings

Directive (EPBD) approved in March 2024 [11]. The directive introduces tools like the renovation passport, offering a roadmap for phased renovations, helping property owners and investors plan interventions [12].

The implementation of bioclimatic principles varies depending on the local climate, building type, and geographical location (Figure 3) [12].

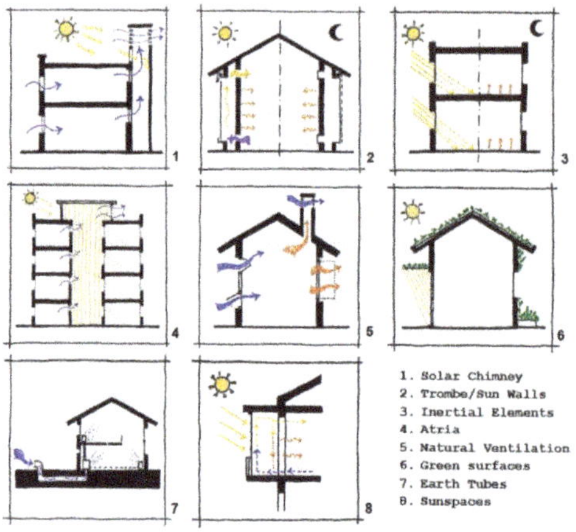


**Figure 3.** Graphic schemes of bioclimatic design strategies [12].

However, the main strategies are broadly applicable in different contexts, as follows.

- Natural Ventilation: Conventional passive cooling strategies, such as shading transparent surfaces and employing cross-ventilation and stack ventilation, facilitate natural cooling and enhance indoor air quality [4,13,14].
- Passive Solar Heating: Optimizing orientation, window placement, and building rotation maximizes winter solar gain, enhances ventilation, balances daylight, and leverages thermal mass for heat storage [7,8].
- Shading and Solar Control: In warm climates, passive shading devices—such as overhangs, louvers, and pergolas-along with vegetative solutions like green walls, are employed to reduce excessive solar exposure and mitigate overheating [15].
- Thermal Insulation: Effective insulation, reflective materials, and high-performance windows, significantly reduce heat transfer—lowering energy consumption by up to 60% and ensuring thermal stability [8].
- Green Roofs and Walls: Vegetation integrated into building designs provide insulation, reduce energy use, mitigate urban heat island effects, and improve biodiversity [7,16].
- Rainwater Harvesting and Greywater Recycling: Rainwater harvesting, greywater recycling, and water features reduce potable water use and contribute to passive cooling [13].
- Integration of Smart and Renewable Technologies: The adoption of on-site renewable energy generation (e.g., photovoltaic panels), local energy storage systems, and intelligent building automation enhances energy autonomy, operational efficiency, and environmental performance [17,18].

## 3. Artificial Intelligence Evolution, Techniques and Application in AEC

Artificial intelligence (AI) is a subdiscipline of computer science focused on the development of systems that are capable of simulating human cognitive processes, such as perception, reasoning, problem-solving, natural language understanding, decision-making and learning from experience. The fundamental principles of AI center on the capacity of machines to replicate or enhance human cognitive functions, encompassing autonomy, learning, reasoning, problem-solving, and adaptation.

The origins of AI date back to the mid-20th century, marked by the Turing Test and the Dartmouth Conference, where the field was formally established [19,20]. Early research focused on symbolic reasoning and rule-based systems but was limited by computational constraints. In the late 20th century, the focus shifted toward machine learning and neural networks, laying the groundwork for deep learning. In the 21st century, growing computational power and the availability of big data have enabled major breakthroughs in pattern recognition, natural language processing, and computer vision. Recent advancements in reinforcement learning and generative models, such as Generative Adversarial Networks (GANs), have further expanded AI's capabilities. Its integration with big data, cloud computing, and the Internet of Things (IoT) has expanded its application areas. Moreover, AI-specific hardware like Graphic Processing Units (GPU) and Tensor Processing Units (TPU) has accelerated AI training and inference [20].

Contemporary research emphasizes explainable AI (XAI) for model transparency, ethical considerations, and applications in healthcare, robotics, and autonomous systems, shaping the future of intelligent technologies [21].

This growing focus on responsible and transparent AI aligns with broader trends in the field, as highlighted by the AI Index Report 2024. The report shows a rapid rise in open-source AI research, with projects on GitHub 3.10 growing from 845 in 2011 to 1.8 million in 2023—a 59.3% increase in 2023 alone—demonstrating both the accelerating pace of innovation and the importance of collaborative development [22].

In parallel with technological progress, ethical considerations and governance frameworks are becoming central to AI deployment. The EU has taken a leading role by introducing regulations for "trustworthy AI" combining legal, ethical, and technical standards [23,24]. These guidelines define seven key requirements: human oversight, safety, privacy, transparency, fairness, societal well-being, and accountability (Figure 4) [23]. This risk-based regulatory model is central to the EU AI Act, ensuring responsible AI development and use [25].

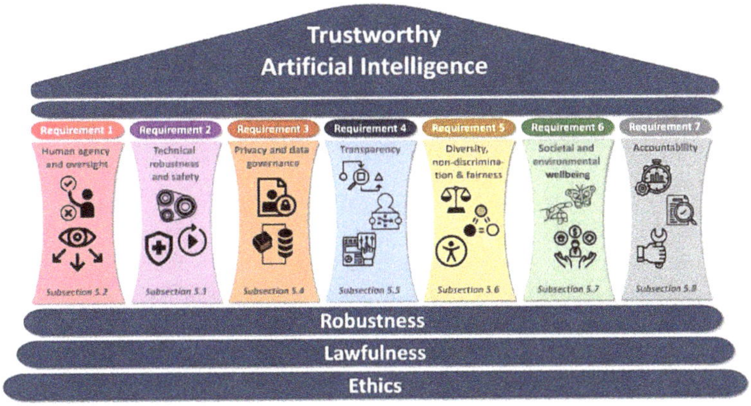


**Figure 4.** Pillars and requirements of trustworthy AI [23].

### 3.1. *Artificial Intelligence Techniques*

AI can be based on several different techniques, depending on the application and the problem being solved. Some of the most commonly used techniques in AI are based on machine learning and deep learning (Figure 5) visually illustrate relationship and the distinction between machine learning and deep learning in terms of feature extraction and learning.

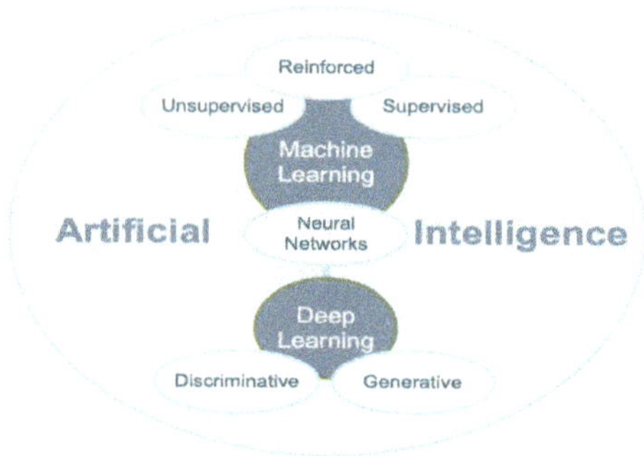


**Figure 5.** AI Taxonomy.

#### 3.1.1. Machine Learning (ML)

This is the backbone of most AI applications. ML involves training algorithms to recognize patterns in data and make predictions based on that data, without being explicitly programmed for specific tasks. ML techniques allow computers to recognize patterns, make predictions, and perform decision-making tasks based on historical data. There are three primary types of machine learning:

- Supervised Learning: The model is trained using labeled data (input-output pairs), where the desired output is known. The model learns to map inputs to correct outputs. This branch of ML includes methods like Linear Regression, Bayesian network, K-nearest neighbors (kNN), Decision Tree, etc.
- Unsupervised Learning: The model is trained using unlabeled data, meaning the system must identify patterns or groupings on its own. The algorithm must find structure or patterns in the input data without guidance. The main goal of unsupervised learning is to explore the data and extract useful insights. This includes Fuzzy C Means, Balanced Iterative Reducing and Clustering using Hierarchies (BIRCH) and K-Means.
- Reinforcement Learning: The system learns by interacting with its environment and receiving feedback in the form of rewards or penalties, optimizing its behavior over time: Q-Learning, Markov decision process [26].

#### 3.1.2. Deep Learning (DL)

Being a subset of machine learning, deep learning uses multi-layered neural networks to model complex patterns in data. These networks are inspired by the human brain's structure and are capable of handling vast amounts of data and performing complex tasks like image recognition, speech processing, and natural language understanding. Deep learning has achieved remarkable success in fields such as autonomous driving, medical image analysis, and Natural Language Processing (NLP) due to its ability to automatically learn features from raw data without needing explicit programming or feature engineering.

Artificial Neural Networks (ANNs) composed of layers of neurons that work together to process input data and produce output predictions. Each neuron in a layer is connected to several neurons in the previous and next layers, creating a network of computational units [27].

- Discriminative (Supervised) models: focus on learning the boundary between different classes in a dataset, rather than modeling data distribution. These supervised deep models typically estimate class probabilities from observable data, enabling accurate classification. In simple terms, discriminative models learn to distinguish between different categories or labels by focusing on the differences in the corresponding data. Common models include Multi-Layer Perceptron (MLP), Convolutional Neural Networks (CNNs), Recurrent Neural Networks (RNN), and their variations. The most powerful class of this type is CNNs, which are extensively used across a range of tasks, such as object detection, speech recognition, computer vision, image classification, and bioinformatics by learning hierarchical features from raw data.

- Generative (Unsupervised) models: The primary objective of generative models (GMs) is to produce data resembling real-world distributions. Despite ongoing research challenges, recent advancements have expanded their applications, particularly in computer vision research. GMs utilize training data from an unknown data-generating distribution to create new samples that match the original distribution. Key models include Auto-Encoder, Generative Adversarial Network (GAN), Restricted Boltzmann Machine (RBM), and Deep Belief Network (DBN) [28].

Artificial intelligence is a rapidly evolving field that allows machines to perform tasks that were traditionally thought to require human intelligence. By leveraging techniques like machine learning, deep learning, natural language processing, and computer vision, AI systems can analyze vast amounts of data, learn from it, and make decisions or predictions autonomously.

Today, AI systems are transforming industries ranging from healthcare to finance, manufacturing, and even architecture, playing a pivotal role in driving Industry 5.0 and Society 5.0-concepts that emphasize the integration of human intelligence, AI, and other emerging technologies for a sustainable future [29]. Its impact spans multiple sectors, including Architecture, Engenieering and Construction (AEC) and Natural Hazards Engineering (NHE) [30,31].

## 4. AI in Bioclimatic Building Design

The application of AI in the AEC sector has evolved significantly over the past few decades from early automated design tools like Computer-Aided Design (CAD) to advanced generative algorithms and sustainable design solutions. In 1976, Cedric Price developed the concept of the "Generator"—a vision of AI-driven architecture, followed by Soddu's artificial DNA model for medieval cities in 1987 [27,32]. By the 2000s, AI integrated data-driven decision-making to enhance energy efficiency, resource management and waste reduction. Innovations like genetic algorithms for ecological design and optimization tools such as GENE-ARCH and OPTIMUS have advanced sustainable architecture [32]. Today, AI, alongside big data and cloud computing, is crucial for smart cities and green building solutions, optimizing environmental performance in real time.

### *4.1. AI, Sustainable Building Design, and Construction 4.0 and 5.0*

The construction industry is undergoing a major transformation driven by digitalization and AI. Construction 4.0 signifies a profound shift within the industry, characterized by the fusion of digital technologies and data-driven solutions, including the widespread adoption of Internet of Things (IoT) devices, Artificial intelligence, robotics, 3D printing, and Building Information Modeling (BIM), which collectively work to optimize processes, reduce costs, and improve building performance. While Construction 4.0 focuses on digital integration and automation, Construction (Industry) 5.0 advances by prioritizing sustainability, environmental stewardship, principles of a circular economy and human needs and well-being (Figure 6) [29,33].

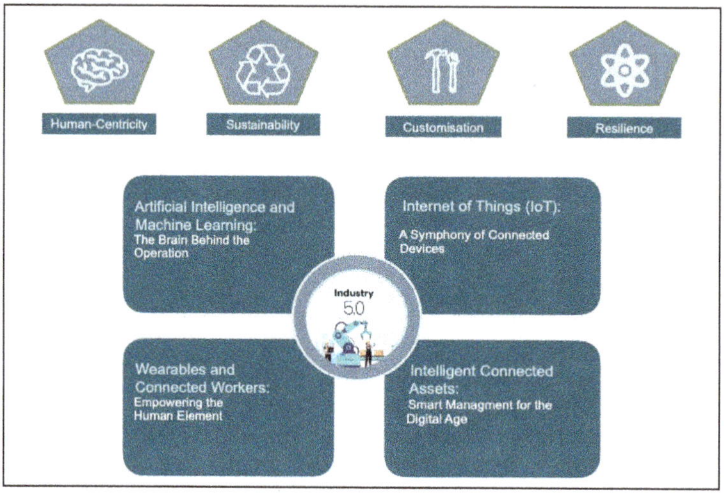


**Figure 6.** Focus and main priorities of the concept of Construction (Industry) 5.0 [29,33].

AI enhances bioclimatic design by automating analysis, predicting performance, and optimizing energy efficiency. Its potential extends to generative design, automation, predictive maintenance, and lifecycle optimization, addressing inefficiencies and reducing environmental impact. The industry faces challenges like excessive material waste (11–15%) and high energy consumption, with buildings accounting for 28% of U.S. commercial energy use and 41.1% in China. AI-driven solutions offer a path toward more sustainable, resource-efficient construction [2].

### *4.2. Recent AI Techniques and Combination Technologies in AEC*

The integration of AI in the AEC industry, particularly in sustainable and bioclimatic design, is advancing rapidly. AI enhances sustainability by optimizing energy efficiency, reducing environmental impact, and improving building performance. Its integration with conventional modeling, simulation, and analytics significantly enhances all phases of building design, construction, and operation (Table 1) [2].

**Table 1.** AI-driven technologies in AEC sector.

| Building Design | Construction | Operation |
|---|---|---|
| Big Data | Intelligent Decision Support Systems (DSS) | Building Performance Monitoring (BPM) |
| Generative design | Radio-Frequency Identification (RFID) | Predictive Analytics |
| Surrogate Modeling | Computer Vision Systems | Structural Health Monitoring (SHM) |
| Augmented Reality<br>Virtual Reality | Iot Devices<br>Construction Robots | Smart Building Systems |

#### 4.2.1. Building Design

At the building design stage, advancements in artificial intelligence provide exceptional opportunities to optimize sustainability objectives across a range of architectural and engineering parameters. Early-stage AI integration enhances decision-making, accelerates design processes, and improves energy efficiency to meet regulatory and cost-saving demands. Techniques such as Knowledge-Based Engineering (KBE), fuzzy logic, neural networks, and genetic algorithms are widely applied in AEC design [34]. Late 20th-century research explored design and morphogenesis, inspiring complex and unconventional forms. Contemporary studies, however, prioritize refining and optimizing existing designs, particularly through the optimization of their shapes [32].

Big Data is transforming architecture by providing architects with vast amounts of information, such as climate patterns, energy use, user behavior, and material performance. This enables data-driven design decisions, balancing esthetics, functionality, and sustainability. For example, data-driven simulations optimize natural light, reducing reliance on artificial lighting, while its use in parametric design allows architects to generate numerous design variations through algorithms. Analyzing material, cost, and environmental impact data refines designs, minimizes waste, and promotes sustainability [29].

Meanwhile, Generative Design—AI-driven generative design algorithms, or Generative Adversarial Network (GAN)—have become integral in the design process, allowing architects and engineers to explore vast numbers of design alternatives based on specific goals (e.g., energy efficiency, structural integrity, material optimization). GANs are applied to optimize building forms, orientations, and window placements to maximize passive heating in winter or minimize solar heat gain in summer, considering local weather patterns. Using the Generator-Discriminator principle, AI autonomously generates options, which are then evaluated and refined [27].

Recent studies highlight the effectiveness of surrogate modeling in optimizing low-energy building design. Surrogate models (SMs) mimic high-fidelity physics-based models by statistically correlating input (design variables) and output (performance targets) data from a sample within the design space. This research underscores the potential advantages of machine learning methods, including ANNs, in handling computationally demanding tasks like building design optimization. Surrogate-assisted building performance optimization models are used to optimize energy performance by improving systems such as HVAC, lighting, and insulation, reducing overall energy consumption. Additionally, surrogate models optimize building shape and layout, enhancing natural lighting, ventilation, and thermal comfort while reducing energy waste, using eco-friendly materials, and improving energy efficiency throughout the building's lifecycle [9].

In addition, Augmented Reality (AR) and Virtual Reality (VR) technologies enhance collaboration, enabling stakeholders to virtually walk-through buildings before they are constructed, adding design modifications and approvals more efficient and accurate [29].

#### 4.2.2. Construction

The construction industry plays a crucial role in global economic development, with an estimated annual value surpassing $10 trillion. However, it often faces challenges such as high costs, project delays, and safety concerns with average cost overruns of 80% leading to significant financial consequences [1]. AI integration with traditional construction techniques, can optimize workflows, streamline processes, and ensure higher-quality construction outcomes.

Intelligent Decision Support Systems (DSS) utilize a vast amount of historical data, real-time information, and predictive analytics to provide accurate insights and recommendations. By automating repetitive tasks, optimizing resource allocation, and identifying potential risks, DSS help decision-makers improve project outcomes, boost productivity, and reduce costs [1].

AI algorithms optimize resource utilization, including materials, labor, and equipment, by processing real-time data from Radio-Frequency Identification (RFID) tags and sensors attached to shipments, ensuring timely delivery, reducing losses, and lowering inventory costs [29]. These systems are capable of dynamically adjusting construction schedules and resource allocations in response to variables such as weather conditions, material availability, or workforce productivity, thereby enhancing operational efficiency and minimizing resource waste. Furthermore, AI-powered computer vision systems conduct real-time inspections of construction materials and components, detecting defects and ensuring compliance with quality standards [3].

Construction sites are inherently complex and hazardous environments. Integration of IoT devices such as wearables, drones, and sensors facilitate real-time monitoring of both worker activities and site conditions. Wearable devices track workers' movements and vital signs, providing safety alerts in the event of accidents, while drones conduct aerial surveys and assess progress, improving project management and decision-making [29]. These systems can trigger alerts or automatically shut down operations in hazardous situations, enhancing safety on the site [3].

Another revolutionary application of AI is the deployment of construction robots, which are capable of performing tasks such as bricklaying, concrete pouring, and 3D printing, improving speed, precision, and reducing reliance on manual labor [29].

#### 4.2.3. Operation

AI integration with IoT and Big Data during the operational phase of bioclimatic design enhances building efficiency and sustainability by continuously monitoring and optimizing performance based on real-time data.

Building Performance Monitoring (BPM) algorithms analyze real-time data from sensors integrated into building systems to detect patterns in energy use and environmental conditions. By processing this data, machine learning algorithms can automatically adjust settings for HVAC systems, lighting, and other equipment, enhancing efficiency while maintaining occupant comfort [2].

Predictive analytics, another subset of AI, revolutionizes maintenance practices by forecasting equipment failures based on variables such as temperature, vibration, and usage patterns, enabling proactive maintenance, minimizing downtime and reducing costly repairs [3]. Structural Health Monitoring (SHM) utilizes IoT sensors to assess the integrity of buildings, bridges, and infrastructure by tracking vibrations, deformation, and stress. Real-time data analysis enables early detection of damage, ensuring timely maintenance, preventing severe failures and ensuring both occupant safety and structural longevity. Additionally, IoT-driven systems optimize resource use, monitor emissions, and support compliance with environmental regulations, promoting sustainable construction [29].

AI-driven machine learning and predictive analytics improve forecasting, maintenance, and performance optimization for renewable energy systems, reducing costs and enhancing efficiency. By analyzing large datasets, AI predicts energy outputs, assists in project planning, and identifies optimal locations for renewable infrastructure [35].

Smart building systems leverage AI to enhance occupant comfort and productivity through personalized environmental control, using sensors and real-time adjustments [2]. Additionally, AI-driven urban planning analyzes infrastructure, transportation, and public services to create inclusive, accessible cities. Smart sensors and algorithms optimize traffic flow, reducing congestion and improving mobility [29].

A comparative synthesis of the most recurrent AI tools, their application stages, corresponding performance metrics, reported impacts and evidence levels is provided in Table 2, offering an integrated view of how digital intelligence is currently quantified across the building lifecycle.

**Table 2.** Comparative summary of AI techniques, their mechanisms and measurable impacts on building performance and sustainability (references).

| AI Technique | Lifecycle Stage (Phase) | Typical Measurable Metrics | Impacts Mechanism | Observed/Reported Impacts | Evidence Level |
|---|---|---|---|---|---|
| Big Data/Data-driven design | Design/Operation | Energy use (kWh/m$^2$·year), daylighting, embodied carbon, waste reduction (%) | Aggregates climate, usage, material and sensor data for parametric/simulation-driven choices. | Energy demand reductions up to 15–25% through climate-responsive layouts and daylighting optimization; waste reduction ~10–15% in case studies [2,29]. | Review/Meta-analysis |
| Generative design (GANs, paremetric search) | Design | Annual energy %, solar gain, WWR, material use | Generates multiple alternatives constrained by performance goals. | Case studies show 20–30% improvement in daylighting; orientation optimization reduced heating/cooling load by up to 22% [2,27,32]. | Simulation/Case study |
| Surogate modeling/surrogate-assisted optimization | Design | Optimization time, predicted energy use, thermal comfort (PMV/PPD) | ML surrogates replicate physics-based models at lower cost. | Reduced simulation time by 70–90% while maintaining <5% error; HVAC energy savings potential 12–18% [9,16]. | Simulation/Case study |
| KBW, fuzzy logic, genetic algorithms and neural networks | Design | Multi-objective scores, iteration time, compliance rate | Encodes expert heuristics and evolutionary optimization for tradeoffs. | Reported 15–20% faster design cycles, compliance with energy codes improved by >10% [26,34]. | Review/Simulation |
| AR/VR | Design/Construction | RFI count, stakeholder approval time, coordination errors | Enables immersive design review and collaboration. | Reduced design change orders by ~20%; decision-making time cut by 30–40% in pilot projects [10,29]. | Simulation/Case study; some empirical pilots |
| Decision Support Systems (DSS) | Construction | Cost overrun %, delay %, resource efficiency | Uses predictive analytics to optimize resource allocation. | Case studies report up to 15% cost savings and 10–12% reduction in project delays [1,29]. | Simulation/Case study |
| Radio-Frequency Identification (RFID) and IoT (logistics) | Construction | Inventory loss %, waste, delivery lateness | Tracks materials/equipment, prevents losses, supports circular flows. | Field tests show 8–12% material waste reduction and improved inventory accuracy >95% [29,36]. | Empirical/Field |
| Compute Vision systems (inspection) | Construction/Operation | Defect detection rate, rework %, safety incidents | Automated defect and safety compliance detection. | Achieved >90% defect detection accuracy; reduced manual inspection time by 40–50% [3,29]. | Empirical/Field |
| Construction robots/automation | Construction | Productivity, labor hours saved, material tolerance, waste % | Automates repetitive/manual tasks. | Increased bricklaying/3D printing productivity by 2–3×; waste reduced by 15–20% [29]. | Empirical/Field |
| Building Performance Modelling and Smart Building Systems | Operation | Energy use, peak demand, IEQ, occupant satisfaction | IoT + ML for real-time system adjustments. | Real buildings show 10–25% operational energy savings, improved occupant comfort (PMV closer to neutral) [2,35]. | Empirical/Field |
| Predictive Analytics | Operation | MTBF, downtime, maintenance savings, forecast error | Forecasts failures and renewable output for proactive planning. | Case studies report 30–40% reduction in unplanned downtime, renewable forecasting error reduced to <10% RMSE [3,35]. | Empirical/Field |
| Structural Helth Monitoring (SHM)/Digital Twins | Operation/Asset management | Damage detection time, RUL accuracy, alerts | Continuous sensing and digital twin analysis for early anomaly detection. | Detected structural anomalies weeks earlier than manual inspection; improved RUL prediction accuracy by 20–25% [37,38]. | Empirical/Field (pilots); Review |

## 5. AI-Driven Digital Twins

Digital twin (DT) was initially implemented in NASA's Apollo program in 2010, and since then, its use has expanded across various industries, including manufacturing, aviation, defense, and healthcare, enhancing automation and efficiency [37]. The convergence of artificial intelligence with digital twin technologies has sparked growing interest in their potential applications within the AEC sector. While digital twins offer a virtual replica of physical assets—such as buildings, infrastructure, or entire urban environments—the integration of AI enhances their functionality, rendering them more dynamic, predictive, and adaptive in response to real-time data and evolving conditions. The potential of a human-centric digital twin (DT) approach is envisioned to create adaptive, data-driven built environments that interact dynamically with users, facilitating more dynamic and personalized interactions. Its ecosystem consists of three core elements: the physical environment, its digital replica, and the system managing their interaction [38].

Modeling technologies create a virtual model that reflects the physical environment's parameters, such as structure, functionality, location, and performance, While IoT technologies enable bi-directional data exchange via communication protocols. High-volume, multi-source data is stored, integrated, and analyzed using advanced analytics, then presented to users through interactive visualizations supported by visualization technologies [37].

### 5.1. Digital Twin System Architecture

The digital twin system is conceptualized in five development layers: data acquisition, data transmission, digital modeling, data/model integration, and service.

1. Data Acquisition: This layer collects dynamic data from the physical environment through IoT sensors that detect changes in physical, chemical, and electrical properties of the surroundings (temperature, humidity, gas concentrations, light intensity, motion, etc.), producing an electrical output in response. Input from various sensors can be collected by the more advanced control systems like Supervisory Control and Data Acquisition (SCADA) system for HVAC plants, Direct Digital Control (DDC) system, etc. These technologies include ultrasonic and gyroscopic sensors, which are used to detect clashes, track machinery locations, and ensure accurate placement of resources on construction sites. While building surveillance systems with video streams detect pedestrians and measure environmental conditions like ambient brightness and surface temperatures using thermal imaging modules [37].
2. Data Transmission: Raw data from the data acquisition layer is transmitted to other system components via wired or wireless technologies (Wi-Fi, Bluetooth, WLAN and Ultra-Wideband (UWB)), following communication protocols like Message Queuing Telemetry Transport (MQTT) or Hypertext Transfer Protocol (HTTP). Building Management Systems (BMS) use the internet and Building Automation and Control Networks (BACnet) protocols for data communication between devices and sensors. Additionally, platforms like SophyAI and Gazebo-ROS are utilized for visualizing and processing sensor data [37].
3. Digital Modeling: Virtual models of the physical environment are created using technologies like laser scanning, photogrammetry, and software tools (Autodesk Revit, Navisworks, Solidworks, etc.) to capture and represent parameters such as geometry, functionality, location, and performance. Additionally, game development software like Unity 3D is employed for creating interactive 3D models, avatars, and virtual environments (e.g., Virtual Reality setups with Oculus devices by Meta Platforms, Inc., Menlo Park, CA, USA) [9,35]. Specifically, Autodesk defines five levels of digital twins in the AEC sector, each with a distinct function: Descriptive, Informative, Predictive, Comprehensive and Autonomous. Currently, most DT applications are at Level 2 (Informative Twin), with some moving toward Level 3 (Predictive Twin) (Figure 7) [38].

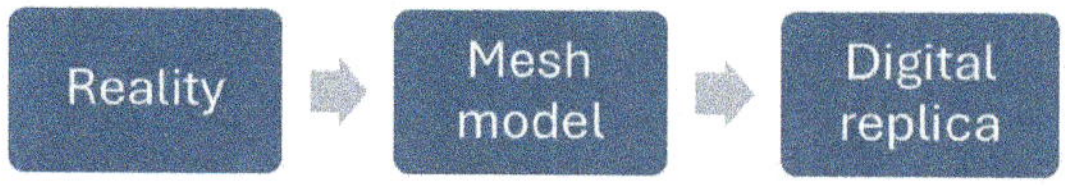


**Figure 7.** Primary data processing: Mesh model, textured model.

4. Data/Model Integration: Multi-source, high-volume data is stored in cloud platforms like Google Cloud Microsoft Azure and Amazon Web service (AWS). Some systems also use mirrored databases to store data from existing building systems like BMS. Data fusion techniques integrate various data types (e.g., sensor, mechanical, and image/video data) into a unified digital twin model, often utilizing customized

Application Programming Interfaces (APIs). Platforms like Autodesk Revit, Unity 3D, and Midas Gen are used for this integration, enabling real-time updates and the fusion of multi-form data into BIM or other virtual environments. Advanced AI technologies process this data for insights and predictive analysis. The processed data then is visualized in digital twin systems through various software platforms, primarily 3D modeling tools like Autodesk Revit, Autodesk Navisworks, Unity 3D, Virtual and Augmented Reality [37].

5. Service: Represents the range of services it provides to users, and these services vary depending on the specific application context. It enables real-time monitoring, predictive analytics, early issue detection, and data visualization, supporting decision-making and operational efficiency. It tracks structural assets, construction activities, and environmental conditions (e.g., temperature, energy consumption, occupancy) while identifying faults in building systems, forecasting failures, and triggering alarms for anomalies. Additionally, digital twins facilitate scenario simulations, robotic control, and smart home management [37].

These layers work together to provide a comprehensive view of the physical and virtual systems in real-time, allowing stakeholders to monitor, analyze, and make data-driven decisions in the built environment.

### 5.2. *The Role of Digital Twins in Bioclimatic Design*

The advent of digital twin technologies has revolutionized the approach to building design, construction, and urban planning, particularly in the context of bioclimatic design. By using digital twins, designers and engineers can create precise virtual models of buildings or urban areas that integrate real-time environmental data, sensor feedback, and predictive simulations. This technology facilitates continuous performance monitoring and optimization, ensuring adaptive responses to climatic conditions. By linking physical assets with virtual counterparts, digital twins incorporate Building Information Modeling, IoT, and sensor networks to implement climate-responsive features such as solar shading, natural ventilation, and energy-efficient building envelopes. Real-time sensor data on indoor and outdoor conditions supports adaptive design modifications, reducing energy consumption and enhancing occupant comfort [10,37].

An increasing interest in optimizing building envelope systems to reduce energy consumption and enhance occupant comfort, driving research into advanced technologies. For example, a data-driven approach to the digital twinning and optimization of a naturally ventilated solar façades with phase-changing materials (PCMs) and double-façade systems integrated with active air conditioning system scan improve thermal performance, energy efficiency of buildings and offer a viable solution for low-income communities facing challenges related to energy costs and indoor air quality. The role of digital twins in monitoring, analyzing, and optimizing the façade system's behavior in real-time under various climate scenarios, providing valuable insights for sustainable building design and reducing energy demand, particularly for low-income communities [36,39].

Monitoring indoor thermal comfort has also gained increasing attention as it plays a key role in optimizing building energy use while maintaining desired indoor conditions. Recent advances integrate BIM, IoT, and immersive VR to provide real-time, interactive visualizations of comfort indicators such as Predicted Mean Vote (PMV) and Predicted Percentage of Dissatisfied (PPD). These systems allow users to navigate virtual building models, access live sensor data, and adjust parameters like metabolic rate or clothing insulation to simulate various scenarios, enabling more intuitive and efficient comfort assessment [10].

In urban environments, digital twins are being explored in the multi-energy system digitalization. By integrating data from IoT sensors, building management systems, and grid infrastructure, DTs provide a dynamic and accurate virtual representation of the physical energy systems, enabling stakeholders to model, simulate, and analyze energy flows and system behavior. By modeling and analyzing energy flows, cities can optimize the integration of renewable sources like solar, wind, and geothermal, reducing reliance on non-renewable energy and enhancing sustainability [40].

AI-driven digital twins also can play a crucial role in stormwater management, addressing urbanization and climate change challenges. Using machine learning, they process vast datasets from sensors, weather forecasts, and monitoring systems to simulate stormwater behavior, predict flooding, and optimize drainage performance, reducing the need for costly infrastructure upgrades [41,42]. By integrating real-time sensor data with hydraulic models, digital twins enhance stormwater depth estimation, improve near-term forecasts, and detect sensor faults with over 99% accuracy, minimizing false flood alarms. Enabled by low-power sensing and wireless communication, they support street-level flood warnings, sewer blockage detection, and active control of valves and gates, ultimately mitigating flooding and pollution risks [42].

Additionally, digital twins aid in fire risk management in urban and wildland-urban areas by integrating temperature sensors, smoke detectors, and satellite imagery to detect and simulate fire behavior. They support bioclimatic design by assessing fire risks and optimizing mitigation strategies, including design of fire-resistant materials, building layouts, and vegetation planning [43].

Finally, the use of digital twins in the field of heritage conservation is emerging as a vital tool for sustainable and effective cultural heritage preservation. Photogrammetry and laser scanning have revolutionized documentation by enabling the rapid acquisition of detailed 3D point clouds. Heritage assets, subject to natural decay and external factors, require ongoing monitoring and maintenance, as emphasized by international organizations like ICOMOS and UNESCO. Heritage Building Information Modeling (HBIM) applies BIM techniques to build heritage, using digital surveys to create accurate models for conservation. Traditionally reliant on visual inspections, monitoring is now enhanced by sensors and scanning technologies. digital twins integrate these advancements with AI and machine learning to predict deterioration and simulate conservation needs, though research in this area remains limited. The Heritage Digital Twin (HDT) framework applies DT concepts to heritage sites, assessing them through four key attributes: Fidelity (model accuracy), Synchronization (real-time interaction), Intelligence (data and AI integration), and Autonomy (automation in conservation tasks) [44].

AI-driven digital restoration is also gaining importance, particularly in detecting and repairing damage to ancient heritage. Deep learning algorithms can identify cracks in murals with high precision, while Generative Adversarial Networks (GANs) simulate and restore original colors and textures. Integrated with 3D modeling and virtual reality, these technologies enable precise digital reconstruction, improving restoration accuracy and efficiency [45].

## 6. Future Perspectives of AI in AEC Industry and Sustainable Building Design

The integration of artificial intelligence into the AEC industry presents significant opportunities for advancing technology and promoting sustainability. AI-powered systems offer innovative solutions to optimize energy use, minimize waste, and streamline resource allocation, ultimately reducing carbon emissions and conserving natural resources. By leveraging AI in BIM, IoT, Big Data, and predictive systems, the industry can improve efficiency, sustainability, and safety while progressing toward Sustainable Development Goals such as SDG 7 (Affordable and Clean Energy), SDG 9 (Industry, Innovation, and Infrastructure), and SDG 11 (Sustainable Cities and Communities). AI-driven innovations in predictive analytics, supply chain optimization, and data-driven communication further foster economic growth and environmental preservation, supporting a holistic approach to sustainable construction [3].

However, despite AI's potential, adoption in the construction sector remains slow due to traditional industry practices, project complexity, and a limited understanding of AI's potential benefits [2]. Key challenges affecting AI integration include:

- High Initial Costs: AI-driven tools such as BIM, generative design, and automation demand substantial upfront investment, posing financial challenges. Adopting tools like BIM integrated with AI, deploying AI-powered energy management systems like urban multi-energy systems (UMES) DTs, or investing in robotic construction technologies involve significant costs for hardware, software, and training. These financial barriers can be especially challenging for small and medium-sized enterprises (SMEs). Financial incentives, government subsidies, and demonstrated long-term return on investment (ROI) could facilitate wider adoption. However, the hidden costs associated with AI adoption, such as system customization, integration with existing infrastructure, and ongoing maintenance, must also be considered [2,29].
- Data Transmission and Security: AI applications in construction heavily rely on data from various sources, including sensor networks, environmental data, construction schedules, and building materials. The data generated in these systems often includes heterogeneous types, such as image data, video data, mechanical data, and environmental data. In terms of data transmission, most studies nowadays have focused on short-range wireless technologies, such as Wi-Fi, Bluetooth, and UWB. Moreover, ensuring secure data transmission is a key consideration. Many construction-related data are confidential, and the transmission of such sensitive information can expose the system to cyber-attacks. A breach in these systems could have far-reaching consequences, including compromising building safety, exposing sensitive data, or disrupting critical infrastructure. Future research should focus on privacy-preserving network models and secure data exchange mechanisms [37].
- Data Integration and Compatibility: The AEC industry struggles with fragmented, inconsistent, and siloed data across various stages of a building's lifecycle. Construction projects typically involve numerous stakeholders using different software platforms, which complicates system interoperability [29]. This data often comes from multiple sources, including design models (e.g., BIM), building materials, construction schedules, sensor networks, and environmental data from IoT devices, which can be incompatible or poorly organized. This lack of integration often leads to delays, cost overruns, and project disruptions. Open-source platforms, standardized interfaces, and semantic web technologies can enhance interoperability. Additionally, collaborative approaches, such as BIM and Integrated Project Delivery (IPD), can streamline workflows, improve coordination, and enhance overall project delivery) [1,2,37].
- Integration of AR and VR for Data Processing and Visualization: In terms of data visualization, the use of 3D modeling platforms, along with immersive technologies like Virtual and Augmented Reality, has advanced the visualization and interaction with digital twin data, driving increasing interest in their integration within Intelligent DSS for construction engineering. These technologies provide interactive and immersive experiences that improve spatial comprehension, collaboration and enhance decision-making but face usability and cost barriers. User-friendly solutions, real-time

synchronization with BIM and cost-effective solutions are needed for widespread adoption in the AEC sector [1,37].

- Scalability and Standardization: While AI-driven solutions show significant potential in individual construction projects, scaling these solutions across entire industries and supply chains presents a distinct challenge. The construction industry's diversity, characterized by varying regulations, building codes, materials, and construction techniques, complicates the widespread implementation of AI solutions without considerable customization. AI adoption is hindered by industry diversity, varying regulations, and a lack of standardized protocols. Unlike industries such as healthcare, which have established broad standards like HL7, or manufacturing with ISO 9000, the construction sector lacks cohesive frameworks for data exchange, interoperability, and quality assurance. Establishing industry-wide frameworks is essential for interoperability and efficiency [1,29].
- Continuous Learning and Adaptation: Studies have shown benefits of DSS in areas like project scheduling, risk management, and material selection. Online learning algorithms in risk management, allowing the system to adjust to evolving risk profiles and enhance decision-making. Meanwhile, machine learning for material selection, enabling the system to learn and refine recommendations based on feedback and new data. DSS enhance project management but require reliable data access and scalable algorithms [1].
- Explainability and Transparency: AI models, particularly those using deep learning, often operate as "black boxes" where their decision-making processes are not transparent or understandable to humans. AI decision-making must be interpretable to ensure trust and accountability. The field of explainable artificial intelligence aims to address this issue by creating models that provide clear, understandable explanations for their decisions. XAI frameworks should balance accuracy and interpretability [21].
- Environmental Impact: As AI systems grow in sophistication, they require massive computational resources, which translates to a substantial environmental impact. AI models require significant computational resources, contributing to a high energy demand and carbon footprint. Sustainable AI practices, data optimization, and governance frameworks should be prioritized [46,47]. In this context, the concept of Net Zero Energy Data Centers (NZEDC) encapsulates key sustainability strategies, defined by the RenewIT project (Deliverable 4.5) as data centers that achieve a net-zero balance by exclusively consuming renewable energy while generating an equivalent amount of electrical and thermal energy over their operational lifespan [18].
- Regulatory and Ethical Considerations: AI systems, which can significantly impact society, the economy, and individual lives, require robust frameworks to ensure they are developed and deployed responsibly. Design teams often lack regulatory support documents with performance benchmarks for non-mechanical solutions, unlike mechanical systems (e.g., HVAC, heat pumps) validated by European standards (Delegated Regulation 2022/759; Commission, 2014). These benchmarks are crucial for assessing energy savings and comfort across climates [12]. Moreover, AI governance must address fairness, transparency, and human oversight. According to the AI Index Report 2024, in 2023, policymakers in both the European Union and the United States made significant strides in AI regulation [22]. The European Union's AI Act and similar frameworks emphasize ethical AI development [25].
- Skill Gaps and Workforce Adaptation: AI adoption in the AEC sector requires significant workforce reskilling to bridge the technical expertise gap. While AI can greatly enhance efficiency, it may reduce the need for certain manual tasks, leading to potential workforce displacement. This raises important ethical concerns about balancing

the benefits of automation with the preservation of employment opportunities [33]. In addition, many professionals in construction lack the technical expertise required to effectively use AI tools, which demand knowledge in areas like machine learning, data analytics, robotics, and programming. Building and deploying AI solutions in construction require specialized knowledge, and there is currently a large gap between the demand for AI talent and the availability of qualified professionals. Talent development and collaborative training programs are essential to prepare professionals for AI-driven roles [2,29].

To fully realize AI's potential in sustainable construction, industry stakeholders must address these challenges through policy support, investment, and technological advancements. Future research should focus on scalable, cost-effective AI solutions that enhance efficiency, safety, and sustainability while ensuring ethical and responsible implementation.

## 7. Conclusions

Amid growing environmental challenges like climate change and resource depletion, bioclimatic design has emerged as a key strategy for creating energy-efficient, climate-responsive buildings, aligning with global sustainability goals. Simultaneously, AI technologies are transforming decision-making across various domains, driving the development of intelligent, eco-friendly infrastructure. This review explores the intersection of bioclimatic design and AI, highlighting their integration as a solution to sustainability and efficiency challenges in the AEC industry.

AI is transforming bioclimatic design across the building lifecycle, enhancing energy efficiency and performance through tools like generative design algorithms and surrogate models. Among these advancements, digital twin technologies stand out, enabling real-time monitoring, simulation, and optimization by integrating AI, IoT, and BIM. Their applications in façade optimization, energy flow analysis, and risk assessment highlight their potential in sustainability and energy efficiency. Looking ahead, DTs are expected to drive autonomous, human-centric design under Construction 5.0 and Industry 5.0, with applications extending to heritage conservation and resource-efficient construction. However, challenges such as computational and energy demands, interoperability, and regulatory constraints must be addressed for broader adoption. One can predict that AI and DT can influence and maybe transform architectural patterns, shifting the design from stylistic or mainstream choices to more optimized choices. Creativity supported by AI can explore varied, innovative design choices. digital twin allows for real-time simulation, reinforcing the design loop. Future research should prioritize scalable, ethical, and transparent implementations to maximize AI's impact. By integrating AI, DTs, and emerging technologies, the AEC industry can set new benchmarks for sustainable, adaptive, and intelligent built environments.

**Author Contributions:** Conceptualization, E.F., T.Z., S.H. and M.M.; methodology, E.F., T.Z., S.H. and M.M.; formal analysis, E.F.; investigation, E.F.; data curation, E.F.; writing—original draft preparation, E.F.; writing—review and editing, M.M.; visualization, E.F.; supervision, M.M. All authors have read and agreed to the published version of the manuscript.

**Funding:** This research received no external funding.

**Data Availability Statement:** The original contributions presented in this study are included in the article. Further inquiries can be directed to the corresponding authors.

**Conflicts of Interest:** The authors declare no conflicts of interest.

## Abbreviations

The following abbreviations are used in this manuscript:

| | |
|---|---|
| AI | Artificial Intelligence |
| AEC | Architecture, Engineering and Construction |
| DT | Digital Twin |
| SDG | Sustainable Development Goals |
| ML | Machine Learning |
| DL | Deep Learning |
| EPBD | Revised Energy Performance of Buildings Directive |
| GAN | Generative Adversal Networks |
| IoT | Internet of Things |
| GPU | Graphic Processing Units |
| TPU | Tensor Processing Units |
| XAI | Explainable Artificial Intelligence |
| EU | Europian Union |
| kNN | k-Nearest Neighbors |
| BIRCH | Balanced Iterative Reducing and Clustering using Hierarchies |
| NLP | Natural Language Processing |
| ANN | Artificial Neural Networks |
| MLP | Multi-Layer Perception |
| CNN | Convolutional Neural Networks |
| RNN | Recurent Neural Networks |
| GM | Generative Models |
| GAN | Generative Adversal Network |
| RBM | Restricted Boltzman Machine |
| DBN | Deep Belief Network |
| NHE | Natural Hazards Engineering |
| CAD | Computer-Aided Design |
| BIM | Building Information Modeling |
| DSS | Decision Support Systems |
| RFID | Radio-Frequency Identification |
| BPM | Building Performance Monitoring |
| SHM | Structural Health Monitoring |
| KBE | Knowledge-Based Engineering |
| SM | Surrogate model |
| HVAC | Heating, Ventilation and Air Conditionig |
| AR | Augmented Reality |
| VR | Vurtual Reality |
| SCADA | Supervisory Control and Data Acquistion |
| DDC | Direct Digital Control |
| WLAN | Wireless Local Area Network |
| UWB | Ultra-Wideband |
| MQTT | Message Queuing Telemetry Transport |
| HTTP | Hypertext Transfer Protocol |
| BMS | Building Management Systems |
| AWS | Amazon Web Service |
| API | Application Programming Interface |
| PCM | Phase-Changing Material |
| PMV | Predicted Mean Vote |
| PPD | Predicted Percentage Dissatisfied |
| HBIM | Heritage Building Information Modeling |
| UMES | Urban Multi-Energy Systems |

| | |
|---|---|
| SME | Medium-Sized Enterprise |
| ROI | Return on Investment |
| IPD | Integrated Project Delivery |
| NZEDC | Net Zero Energy Data Center |

## References


1. Waqar, A. Intelligent decision support systems in construction engineering: An artificial intelligence and machine learning approach. *Expert Syst. Appl.* **2024**, *249*, 123503. [CrossRef]
2. Adewale, B.A.; Ene, V.O.; Ogunbayo, B.F.; Aigbavboa, C.O. A Systematic Review of the Applications of AI in a Sustainable Building's Lifecycle. *Buildings* **2024**, *14*, 2137. [CrossRef]
3. Regona, M.; Yigitcanlar, T.; Hon, C.; Teo, M. Artificial intelligence and sustainable development goals: Systematic literature review of the construction industry. *Sustain. Cities Soc.* **2024**, *108*, 105499. [CrossRef]
4. Mejía-Parada, C.; Mora-Ruiz, V.; Attia, S. Bioclimatic design recommendations for novel cluster analysis-based mapping for humid climates with altitudinal gradient variations. *J. Build. Eng.* **2024**, *82*, 108262. [CrossRef]
5. Elaouzy, Y.; El Fadar, A. Impact of key bioclimatic design strategies on buildings' performance in dominant climates worldwide. *Energy Sustain. Dev.* **2022**, *68*, 532–549. [CrossRef]
6. Global Alliance for Buildings and Construction. *Global Status Report for Buildings and Construction 2020*; United Nations Environment Programme (UNEP): Nairobi, Kenya, 2020; Available online: https://globalabc.org/resources/publications/2020-global-status-report-buildings-and-construction (accessed on 5 November 2024).
7. Barea, G.; Mercado, M.V.; Filippín, C.; Monteoliva, J.M.; Villalba, A. New paradigms in bioclimatic design toward climatic change in arid environments. *Energy Build.* **2022**, *266*, 112100. [CrossRef]
8. Elaouzy, Y.; El Fadar, A. Sustainability of building-integrated bioclimatic design strategies depending on energy affordability. *Renew. Sustain. Energy Rev.* **2023**, *179*, 113295. [CrossRef]
9. Elwy, I.; Hagishima, A. The artificial intelligence reformation of sustainable building design approach: A systematic review on building design optimization methods using surrogate models. *Energy Build.* **2024**, *323*, 114769. [CrossRef]
10. Shahinmoghadam, M.; Natephra, W.; Motamedi, A. BIM- and IoT-based virtual reality tool for real-time thermal comfort assessment in building enclosures. *Build. Environ.* **2021**, *199*, 107905. [CrossRef]
11. European Union. *Directive (EU) 2024/XXX of the European Parliament and of the Council on the Energy Performance of Buildings (Recast)*; Publications Office of the European Union: Luxembourg, 2024; Available online: https://energy.ec.europa.eu/topics/energy-efficiency/energy-efficient-buildings/energy-performance-buildings-directive_en (accessed on 5 November 2024).
12. López Plazas, F.; Crespo Sánchez, E.; González Ruiz, P. Bioclimatic strategies in existing multifamily buildings to achieve cities decarbonization goals: Potential and relevance for Catalonia climates. *Cities* **2024**, *154*, 105335. [CrossRef]
13. Ozarisoy, B.; Altan, H. Systematic literature review of bioclimatic design elements: Theories, methodologies, and cases in the South-Eastern Mediterranean climate. *Energy Build.* **2021**, *250*, 111281. [CrossRef]
14. Couvelas, A. Bioclimatic building design theory and application. *Procedia Manuf.* **2020**, *44*, 326–333. [CrossRef]
15. Xhexhi, K.; Aliaj, B. Modular construction and bioclimatic strategies: A sustainable approach to building design. In Proceedings of the 4th International Conference on Scientific and Academic Research, Konya, Turkey, 19–20 July 2024; All Sciences Academy: Konya, Turkey, 2024. Available online: https://www.researchgate.net/publication/382561521 (accessed on 5 November 2024).
16. Mazzeo, D.; Matera, N.; Peri, G.; Scaccianoce, G. Forecasting green roofs' potential in improving building thermal performance and mitigating urban heat island in the Mediterranean area: An artificial intelligence-based approach. *Appl. Therm. Eng.* **2023**, *222*, 119879. [CrossRef]
17. Skandalos, N.; Karamanis, D. Decarbonizing Operational Emissions in Urban Neighborhoods with the Integration of Rooftop Photovoltaics and Green Infrastructure under Current and Future Climate Conditions. *Energy Build.* **2025**, *329*, 115306. [CrossRef]
18. Manganelli, M.; Soldati, A.; Martirano, L.; Ramakrishna, S. Strategies for Improving the Sustainability of Data Centers via Energy Mix, Energy Conservation, and Circular Energy. *Sustainability* **2021**, *13*, 6114. [CrossRef]
19. Hoffmann, C.H. Is AI intelligent? An assessment of artificial intelligence, 70 years after Turing. *Technol. Soc.* **2022**, *68*, 101893. [CrossRef]
20. Padmaja-Chinimilli, V.R.; Sadasivuni, L.N.; Gouthami, L.A.; Priyanka, K.B. The rise of AI: A comprehensive research review. IAES Int. *J. Artif. Intell.* **2024**, *13*, 2226–2235. [CrossRef]
21. Longo, L.; Brcic, M.; Cabitza, F.; Choi, J.; Confalonieri, R.; Del Ser, J.; Guidotti, R.; Hayashi, Y.; Herrera, F.; Holzinger, A.; et al. Explainable Artificial Intelligence (XAI) 2.0: A manifesto of open challenges and interdisciplinary research directions. *Inf. Fusion* **2024**, *106*, 102301. [CrossRef]
22. HAI, Stanford University. *Artificial Intelligence Index Report*; Stanford University: Palo Alto, CA, USA, 2024; Available online: https://aiindex.stanford.edu/report/ (accessed on 2 December 2024).

23. European Commission. *Ethics Guidelines for Trustworthy AI*; European Commission: Brussels, Belgium, 2019; Available online: https://digital-strategy.ec.europa.eu/en/library/ethics-guidelines-trustworthy-ai (accessed on 10 December 2024).
24. European Commission. *Artificial Intelligence Act 2024*; European Commission: Brussels, Belgium, 2024; Available online: https://eur-lex.europa.eu/legal-content/EN/TXT/?uri=CELEX%3A52021PC0206 (accessed on 10 December 2024).
25. Díaz-Rodríguez, N.; Del Ser, J.; Coeckelbergh, M.; López de Prado, M.; Herrera-Viedma, E.; Herrera, F. Connecting the dots in trustworthy Artificial Intelligence: From AI principles, ethics, and key requirements to responsible AI systems and regulation. *Inf. Fusion.* **2023**, *99*, 101896. [CrossRef]
26. Bellini, V.; Cascella, M.; Cutugno, F.; Russo, M.; Lanza, R.; Compagnone, C.; Bignami, E. Understanding basic principles of Artificial Intelligence: A practical guide for intensivists. *Acta Bio. Med.* **2022**, *93*, e2022297. [CrossRef]
27. Krausková, V.; Pifko, H. Use of Artificial Intelligence in the Field of Sustainable Architecture: Current Knowledge. *Archit. Pap. Fac. Archit. Des. STU* **2021**, *26*, 20–29. [CrossRef]
28. Mortezapour Shiri, F.; Perumal, T.; Mustapha, N.; Mohamed, R. A comprehensive overview and comparative analysis on deep learning models. *J. Artif. Intell.* **2024**, *6*, 301–360. [CrossRef]
29. Rane, N. Integrating Leading-Edge Artificial Intelligence (AI), Internet of Things (IoT), and Big Data Technologies for Smart and Sustainable Architecture, Engineering and Construction (AEC) Industry: Challenges and Future Directions. *Int. J. Data Sci. Big Data Anal.* **2023**, *3*, 73–95. [CrossRef]
30. Deierlein, G.G.; Zsarnóczay, Á. (Eds.) *State of the Art in Computational Simulation for Natural Hazards Engineering*, 2nd ed.; Report Number 2021-01; Zenodo: Genève, Switzerland, 2021. [CrossRef]
31. Olawade, D.B.; Wada, O.Z.; Ige, A.O.; Egbewole, B.I.; Olajo, A.; Oladapo, B.I. Artificial intelligence in environmental monitoring: Advancements, challenges, and future directions. *Hyg. Environ. Health Adv.* **2024**, *12*, 100114. [CrossRef]
32. Castro Pena, M.L.; Carballal, A.; Rodríguez-Fernández, N.; Santos, I.; Romero, J. Artificial intelligence applied to conceptual design: A review of its use in architecture. *Autom. Constr.* **2021**, *124*, 103550. [CrossRef]
33. Huang, S.; Wang, B.; Li, X.; Zheng, P.; Mourtzis, D.; Wang, L. Industry 5.0 and Society 5.0—Comparison, complementation and co-evolution. *J. Manuf. Syst.* **2022**, *64*, 424–428. [CrossRef]
34. Gilner, E.; Galuszka, A.; Grychowski, T. Application of Artificial Intelligence in Sustainable Building Design—Optimisation Methods. In Proceedings of the 24th International Conference on Methods and Models in Automation and Robotics (MMAR), Międzyzdroje, Poland, 26–29 August 2019. [CrossRef]
35. Ziarati, T.; Hedayat, S.; Moscatiello, C.; Sappa, G.; Manganelli, M. Overview of the Impact of Artificial Intelligence on the Future of Renewable Energy. In Proceedings of the IEEE International Conference on Environment and Electrical Engineering and 2024 IEEE Industrial and Commercial Power Systems Europe (EEEIC/I&CPS Europe), Rome, Italy, 17–20 June 2024. [CrossRef]
36. Tariq, R.; Torres-Aguilar, C.E.; Sheikh, N.A.; Ahmad, T.; Xamán, J.; Bassam, A. Data engineering for digital twining and optimization of naturally ventilated solar façade with phase changing material under global projection scenarios. *Renew. Energy* **2022**, *187*, 1184–1203. [CrossRef]
37. Tuhaise, V.V.; Tah, J.H.M.; Abanda, F.H. Technologies for digital twin applications in construction. *Autom. Constr.* **2023**, *152*, 104931. [CrossRef]
38. Osama, Z. The digital twin framework: A roadmap to the development of user-centred digital twin in the built environment. *J. Build. Eng.* **2024**, *98*, 111081. [CrossRef]
39. Tariq, R.; Bassam, A.; Orozco-del-Castillo, M.G.; Ricalde, L.J.; Carvente, O. Sustainability framework of intelligent social houses with a synergy of double-façade architecture and active air conditioning systems. *Energy Convers. Manag.* **2023**, *288*, 117120. [CrossRef]
40. Koirala, B.; Cai, H.; Khayatian, F.; Munoz, E.; An, J.G.; Mutschler, R.; Sulzer, M.; De Wolf, C.; Orehounig, K. Digitalization of urban multi-energy systems—Advances in digital twin applications across life-cycle phases. *Adv. Appl. Energy* **2024**, *16*, 100196. [CrossRef]
41. Kim, Y.; Oh, J.; Bartos, M. Stormwater digital twin with online quality control detects urban flood hazards under uncertainty. *Sustain. Cities Soc.* **2025**, *118*, 105982. [CrossRef]
42. Sharifi, A.; Tarlani Beris, A.; Sharifzadeh Javidi, A.; Nouri, M.; Gholizadeh Lonbar, A.; Ahmadi, M. Application of artificial intelligence in digital twin models for stormwater infrastructure systems in smart cities. *Adv. Eng. Inform.* **2024**, *61*, 102485. [CrossRef]
43. Yang, S.; Huang, Q.; Yu, M. Advancements in remote sensing for active fire detection: A review of datasets and methods. *Sci. Total Environ.* **2024**, *943*, 173273. [CrossRef]
44. Sugiyama, G.; Bourgeois, I.; Rodrigues, H. A holistic methodology for the assessment of Heritage Digital Twin applied to Portuguese case studies. *Digit. Appl. Archaeol. Cult. Herit.* **2025**, *36*, e00390. [CrossRef]
45. Yan, Y.; Chai, B.; Li, J. ArtDiff: Integrating IoT and AI to enhance precision in ancient mural restoration. *Alex. Eng. J.* **2025**, *111*, 511–520. [CrossRef]

46. Van Wynsberghe, A. Sustainable AI: AI for sustainability and the sustainability of AI. *AI Ethics* **2021**, *1*, 213–218. [CrossRef]
47. Wu, C.J.; Raghavendra, R.; Gupta, U.; Acun, B.; Maeng, K.; Chang, G.; Aga, F.; Huang, J.; Bai, C.; Gschwind, M.; et al. Sustainable AI: Environmental Implications, Challenges and Opportunities. *Proc. Mach. Learn. Syst.* **2022**, *4*, 795–813.

*energies*

*Review*

# Reinforcement Learning for Optimizing Renewable Energy Utilization in Buildings: A Review on Applications and Innovations

**Panagiotis Michailidis [1,2,*,†], Iakovos Michailidis [1,2,*,†] and Elias Kosmatopoulos [1,2,†]**

1 Center for Research and Technology Hellas, 57001 Thessaloniki, Greece; kosmatop@iti.gr
2 Department of Electrical and Computer Engineering, Democritus University of Thrace, 67100 Xanthi, Greece
* Correspondence: panosmih@iti.gr (P.M.); michaild@iti.gr (I.M.); Tel.: +30-2310-464160 (P.M.)
† These authors contributed equally to this work.

**Abstract:** The integration of renewable energy systems into modern buildings is essential for enhancing energy efficiency, reducing carbon footprints, and advancing intelligent energy management. However, optimizing RES operations within building energy management systems introduces significant complexity, requiring advanced control strategies. One significant branch of modern control algorithms concerns reinforcement learning, a data-driven strategy capable of dynamically managing renewable energy sources and other energy subsystems under uncertainty and real-time constraints. The current review systematically examines RL-based control strategies applied in BEMS frameworks integrating RES technologies between 2015 and 2025, classifying them by algorithmic approach and evaluating the role of multi-agent and hybrid methods in improving real-time adaptability and occupant comfort. Following a thorough explanation of a rigorous selection process—which targeted the most impactful peer-reviewed publications from the last decade, the paper presents the mathematical concepts of RL and multi-agent RL, along with detailed summaries and summary tables of the integrated works to facilitate quick reference to key findings. For evaluation, the paper examines and outlines the different attributes in the field considering the following: methodologies of RL; agent types; value-action networks; reward functions; baseline control approaches; RES types; BEMS types; and building typologies. Grounded on the findings presented in the evaluation section, the paper offers a structured synthesis of emerging research trends and future directions, identifying the strengths and limitations of RL in energy management.

**Keywords:** reinforcement learning; renewable energy; building energy management; adaptive control; model-free control; smart buildings

## 1. Introduction

### 1.1. Motivation

Energy efficiency has become a global priority, driven by rising energy costs, environmental concerns, and the urgent need to curb greenhouse gas emissions [1,2]. Buildings, among the largest energy consumers worldwide, are responsible for approximately 34% of global energy use and carbon emissions, particularly associated with urban areas [3]. Given their significant environmental impact and operational costs, improving building energy efficiency has become a cornerstone of sustainability efforts. A key strategy in achieving this goal concerns the integration of renewable energy systems (RES), such as solar photovoltaic panels (PVs) and wind turbines (WTs), within building infrastructures [4]. By leveraging on-site renewable energy sources, buildings may decrease reliance on conventional power grids [5], reduce energy costs [6], and contribute to broader sustainability objectives [7].

The concept of integrating RES into buildings gained traction in the late 20th century, fundamentally reshaping how energy is generated and managed in the built environment. Early initiatives primarily focused on solar thermal and PV systems, which initially emerged in pilot projects during the 1970s in response to the global oil crisis [8,9]. As renewable technologies advanced and the environmental consequences of fossil fuels became more apparent, governments and institutions worldwide increased their investment in RES for buildings [10]. By the 1990s, improvements in PV efficiency and cost reductions enabled more widespread adoption, which led early efforts in promoting renewable energy integration across residential and commercial sectors [9,11]. In more recent years advanced policies, such as the European Union's Green Deal [12], played a pivotal role in promoting RES integration by setting ambitious climate targets and providing financial mechanisms like the REPowerEU [13], Renewable Energy Directive [14], and LIFE Programme [15] to support regions transitioning from fossil fuels. As is evident, such initiatives inevitably paved the way for the broader adoption of RES in buildings by providing a strong regulatory framework, financial support, and streamlined integration pathways. As RES penetration increased, the role of building energy management systems (BEMS) became indispensable in optimizing energy consumption, demand response (DR), and grid interaction.

While integrating RES in buildings offers significant benefits, including reducing reliance on conventional energy, lowering costs, and enhancing sustainability, it also poses significant challenges, including policy constraints, high costs, maintenance demands, installation disruptions, and the need for increased energy flows optimization, demand forecasting, and system management, alongside stability concerns associated with renewable energy fluctuations, grid integration, and real-time control [16–18]. Unlike conventional energy systems, RES are inherently variable, with outputs fluctuating based on external factors such as weather conditions and time of day. For example, solar energy generation is directly affected by sunlight availability, which varies with daily and seasonal cycles. This variability complicates energy management, requiring buildings to adapt their energy consumption patterns to align with high RES output periods while relying on storage solutions or grid support during low-production intervals [19,20]. Moreover, additional dynamic factors, including occupancy levels, energy pricing, and climate conditions, further increase the complexity of managing RES-equipped buildings [21]. Traditional rule-based control (RBC) approaches often fall short in addressing these complexities, as they lack the flexibility and adaptability required to respond effectively to fluctuating conditions [22–24]. Consequently, there is a growing need for advanced, intelligent control methodologies that can optimize energy use and enhance efficiency in RES-integrated building environments.

To overcome these challenges, Artificial Intelligence (AI) and machine learning-based control frameworks have emerged as transformative tools in BEMS [25–28]. Such intelligent systems are adequate to dynamically adjust to real-time fluctuations in occupancy, RES output, and energy prices, optimizing the operation of integrated energy subsystems, such as heating, ventilation, and air conditioning (HVAC), lighting systems (LS), and energy storage systems (ESS) or thermal storage systems (TSS) [29,30]. BEMS control strategies typically fall into two primary categories: model-based and model-free approaches [31]. Model-based methods rely on predefined mathematical representations to simulate system behavior, enabling predictive and precise control [32–34]. The efficiency of such frameworks is highly dependent on the accuracy of these models, making them complex and often expensive to implement, particularly in highly variable systems with uncertain parameters. On the other hand, model-free approaches do not require prior system modeling; instead, they learn optimal control strategies by interacting directly with the environment [31,35–37].

According to the literature, reinforcement learning (RL) stands out as one of the most prominent model-free approach for managing building energy systems, including HVAC systems and similar frameworks (see Figure 1) [38]. In model-free RL, an agent learns optimal control policies through trial and error, continuously refining its strategy by exploring various actions and maximizing rewards [39–42]. The application of RL in managing RES within buildings dates back to the early 2000s, when researchers began leveraging its adaptive learning capabilities to address the inherent variability in RES output. Until then, a significant number of studies demonstrated RL's effectiveness in dynamically optimizing building energy management, particularly in HVAC operations under fluctuating environmental conditions [43–45].

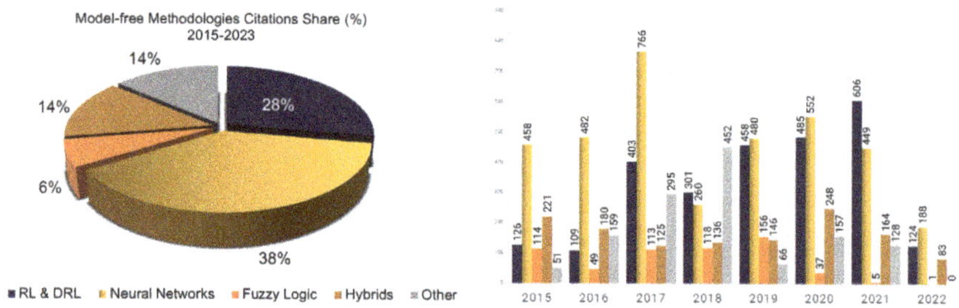


**Figure 1.** Citations share between the different model-free methodologies for HVAC frameworks: Overall share (%) (**Left**) and occurrence per year (**Right**) for 2014–2022 period.

Over the past decade, RL has firmly established itself as a promising model-free control strategy, tackling the complexities of RES integration in smart buildings and paving the way for more sophisticated applications in the years ahead. Its growing prominence in building energy management is largely attributed to its adaptability and capacity to handle the stochastic nature of RES [46,47]. Unlike model-based control methods, RL-based approaches are adequate to dynamically adjust to real-time system and environmental variations, making them particularly well suited for highly variable energy systems, such as those incorporating RES. A key advantage of RL relies in its ability to learn directly from interactions with the environment, refining its control policy over time without requiring a predefined system model—a feature that has proven especially valuable for complex energy management scenarios [47]. To this end, over the past years, the RL approach has demonstrated its transformative potential in building energy efficiency by optimizing energy consumption, reducing operational costs, and improving occupant comfort [48]. Recent advancements in algorithmic design, function approximation, and real-time adaptability have significantly enhanced RL's ability to manage complex building energy systems. As such, numerous studies highlight improvements in reward shaping, multi-agent coordination, and self-learning mechanisms, allowing RL-based controllers to dynamically adjust HVAC, lighting, and renewable energy storage, leading to measurable energy savings and benefits [49–51]. Moreover, it is evident that RL introduced a significant number of hybridized approaches for optimizing energy systems, moving beyond traditional control methods, shaping next-generation energy management solutions. For instance, in [52], scientists proposed a Deep Q-Network RL approach with fuzzified reward mechanisms for PV Maximum Power Point Tracking (MPPT), improving performance in dynamic conditions without relying on explicit system models. Moreover, in [53], researchers applied a Twin Delayed Deep Deterministic Policy Gradient to pulverized coal boiler combustion, integrating predictive modeling for real-time optimization.

Recognizing the pivotal role of RES in sustainable building practices, this review focuses specifically on RL-based control strategies for optimizing RES operation in buildings, alongside other integrated energy systems. By analyzing recent RL applications within BEMS that incorporate RES into the energy mix, this work examines high-value research contributions to highlight how RL enhances energy efficiency and operational intelligence in building environments. The analysis spans various building types, control frameworks, and both real-world and simulated implementations, offering a comprehensive perspective on the current state of RL-based control in the field. The findings of this review provide valuable insights for researchers and practitioners aiming to implement intelligent, adaptive control strategies in RES-integrated buildings, ultimately supporting the advancement of sustainable and resilient building technologies.

### 1.2. Literature Analysis Approach

This review aims to provide a comprehensive exploration of key studies on the application of RL for controlling RES in buildings, highlighting significant trends and critical findings in the field. By systematically analyzing research from the past decade, this work evaluates fundamental concepts, control methodologies, RL algorithms, and their specific applications in managing RES within building environments. To ensure a structured and in-depth examination, studies are categorized based on RL control approaches—value-based, policy-based, and actor-critic—along with training methodologies and different RES-integrated BEMS. Additionally, details regarding testbed characteristics are included to offer a detailed overview of the current state of RL control, ensuring a well-rounded evaluation of each selected study.

- **Article Selection:** A rigorous selection process was conducted using academic databases such as Scopus- and Web of Science (WoS)-indexed peer-reviewed journals and conferences to ensure quality and reliability. An initial pool of over 300 papers was reviewed based on abstracts, from which the most relevant studies were shortlisted for detailed analysis. More specifically, the work followed a multi-step quality assessment process considering the following: **Citation Impact** (studies with at least 10 citations—excluding self-citations—were selected to ensure academic influence, verified via Scopus at the time of selection); **Relevance** (only papers explicitly addressing RL-based control for RES-integrated BEMS were included, excluding studies on fault detection, generic demand-side management, or isolated RES control without RL); Peer Review (only peer-reviewed journal articles and high-quality conference proceedings—IEEE, Elsevier, Springer, MDPI—were considered, excluding preprints and non-peer-reviewed reports); **Rigor** (selected studies had to clearly describe RL implementation, optimization objectives, and evaluation frameworks, with benchmark comparisons and experimental validation); **Diversity** (a balanced selection of value-based, policy-based, actor-critic, and hybrid RL methods was ensured, covering various building types—residential, commercial, university, hospital—and RES-Integrated BEMS setups).
- **Keyword Research:** A comprehensive keyword analysis was performed, incorporating terms such as *"Reinforcement Learning in RES for buildings"*, *"RL control in BEMS"*, *"RL-based energy management"* and other specific phrases related to RES integration. This approach ensured a broad yet precise capture of the challenges and advancements in RL-based RES management.
- **Data Collection:** Each publication was systematically categorized based on the RL techniques applied to RES control, the integration of additional energy systems, the specific application context, and evaluations of the advantages, limitations, and practical implications within building energy management.

- **Quality Assessment:** A structured quality assessment was carried out, considering citation count, the academic contributions of authors, and the methodological rigor of each study. More specifically, the citation-based selection concerned a total number of more than ten citations acting as a measure of academic impact, ensuring the inclusion of studies with established influence. Citation counts, retrieved from Scopus and excluding self-citations, provided reliability. Preference was given to publications in high-impact journals and conferences. Moreover the research background of authors in RL, RES, BEMS was evaluated based on their publications in top-tier journals, contributions to widely adopted RL algorithms or methodological advancements in energy management, and affiliations with leading research institutions or energy-focused labs. Studies from authors with a strong track record in RL-based energy optimization were prioritized to ensure credibility.
  This evaluation helped determine the relative significance of each research work's contribution in the field.
- **Data Synthesis:** Findings were synthesized into distinct categories, enabling clear comparisons across studies and facilitating a holistic understanding of the evolving landscape of RL-based control for RES in buildings.

### 1.3. Previous Work

The literature presents a wealth of studies exploring the application of RL in managing various energy systems. Among these, Wang et al. [46] conducted a comprehensive review of RL-based building control strategies, emphasizing their potential to enhance energy efficiency and adaptability. The study identified key challenges, including data-intensive training, security vulnerabilities, limited real-world adoption, and poor generalization. To address these issues, the authors highlighted the importance of transfer learning and the development of open-source testbeds to advance RL controller design and benchmarking. Al et al. [45] focused specifically on RL applications for HVAC control in intelligent buildings, underscoring its effectiveness in optimizing energy consumption while maintaining occupant comfort. By comparing RL-based approaches with conventional HVAC control methods, the study demonstrated RL's superiority in overcoming limitations such as high computational demands and a lack of adaptive learning capabilities. A significant contribution by Perera et al. [54] examined RL applications across broader energy systems, classifying existing research into seven key categories, with a strong emphasis on building energy management and dispatch optimization. The study reinforced RL's potential to enhance energy system efficiency while also identifying critical research gaps, including the limited integration of deep learning techniques, benchmarking inconsistencies, and challenges related to reproducibility.

Gaviria et al. [55] extended the discussion to machine learning frameworks for optimizing PV system operations. Their review highlighted gaps in benchmarking, real-world validation, and the integration of advanced machine learning models. The study proposed future directions, such as leveraging RL for Maximum Power Point Tracking (MPPT) and exploring the use of transformer-based models for improved PV performance forecasting. Fu et al. [56] provided a focused review of RL applications in building energy efficiency, particularly in HVAC optimization. The study systematically classified RL algorithms, distinguishing value-based methods, e.g., Q-learning (QL), Deep-Q-Networks (DQN), for discrete action spaces and policy-based methods, e.g., DDPG, proximal policy optimization (PPO), for continuous control tasks. While the research underscored RL's advantages over traditional control techniques, it also highlighted persistent challenges, such as prolonged training times and the high costs associated with real-world deployment.

### 1.4. Contribution and Novelty

This review distinguishes itself from the existing literature on RL control in RES-integrated BEMS through several key contributions. Firstly, it is among the few studies that specifically focus on RL control frameworks designed for the optimal operation of RES—including, PV panels, WT, solar water heaters (SWHs), ground heat pumps (GHPs), and biomass energy systems (BIO)—alongside a diverse set of energy management components, such as HVAC, domestic hot water (DHW, LS, ESS), and electric vehicles (EVs) within building environments.

Unlike previous reviews, this work conducts a large-scale, in-depth analysis of RL applications, systematically evaluating a substantial number of influential research studies published between 2015 and 2025. By encompassing nearly all RL-focused studies in this domain that have received more than ten citations (>10), this review ensures that only high-impact contributions inform the conclusions drawn in subsequent sections. Through the synthesis of key findings from the past decade, this study highlights the adaptability and effectiveness of RL-based control strategies in optimizing RES-integrated BEMS.

Additionally, this review presents detailed summary tables that highlight the most influential studies, enabling readers to efficiently identify relevant research and compare different RL approaches. Beyond summarizing existing works, this study provides a thorough evaluation of critical research elements, including the specific RL methodology employed, the agent-based algorithmic framework, the target optimization outputs, baseline control strategies, RES configurations, and building typologies. By adopting this comprehensive approach, the review not only identifies valuable insights and emerging trends but also lays the groundwork for future advancements in RL-based control for RES management.

### 1.5. Paper Structure

The structure of this paper is as follows (Figure 2): Section 1 introduces the motivation behind this review, outlines the literature analysis approach, examines previous studies, and highlights the paper's key contributions and novelty. Section 2 provides an overview of commonly integrated RES technologies in buildings and reviews general RL-based control strategies for managing RES-integrated BEMS. Section 3 delves into the primary RL control methodologies, presenting the generalized mathematical framework. Section 4 examines the most highly cited research works from 2014 to 2024, summarizing key characteristics in tabular format for comparative analysis. Section 5 evaluates various research dimensions, comparing prevalent RL methodologies and their effectiveness. Section 6 identifies emerging trends based on the evaluation and outlines potential future directions for RL-based control in RES-integrated systems. Finally, Section 7 concludes with a summary of the key insights and findings presented in this review.

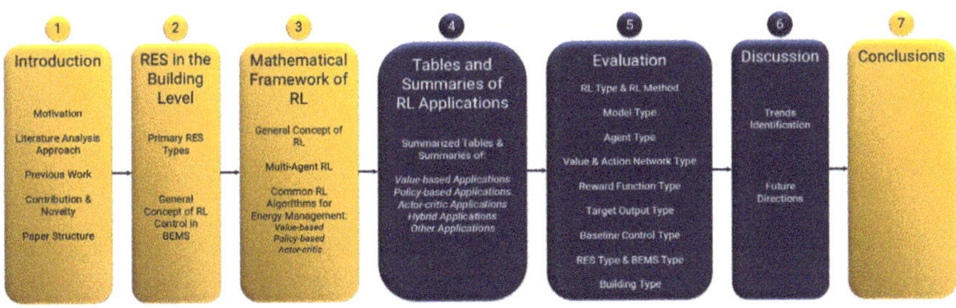


**Figure 2.** Paper structure.

## 2. Renewable Energy Systems in the Building Level

### 2.1. Primary RES Types

RES commonly integrated into buildings include various technologies designed to enhance energy efficiency and sustainability. Commonly integrated RES in buildings may include: [57,58]:

- **Solar Photovoltaic Systems:** PV systems generate electricity by converting sunlight directly into electrical energy through rooftop or facade-mounted solar panels. Their widespread adoption is driven by decreasing costs and ease of installation.
- **Solar Water Heating Systems:** Utilizing solar energy to produce heat, SWH systems are commonly used for water and space heating. They operate through solar collectors that absorb sunlight and transfer heat to a working fluid, which then supplies thermal energy to the building's water heating or HVAC system.
- **Wind Turbine Systems:** Small-scale WT can be installed near buildings to harness wind energy for electricity generation. However, their feasibility is highly dependent on local wind conditions, zoning restrictions, and structural integration.
- **Geothermal Heat Pumps:** Also known as ground-source heat pumps, leverage the earth's stable underground temperature for heating and cooling. By circulating a heat-exchange fluid through buried pipes, they provide an energy-efficient alternative to conventional HVAC systems.
- **Biomass Energy Systems:** Biomass systems convert organic materials—such as wood pellets, agricultural residues, or other bio-based fuels—into heat or electricity. They are particularly advantageous in regions with abundant biomass resources.

The adoption of RES in buildings is influenced by multiple factors, including energy demand, available space, budget constraints, climate conditions, government incentives, and grid connectivity. Residential buildings predominantly incorporate solar PV and solar thermal systems, benefiting from decreasing solar panel costs and sufficient rooftop space, enabling self-consumption and reduced carbon footprints [59]. Commercial buildings, characterized by larger roof areas and substantial energy requirements, frequently deploy solar PV and may integrate GHP systems for efficient heating and cooling solutions [60].

Public buildings, including government and municipal structures, are increasingly adopting solar PV and other RES technologies to align with sustainability policies and regulatory targets [61]. Educational institutions, such as universities and schools, are expanding PV adoption to lower operational costs while incorporating sustainability initiatives into their curricula [62]. Meanwhile, hospitals—given their high and continuous energy demands—exhibit a lower reliance on onsite RES, typically generating only about 7% of their electricity needs from renewable sources. Due to their critical power reliability requirements, hospitals often supplement energy generation with backup systems and grid connectivity [59,62].

One of the most significant parameters for RES adoption is the efficient architectural integration of such technologies in buildings representing a crucial aspect of sustainable urban development, ensuring both functional efficiency and aesthetic harmony [63]. For instance building-integrated photovoltaics (BIPVs), where solar panels are seamlessly embedded into façades, roofs, and windows, exemplify how RES can be integrated without compromising architectural design [64]. Similarly, building-integrated WT, solar thermal collectors, and green roofs contribute to on-site energy generation while maintaining visual appeal. Advanced materials, such as transparent solar glass and PV shingles, further enhance the feasibility of RES integration in modern architecture [65]. Such innovations, coupled with digital design tools and parametric modeling, allow architects to optimize energy performance while preserving design flexibility.

Overall, RES adoption varies across building types, shaped by energy needs, infrastructure constraints, and economic feasibility, reinforcing the importance of tailored integration strategies for maximizing renewable energy utilization.

### 2.2. General Concept of Reinforcement Learning Control in BEMS

In an RL-based building energy management system, four key elements define the control process: environment, sensors, environment update, and reward computation. The overall operation of RL control considering a building energy management system integrated with RES and other various subsystems may be described in the following steps shown in Figure 3:

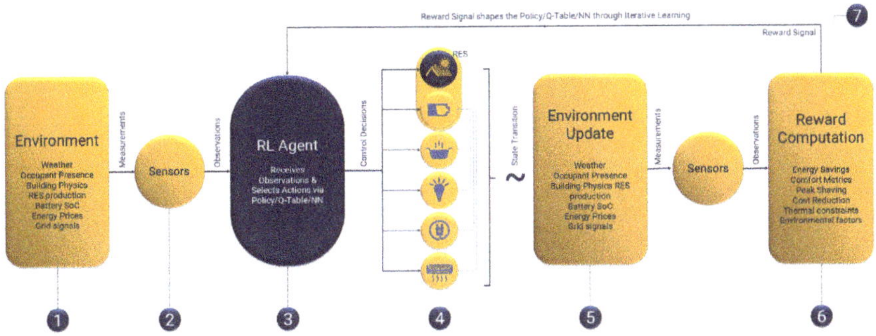


**Figure 3.** General concept of RL control in BEMS.

### 2.3. General Concept of Reinforcement Learning Control in BEMS

The fundamental operation of RL control within BEMS (see Figure 3) may be described through the following key steps:

1. **Environment:** The environment represents all external factors that the RL agent does not directly control but must respond to and account for. This includes both indoor and outdoor conditions, such as weather, occupant presence, building physics, PV/wind energy production, and battery states, as well as external signals like electricity prices and demand-response requests. Additionally, it encompasses constraints imposed by the power grid or regulatory frameworks. In essence, the environment encapsulates the entire building system and its dynamic interactions with external influences over time.
2. **Sensors:** Sensors play a crucial role in collecting real-time data on the state of the environment, providing the RL agent with the necessary observations for decision-making. In a building context, this typically includes measurements of indoor temperature, occupancy, energy consumption, PV/wind energy output, and battery state of charge. Sensor inputs can originate from physical devices, such as temperature sensors, occupancy detectors, and power meters, or from virtual data streams, including electricity price signals and weather forecasts.
3. **RL Agent:** The RL agent serves as the intelligent decision-making entity, processing environmental observations and selecting optimal control actions to enhance energy efficiency, reduce costs, and maintain occupant comfort. Depending on the RL approach, the agent's learned policy may be represented using a Q-table (in traditional RL) or an artificial neural network (ANN) in DRL. The agent makes key operational decisions, such as adjusting HVAC setpoints, scheduling battery storage operations, managing EV charging, and optimizing renewable energy dispatch.
4. **Control Decision Application:** Based on the learned policy, the RL agent applies its selected control actions in real time to optimize the building's energy management.

These decisions might involve regulating HVAC settings to maintain indoor comfort while minimizing energy consumption, scheduling battery charging and discharging to maximize renewable self-consumption and reduce peak loads, or managing PV output, determining whether to store excess solar energy, use it immediately, or feed it into the grid based on real-time electricity prices and demand.

5. **Environment Update:** After the RL agent implements its decisions, the environment undergoes an update, reflecting the impact of the chosen actions. This includes changes in building physics, occupant behavior, weather conditions, and grid interactions, which collectively determine the new environmental state. Mathematically, this transition captures how the building and its systems evolve over time in response to both internal control strategies and external dynamics.
6. **Reward Computation:** Following the environment update, a numerical reward is calculated to assess the effectiveness of the RL agent's actions in the given timestep. The reward function may consider multiple factors, such as energy cost savings, occupant comfort levels, peak load reduction, adherence to thermal constraints, and environmental impact.
7. **Reward Signal Feedback:** Finally, the computed reward is fed back into the RL algorithm, enabling the agent to refine its policy (whether through QL, an ANN, or another RL approach). By continuously interacting with the environment, learning from past actions, and adjusting its strategy accordingly, the RL agent progressively improves its decision-making to better achieve objectives such as minimizing costs, maintaining thermal comfort, and maximizing on-site renewable energy utilization.

## 3. Mathematical Framework of Reinforcement Learning

The integration of RES in buildings has become a critical strategy for reducing carbon footprints and enhancing overall energy efficiency. RL has emerged as a powerful solution for managing the inherent complexity and stochastic nature of RES, enabling dynamic optimization of energy generation, storage, and consumption. Unlike conventional model-based control methods, RL is inherently adaptive, allowing it to respond to environmental variations and unforeseen disturbances. This makes it particularly well suited for addressing the intermittency of renewable energy sources such as solar and wind power.

RL formulates the control problem as a Markov Decision Process (MDP), wherein the objective is to learn optimal policies through iterative trial-and-error interactions with the environment [66,67]. This adaptive learning framework is especially valuable in modern smart buildings, where effective RES integration requires dynamic load balancing, intelligent ESS management, and real-time responsiveness to fluctuating demand. By continuously refining its control strategies, RL enables more efficient and resilient building energy management, ensuring that renewable resources are utilized optimally in diverse and unpredictable conditions.

### 3.1. The General Concept of RL

The mathematical framework of RL may be summarized as follows: RL can be formalized through the MDP, defined by the tuple $(S, A, P, R, \gamma)$. Here, $S$ is the set of possible states representing the environment; $A$ is the set of actions available to the agent; $P(s'|s, a)$ denotes the transition probability from state $s$ to $s'$ under action $a$; $R(s, a)$ is the reward function, providing feedback based on the action taken; and $\gamma \in [0, 1]$ is the discount factor that balances immediate and future rewards.

The objective is to maximize the cumulative reward $G_t = \sum_{k=0}^{\infty} \gamma^k R_{t+k}$ by learning a policy $\pi(a|s)$ that determines the best action to take in a given state. The optimal policy $\pi^*$

maximizes the expected return $V^{\pi}(s) = \mathbb{E}[G_t|s_t = s]$. The action-value function $Q^{\pi}(s,a)$ may be defined as:

$$Q^{\pi}(s,a) = \mathbb{E}[G_t|s_t = s, a_t = a] \tag{1}$$

This function is iteratively updated through algorithms such as QL, where:

$$Q(s,a) \leftarrow Q(s,a) + \alpha[R + \gamma \max_{a'} Q(s',a') - Q(s,a)] \tag{2}$$

where $\alpha$ is the learning rate.

### 3.2. Multi-Agent Reinforcement Learning

In a multi-agent extension of RL, a set of $N$ agents interacts with the environment under the framework of a Markov game (or stochastic game). Let $\mathbf{s} \in \mathcal{S}$ denote the shared or partially shared state, and let $a_i \in \mathcal{A}_i$ be the action of the $i$th agent [20,68]. The transition dynamics are given by:

$$P(\mathbf{s}' \mid \mathbf{s}, a_1, \ldots, a_N), \tag{3}$$

which specifies the probability of transitioning to state $\mathbf{s}'$ when agents choose actions $a_1, \ldots, a_N$. In a cooperative building energy management scenario, agents often share a global reward function $R(\mathbf{s}, a_1, \ldots, a_N)$ and collectively aim to maximize the expected return [68]. Each agent $i$ may maintain a policy $\pi_i(a_i \mid \mathbf{s})$, which in value-based methods is refined by approximating a state-action value function $Q_i(\mathbf{s}, a_i)$.

In practice, these functions are often learned via ANNs called *Value networks* (critics) and *Action networks* (policies/actors) [69]. The value network estimates the expected return, for example, $Q_i^{\pi}(\mathbf{s}, a_i)$, while the action network outputs the probability distribution over possible actions (or a direct action map) $\pi_i(a_i \mid \mathbf{s})$. Each agent updates these networks iteratively via algorithms, coordinating to optimize energy generation, storage, and consumption in the building environment [70].

We might say that value networks (critics) and action networks (policies) operation may be organized around three core dimensions [71]: *Structure*, covering how critics and policies are shared among agents; *Training*, describing whether learning is centralized, decentralized, or mixed; and *Coordination*, which ranges from fully implicit interactions to explicit communication or hierarchical control. Each dimension influences scalability, data requirements, and overall system performance under varying building and renewable integration scenarios [71–73]. More specifically, the subtypes of such elements may be described as follows:

**Structure:** Determines how value and policy networks are shared among agents, ranging from centralized (shared networks) to decentralized (individual networks), impacting coordination and scalability. They may be categorized as follows:

- *One centralized critic with separate policy networks*: A single critic processes global or aggregated information, while each agent retains its own policy.
- *Fully separate critics and policies*: Each agent independently learns both value functions and policies, with no shared parameters or central critic.
- *A shared critic and policy across all agents*: All agents operate under the same critic and policy, effectively functioning as a unified controller.
- *Hybrid/partially shared*: Certain components (e.g., parts of the critic or specific layers in the policy networks) are shared, while others remain agent-specific.

**Training:** Refers to the learning paradigm—centralized, decentralized, or hybrid—that dictates how agents update their networks, influencing learning efficiency and policy development. They may be categorized as follows:

- *Centralized Training, Decentralized Execution (CTDE)*: Agents leverage global or centralized information during learning but act independently in real-time.
- *Fully Decentralized Training*: Each agent learns solely from its own local observations and experiences, with no shared critic or global oversight.
- *Fully Centralized Training*: A single entity holds and updates all agent parameters, effectively treating the multi-agent environment as one large system.
- *Mixed/Hybrid Training*: Combines centralized feedback with decentralized updates, or vice versa, to balance local autonomy and global coordination.

**Coordination:** Encompasses the mechanisms through which agents exchange information or rewards, affecting the convergence and effectiveness of their collective policies. Coordination or communication schemes among agents may be categorized as follows:

- *Implicit Coordination*: Agents rely on shared objectives, rewards, or a global critic but do not communicate directly.
- *Explicit Coordination*: Agents exchange information or messages, enabling direct negotiation or data sharing for coordinated decision-making.
- *Emergent Coordination*: Cooperative behavior arises through repeated interactions within the environment, without any explicit mechanism or communication.
- *Hierarchical Coordination*: Leader-follower roles or multi-level decision architectures establish structured control and optimize collaborative objectives.

Structurally, a design with one centralized critic means all agents share a value network, while fully separate critics means each agent trains its own—both choices also dictate whether policy (action) networks are similarly shared or independent. From a training standpoint, even if the networks are centralized or partially shared, agents may still learn under a mixed or decentralized scheme, each updating its own parameters with only limited global oversight. Finally, coordination mechanisms—implicit or explicit—determine how agents exchange information or rewards and thus whether their value or action networks converge toward cohesive multi-agent policies or evolve more independently.

### 3.3. Common RL Algorithms for Energy Management

According to the literature, RL for energy systems management may be categorized into three main approaches [40]: value-based, policy-based, and actor-critic methods, each operating uniquely to optimize energy systems. More specifically, these are characterized as follows:

#### 3.3.1. Value-Based Algorithms

Such RL methodologies focus on estimating the value of actions in a given state. In order to do so, a value function is utilized in order to guide decision-making by selecting the action with the highest expected reward. According to the literature, value-based RL approaches are particularly effective for discrete control tasks, such as determining optimal time slots for battery charging or scheduling energy loads. However, they struggle with large or continuous action spaces due to the curse of dimensionality. Common value-based algorithms found in the literature considering energy management concern QL and DQN [74]. Their mathematical conceptualization may be expressed as follows [40]:

- **Q-Learning:** QL is an off-policy algorithm that learns the optimal action-value function regardless of the policy being followed. This algorithm is widely used for RES control due to its simplicity and ability to converge to the optimal policy without requiring a model of the environment [75]. However, QL struggles with high-dimensional state spaces, leading to slow convergence [76]. It is primarily applied in small-scale RES systems, such as individual building HVAC control or battery storage management.

QL learns the value of state-action pairs, $Q(s,a)$, through iterative updates using the Bellman equation:

$$Q(s_t,a_t) \leftarrow Q(s_t,a_t) + \alpha\left[R(s_t,a_t) + \gamma \max_a Q(s_{t+1},a) - Q(s_t,a_t)\right] \tag{4}$$

where $\alpha$ is the learning rate. QL performs well in discrete action spaces and has been applied to HVAC optimization in buildings [77].

- **Deep Q-Networks:** DQN addresses the limitations of QL by utilizing deep ANNs to approximate the $Q$-function, enabling the handling of complex, high-dimensional environments. DQN leverages experience replay and target networks to stabilize learning [66]. This makes DQN suitable for larger RES configurations where multi-variable control is required, such as managing distributed energy resources in commercial buildings. However, DQN is computationally intensive and requires significant tuning, making deployment in real-time systems challenging. DQN extends the QL approach by using deep ANNs to approximate $Q(s,a)$ values, enabling the handling of large state spaces [78]. The update rule may be expressed by the following equation:

$$Q(s_t,a_t) \leftarrow Q(s_t,a_t) + \alpha\left[R(s_t,a_t) + \gamma \max_a Q(s_{t+1},a;\theta) - Q(s_t,a_t;\theta)\right] \tag{5}$$

3.3.2. Policy-Based Algorithms

In contrast to value-based approaches, the policy-based methodologies bypass the value function and directly optimize a policy to determine actions. Such approaches are well suited for continuous action spaces, enabling smooth control of solar trackers, or inverters. However, while they perform better in high-dimensional action spaces, they may converge to suboptimal solutions without careful tuning of hyperparameters [40]. Common policy-based algorithms found in the literature considering energy management concern primarily the PPO algorithm. The mathematical concepualizationt of the algorithm may be expressed as follows:

- **Proximal Policy Optimization (PPO):** PPO entails primarily a policy-based approach because it directly optimizes the policy (the actor) by maximizing an objective function related to expected rewards. The policy is typically represented by a neural network that outputs action probabilities (for discrete actions) or parameters of a distribution (for continuous actions). PPO improves stability by limiting the policy update step, thereby preventing drastic policy changes [79]. PPO has been effectively applied to BEMS, particularly for optimizing HVAC systems and renewable energy dispatch. PPO strikes a balance between performance and computational efficiency, making it popular for large-scale applications [80].

$$L^{CLIP}(\theta) = \mathbb{E}[\min(r_t(\theta)A_t, clip(r_t(\theta), 1-\epsilon, 1+\epsilon)A_t)] \tag{6}$$

where $r_t(\theta) = \frac{\pi(a_t|s_t)}{\pi_{old}(a_t|s_t)}$ and $A_t$ concern the advantage function.

3.3.3. Actor-Critic Algorithms

Actor-critic methods combine the strengths of both value-based and policy-based approaches. They consist of two components: the actor, which selects actions, and the critic, which evaluates the actions by estimating a value function. This hybrid design improves training stability and learning efficiency. Such algorithms are frequently used in complex building systems, coordinating multiple RES components to optimize ESS, consumption, and grid interactions [40]. Common actor-critic algorithms found in the literature considering energy management concern the following approaches: Deep Deterministic Policy Gradient (DDPG), Soft Actor-Critic (SAC), and Advantage Actor-Critic (A2C). Their mathematical conceptualization may be expressed as follows:

- **Deep Deterministic Policy Gradient:** DDPG presents an actor-critic algorithm that combines the strengths of policy gradient methods and QL. It is well suited for continuous action spaces, making it advantageous for applications such as battery energy storage control, where actions like charge/discharge rates are continuous [81]. DDPG's major advantage is its ability to handle high-dimensional, continuous control tasks. However, it is sensitive to hyperparameters and prone to instability during training. DDPG portrays an off-policy algorithm suited for continuous action spaces, combining actor-critic architectures [81]. The policy (actor) updates based on gradients derived from the critic's Q-function.

$$CriticUpdate : Q(s_t, a_t) \leftarrow R(s_t, a_t) + \gamma Q(s_{t+1}, \mu(s_{t+1})) \tag{7}$$

$$ActorUpdate : \nabla_\theta J(\theta) = \mathbb{E}[\nabla_\theta Q(s_t, \mu(s_t;\theta))\nabla_\theta \mu(s_t;\theta)] \tag{8}$$

- **Soft Actor-Critic (SAC):** SAC maximizes entropy, encouraging exploration and preventing premature convergence to suboptimal policies. Such algorithms have proven effective in RES control for smart grids and microgrids, where the environment is highly dynamic and stochastic [82]. SAC's primary benefit lies in its robustness to environmental noise and ability to achieve high performance across varying conditions. However, it requires substantial computational resources. SAC achieves sample efficiency and is well suited for real-time energy management applications but requires careful balance of the entropy coefficient $\alpha$.

$$J(\pi) = \sum_t \mathbb{E}[R(s_t, a_t) + \alpha H(\pi(\cdot|s_t))] \tag{9}$$

- **Advantage Actor-Critic:** A2C concerns a synchronous, deterministic version of the asynchronous actor-critic framework. It uses both a policy network (actor) to choose actions and a value network (critic) to evaluate them, optimizing both simultaneously to improve learning efficiency. In the context of RES control for buildings, A2C is particularly useful for tasks that involve sequential decision-making under uncertainty, such as demand-side energy management and dynamic load scheduling [83]. A2C's key advantage lies in training stability, in comparison to purely policy-based or value-based methods. A2C leverages the advantage function to reduce variance in policy gradient estimation, which accelerates convergence. However, its performance may be sensitive to hyperparameter tuning and might struggle with very high-dimensional action spaces without additional enhancements.
The advantage function $A(s_t, a_t)$ is calculated as the difference between the value of taking an action in a given state and the baseline value of the state:

$$A(s_t, a_t) = R(s_t, a_t) - V(s_t) \tag{10}$$

The policy gradient is updated to maximize the expected reward with reduced variance:

$$\nabla_\theta J(\pi_\theta) = \mathbb{E}_{\pi_\theta}[\nabla_\theta \log \pi_\theta(a_t|s_t) A(s_t, a_t)] \tag{11}$$

A2C is well suited for moderately complex RES applications but can require enhancements like parallel training for scaling to larger and more complex environments.

In general, RL presents a transformative approach to controlling RES in buildings by adapting to the stochastic nature of renewable sources and optimizing energy consumption patterns dynamically. The choice of an RL algorithm depends on the specific application, with QL and DQN being suitable for discrete environments, while DDPG, PPO, and SAC excel in continuous, large-scale RES management. Future advancements in RL will further enable autonomous, resilient, and efficient energy systems in smart buildings.

## 4. Tables and Summaries of RL Applications

This section provides a structured analysis of RL applications in RES-integrated BEMS from 2015 to 2025. The Summarized Tables subsection offers a high-level overview, highlighting key attributes of each study, while the Summaries of Applications subsection provides a detailed description of the research scope and outcomes. This approach allows readers to quickly identify relevant applications in the tables and refer to the detailed summaries for deeper insights into their methodologies and findings.

### 4.1. Summarized Tables

Before proceeding to the actual summaries on the different applications, this section presents five summarized tables categorizing the studies based on the type of RL type employed (namely, value-based, policy-based, actor-critic, hybrid, and other applications—see Tables 1–5). Such tables systematically classify each application according to key characteristics (e.g., reference, year of publication, RL method, agent type, BEMS combination, building type (residential/commercial), application type (simulation/real-life), and number of citations), ensuring a comprehensive understanding of the methodologies used in the field. More specifically the Summarized Table columns identify:

- **Ref.:** illustrating the reference application in the first column;
- **Year:** illustrating the publication year for each research application;
- **Method:** illustrating the specific RL algorithmic methodology applied in each work;
- **Agent:** illustrating the agent type of the concerned methodology (single-agent or multi-agent RL approach)
- **BEMS:** illustrating the combination of energy systems within BEMS RES technology integrated in the energy mix (e.g. PV, ESS, TSS, SHW, WT, HVAC, HP, GHP, BIO, etc.)
- **Residential:** defining if the testbed application concerns a residential building control application with an "x";
- **Commercial:** defining if the testbed application concerns a commercial building control application with an "x";
- **Simulation:** defining if the testbed application concerns a simulative building control application with an "x";
- **Real-life:** defining if the testbed application concerns a real-life building control application with an "x";
- **Citations:** portrays the number of citations—according to Scopus—of each work.

The abbreviation "-" represents the "not identified" elements in the tables and figures.

**Table 1.** Summarized table of value-based RL applications for RES-integrated BEMS (2015–2025).

| Ref. | Year | Method | Agent | BEMS | Residential | Commercial | Simulation | Real-Life | Citations |
|---|---|---|---|---|---|---|---|---|---|
| [84] | 2015 | QL | Multi | PV/GHP/HVAC | x | | x | | 150 |
| [85] | 2015 | QL | Multi | PV/ESS | x | | x | | 37 |
| [86] | 2017 | FQI | Single | PV/HVAC/DWH | x | | | x | 19 |
| [87] | 2018 | SARSA | Single | PV/ESS | x | | x | | 16 |
| [88] | 2019 | QL | Single | PV/ESS/EVs | x | | x | | 84 |
| [89] | 2018 | QL | Single | PV | | x | x | | 128 |
| [90] | 2019 | DQN | Multi | PV/ESS | | | x | | 39 |
| [91] | 2020 | QL | Multi | PV/HVAC/LS/EVs | x | | x | | 283 |
| [92] | 2020 | QL | Single | SWH/HVAC | x | | x | | 43 |
| [93] | 2021 | QL | Single | PV/ESS/Other | x | | x | | 24 |
| [94] | 2021 | QL | Single | PV/HVAC/DHW | x | | x | | 132 |
| [95] | 2021 | QL | Single | PV/ESS/Other | x | | x | | 11 |
| [96] | 2022 | DQN | Single | PV/SWH/HVAC | x | | x | | 33 |
| [97] | 2022 | DQN | Multi | PV/Other | x | | x | | 18 |
| [98] | 2022 | DQN | Single | PV/WT/GHP | | x | x | | 64 |
| [99] | 2022 | QL | Single | PV/WT/SWH/BIO | | x | x | | 15 |
| [100] | 2023 | DQN | Single | PV/ESS | x | | x | | 14 |

**Table 2.** Summarized table of policy-based RL applications for RES-integrated BEMS (2015–2025).

| Ref. | Year | Method | Agent | BEMS | Residential | Commercial | Simulation | Real-Life | Citations |
|---|---|---|---|---|---|---|---|---|---|
| [101] | 2019 | DPG | Single | PV/EVs/Other | x | | x | | 464 |
| [102] | 2020 | MARLISA | Multi | PV/ESS/HVAC | x | | x | | 51 |
| [103] | 2021 | PPO | Single | PV/HVAC | | x | x | | 26 |
| [104] | 2021 | PPO | Single | PV | x | | x | | 40 |

**Table 3.** Summarized table of actor-critic RL applications for RES-integrated BEMS (2015–2025).

| Ref. | Year | Method | Agent | BEMS | Residential | Commercial | Simulation | Real-Life | Citations |
|---|---|---|---|---|---|---|---|---|---|
| [105] | 2019 | DDPG | Single | PV/ESS/HVAC | x | | x | | 304 |
| [106] | 2020 | A2C | Multi | PV/ESS/HVAC/Other | x | | x | | 112 |
| [107] | 2021 | TD3 | Single | PV/ESS/HVAC/EVs | x | | x | | 39 |
| [108] | 2021 | DDPG | Single | PV/ESS/HVAC | | x | | x | 56 |
| [109] | 2021 | - | Multi | PV/ESS/HVAC | | x | x | | 16 |
| [110] | 2021 | SAC | Single | PV/ESS/TSS/HVAC/DHW | | x | x | | 69 |
| [111] | 2021 | SAC | Single | PV/HVAC/TSS | | x | x | | 72 |
| [112] | 2022 | TD3 | Single | PV/ESS/BIO | | x | x | | 46 |
| [113] | 2022 | DDPG | Single | PV/SWH/ESS/HVAC/TSS | x | | x | | 25 |
| [114] | 2022 | SAC | Multi | PV/SWH/ESS | x | x | x | | 26 |
| [115] | 2023 | SAC | Multi | PV/ESS | x | | x | | 21 |
| [116] | 2023 | DDPG | Multi | PV/ESS/HVAC | x | x | x | | 46 |
| [117] | 2023 | SAC | Multi | PV/ESS/DHW/HVAC | | x | x | x | 35 |
| [118] | 2023 | TD3 | Single | PV/WT/ESS | | x | x | | 10 |
| [119] | 2024 | SAC | Single | PV/ESS/EVs | x | x | x | | 12 |

Table 4. Summarized table of hybrid RL applications for RES-integrated BEMS (2015–2025).

| Ref. | Year | Method | Agent | BEMS | Residential | Commercial | Simulation | Real-Life | Citations |
|---|---|---|---|---|---|---|---|---|---|
| [120] | 2016 | HRL/ACO | Single | PV/DHW/HVAC | x | | x | x | 69 |
| [121] | 2018 | QL/FLC | Multi | PV/ESS/Other | x | | x | | 179 |
| [122] | 2019 | Monte Carlo/ADP | Multi | PV/ESS | x | | x | | 10 |
| [123] | 2020 | DQN/TD3 | Multi | PV/ESS/BIO | | x | | x | 21 |
| [124] | 2021 | QL/ANN | Single | PV/ESS/Other | x | | x | | 14 |
| [125] | 2021 | SAC/RBC | Single | PV/TSS | | x | x | | 24 |
| [126] | 2022 | DQN/DDPG | Single | PV/ESS/HVAC/Other | x | | x | | 45 |
| [127] | 2022 | TD3/MPC | Single | PV/Other | x | | x | | 11 |
| [128] | 2023 | QL/FLC | Single | PV/ESS/EVs/Other | x | | x | | 12 |
| [129] | 2023 | DDPG/RBC | Single | PV/ESS | | x | x | | 26 |
| [130] | 2024 | QL/ANN | Multi | PV/ESS/EVs | x | | x | | 12 |
| [131] | 2024 | PPO/IL | Multi | PV/ESS/HVAC | x | | x | | 15 |

Table 5. Summarized table of other RL applications for RES-integrated BEMS (2015–2025).

| Ref. | Year | Method | Agent | BEMS | Residential | Commercial | Simulation | Real-Life | Citations |
|---|---|---|---|---|---|---|---|---|---|
| [132] | 2017 | Bayesian RL | Multi | PV/ESS/WT/SWH/HVAC | x | | x | | 37 |
| [133] | 2022 | MCTS | Multi | PV/ESS/WT/GHP | x | | x | | 23 |
| [134] | 2022 | MCTS | Multi | PV/ESS/WT/GHP | x | | x | | 116 |

### 4.2. Summaries of Applications

A more in-depth analysis of the aforementioned applications is illustrated in Tables 6–10 in the same order:

Table 6. Summaries of Value-based applications.

| Author (Year) | Summary |
|---|---|
| *Yang et al. [84]* | MARL QL optimized a PV/T-GHP system with floor heating in a Swiss resident. RL learned online to maximize net thermal and electrical output, ensuring GHP compensation and optimal operation. Achieved 100% heat demand coverage (vs. 97% RBC) and 10% higher PV/T energy capture |
| *Raju et al. [85]* | Multi-agent QL scheduled PV/ESS operations in a university microgrid, coordinating battery storage for optimal grid interactions. The approach reduced grid reliance by 15–20%, improving energy balancing, and enhancing battery utilization over a 10-year horizon |
| *De Somer et al. [86]* | Model-based RL (FQI with ERT regression) optimized DR in DHW buffers of PV-integrated homes. Real-life tests in 6 smart homes showed a 20% increase in PV self-consumption, reducing grid dependence by dynamically scheduling heating cycles. |
| *Ebell et al. [87]* | SARSA-ANN-based RL managed PV/ESS providing frequency containment reserve. The agent minimized grid imports while ensuring real-time compliance with demand fluctuations. Simulations showed a 7.8% reduction in grid reliance compared to RBC |
| *Kim et al. [89]* | QL single-agent control optimized PV/ESS/V2G scheduling in a university building, minimizing costs through ToU pricing strategies. Reduced daily energy costs by 20–25% and grid reliance by 15–30% while dynamically shifting charging/discharging cycles. |
| *Remani et al. [88]* | QL-based RL optimized real-time load scheduling for residential PV-powered homes. Stochastic PV modeling ensured adaptive scheduling, reducing energy costs from 735 units to 16.25 units, enhancing self-sufficiency |

**Table 6.** *Cont.*

| Author (Year) | Summary |
| --- | --- |
| *Prasad et al. [90]* | Multi-agent DQN optimized peer-to-peer energy sharing in a PV/ESS residential community. Agents learn when to store, lend, or borrow power. Simulations showed 40–60 kWh lower grid reliance per household, achieving near-zero energy status |
| *Xu et al. [91]* | Multi-agent QL-based HEMS integrates PV, HVAC, LS, EVs, and shiftable loads. Extreme Learning Machine predicted PV output and prices. Cost reductions reached 44.6%, and dynamic scheduling enhanced system flexibility |
| *Correa et al. [92]* | Tabular QL optimized SWH scheduling in a university with solar thermal collectors and heat pumps (HP). TRNSYS simulations revealed a 21% efficiency boost in low solar conditions, optimizing demand-side response |
| *Chen et al. [93]* | Multi-objective QL optimized DR for PV/ESS smart homes, balancing cost vs. user satisfaction via dual Q-tables. Achieved 8.44% cost savings, with a 1.37% increase in user satisfaction |
| *Lissa et al. [94]* | QL-based HEMS with ANN-approximated Q-values optimized a PV/ESS/DHW system. Dynamic setpoint adjustments improved PV self-consumption (+9.5%), energy savings (+16%), and load shifting (+10.2%) |
| *Raman et al. [95]* | Zap QL optimized PV/ESS scheduling during grid outages, ensuring resilience. Computation time reduced (0.14 ms RL vs. 62.47 s MPC), maintaining 100% reliability for critical loads in disaster scenarios |
| *Heidari et al. [96]* | DDQN optimized a PV/HP/DHW framework, incorporating Legionella control and user comfort constraints. Swiss home trials reported 7–60% energy savings over RBC while maintaining hygiene and adaptability |
| *Lu et al. [97]* | MARL DQN optimized residential appliance scheduling with PV. Individual agents learnt cooperative control policies, reducing costs by 30%, peak demand by 18.2%, and improving punctuality by 37.3%. |
| *Shen et al. [98]* | D3QN-based MARL optimized PV/WT/GHP systems in office buildings. Prioritized experience replay and feasible action screening reduced discomfort by 84%, minimized unconsumed renewable energy by 43%, and cut energy costs by 8% |
| *Wang et al. [99]* | QL-based BEEL method optimizes DHW/PV/BIO interactions in a multi-zone building. Enhanced control over heating, cooling, and storage, reducing heating demand by 26% and cooling energy by 15% |
| *Cordeiro et al. [100]* | DQN/SNN control optimized a PV/ESS framework in a university setting. Boosted self-sufficiency (41.39% → 52.92%), reduced grid reliance (−17.85%), and achieved EUR 25,000 annual savings with an 8.56 kg $CO_2$ cut |

**Table 7.** Summaries of Policy-based applications.

| Author (Year) | Summary |
| --- | --- |
| *Mocanu et al. [101]* | Proposed energy management for PV-integrated residents using DPG and DQN. The model was trained on Pecan Street data, optimizing real-time energy consumption and costs, achieving 26.3% peak demand reduction, 27.4% energy cost savings, and improved PV utilization |
| *Vasquez et al. [102]* | MARLISA MARL optimized urban energy systems with PV, TSS, HPs, and heaters. MARLISA reduced daily peak demand by 15%, ramping by 35%, and increased the load factor by 10%. Combining MARLISA with RBC accelerated convergence within 2 years |
| *Chen et al. [103]* | PPO-based PROF method integrated ANN projection layers to enforce convex constraints in energy optimization. Applied to radiant heating in Carnegie Mellon's campus, enhancing thermal comfort and reducing energy use by 4% |
| *Jung et al. [104]* | GIS-PPO RL model for optimal rooftop PV planning in South Korea. Simulates 10,000 economic scenarios, optimizing panel allocation under future uncertainties. Achieved USD 539,197 profit, outperforming RBC and GA by 4.4%, while reducing global warming potential by 91.8% |

Table 8. Summaries of Actor-critic applications.

| Author (Year) | Summary |
|---|---|
| *Yu et al. [105]* | DDPG optimized HVAC scheduling and ESS management in a resident, integrating PV. The method handled thermal inertia and stochastic DR to minimize energy costs ensuring comfort. Simulations using real-world data, showed 8.10–15.21% less costs compared to RBC |
| *Lee et al. [106]* | The MARL A2C optimized PV-ESS and appliance scheduling across multiple residents. Each household trained locally, periodically updating a global model to improve efficiency. The approach reduced electricity costs while maintaining comfort, achieving learning accelerating convergence |
| *Ye et al. [107]* | The TD3 DR optimized BEMS integrating PV, ESS, HVAC, and EVs with V2G. The algorithm addressed uncertainties in solar power generation and non-shiftable loads, achieving 12.45% and 5.93% lower energy costs compared to DQN and DPG using real-world datasets |
| *Touzani et al. [108]* | Developed a DDPG-based controller for DER management in a commercial building equipped with PV, ESS, and HVAC. The system was trained on synthetic data and deployed in FLEXLAB, demonstrating energy savings through optimized scheduling of RES and load shifting |
| *Lee et al. [109]* | Proposed a hierarchical framework with federated learning to optimize shared ESS and HVAC scheduling in smart buildings. The approach preserved privacy by sharing only abstracted model parameters while reducing HVAC consumption by 24–32% and electricity costs by 18.6–20.6% |
| *Pinto et al. [110]* | Proposed a centralized SAC for HVAC and TSS management in commercial buildings (offices, retail, and restaurant). The model balanced peak demand reduction, operational cost savings (4% reduction), and load profile smoothing (12% peak decrease, 6% improved load uniformity) |
| *Pinto et al. [111]* | Utilized SAC for HP and TSS coordination, integrating LSTM-based indoor temperature predictions. The approach optimized grid interaction and storage utilization, reducing peak demand by 23%, Peak-to-Average Ratio (PAR) by 20%, and having 20% faster training |
| *Gao et al. [112]* | Investigated TD3 and DDPG for optimizing off-grid hybrid RES systems in a Japanese office building. The model ensured safe battery operation and grid stability, with TD3 outperforming DDPG by reducing hourly grid power error below 2 kWh and increasing battery safety by 7.72 h |
| *Langer et al. [113]* | Examined DDPG for home energy management with PV, modulating air-to-water HP, ESS, and TSS. The model transitioned from MILP to DRL, leveraging domain knowledge for stable learning. Simulations in Germany demonstrated 75% self-sufficiency, 39.6% cost savings, over RBC |
| *Pinto et al. [114]* | Developed a MARL-based SAC model for district energy management (three residential buildings and a restaurant). The decentralized approach optimized PV self-consumption and thermal storage, reducing energy costs by 7% and peak demand by 14% compared to RBC |
| *Qiu et al. [116]* | Proposed a federated multi-agent DDPG approach for joint peer-to-peer energy and carbon trading in mixed-use buildings (residential, commercial, industrial). The Fed-JPC model reduced total energy and environmental costs by 5.87% and 8.02% while ensuring data privacy |
| *Nweye et al. [115]* | Introduced the MERLIN framework utilizing SAC for distributed PV-battery optimization in 17 zero-net-energy homes. Combined real-world smart meter data and simulations, reducing training data needs by 50%, improving ramping by 60%, and lowering peak load by 35% |
| *Xie et al. [117]* | Applied a MARL SAC framework with an attention mechanism for DR in grid-interactive buildings. The model controlled PV, ESS, and DHW in residential, office, and academic buildings. Achieved a 6–8% reduction in net demand, USD 92,158 annual electricity savings at Cornell University |
| *Qiu et al. [118]* | Proposed a TD3-based RL approach with dynamic noise balancing to optimize PV, wind, ESS, and hydrogen-based energy systems in commercial buildings. The model reduced operating costs by 18.46%, eliminated RES curtailment, and achieved superior accuracy over traditional methods |
| *Deng et al. [119]* | Developed the DC-RL model integrating a distributed DC energy system with SAC. Optimized PV, ESS, EVs, and flexible loads for residential and office buildings. Increased PV self-consumption by 38%, satisfaction by 9%, and PV self-sufficiency to 93%, while reducing ESS reliance by 33% |

**Table 9.** Summaries of Hybrid applications.

| Author (Year) | Summary |
|---|---|
| *Kazmi et al. [120]* | The hybrid RL-combining RL, heuristics and ACO-optimized DHW production in 46 NZEBs in Holland, integrating PV and ASHPs. Simulations showed 27% energy savings, while real-world tests yielded 61 kWh savings over 3.5 months |
| *Kofinas et al. [121]* | Proposed a MARL system using fuzzy QL to control a microgrid with PV, diesel generator, fuel cell, ESS, and electrolyzer. Each device RL agent, optimized energy balance independently. System achieved 1.54% uncovered energy, less diesel usage (0.87%), and reduced ESS discharges |
| *Chasparis et al. [122]* | Developed a hierarchical ADP-based RL model where an aggregator controlled residential PV/ESS systems for energy bidding. Used Monte Carlo Least Squares ADP for flexibility forecasting. Real-world simulations in Austria (30 homes) improved revenues over 10 test days |
| *Gao et al. [123]* | Employed a MARL RL framework using TD3 (continuous actions) and DQN (discrete actions) for off-grid energy optimization in a Japanese office building with PV, BIO, and ESS. Achieved 64.93% improvement in off-grid operation and an 84% reduction in unsafe battery runtime |
| *Ashenov et al. [124]* | Integrated ANNs for forecasting and QL for load scheduling in a HEMS with PV, ESS, and smart plugs. Distinguished non-shiftable and shiftable loads, dynamically optimizing scheduling. Achieved a 24% cost reduction and a 15% profit increase in a single-home case study |
| *Deltetto et al. [125]* | Combined SAC with RBC for demand response in small commercial buildings (office, retail, restaurant) with PV. SAC-alone reduced costs by 9%, energy use by 7%, and improved peak shaving by 4%, but violated DR constraints. Hybrid SAC-RBC balanced cost and DR compliance. |
| *Huang et al. [126]* | Proposed a mixed DRL model integrating DQN and DDPG for discrete-continuous action spaces in HEMS (PV, ESS, HVAC, appliances). A safe-mixed DRL version mitigated comfort violations. Achieved 25.8% cost reduction and significant thermal comfort improvements. |
| *Nicola et al. [127]* | Combined Synergetic (SYN) and Sliding Mode Control (SMC) with a TD3 RL agent for MPPT in a 100 kW PV-grid system. The RL agent stabilized DC-link voltage, reducing steady-state error (<0.02%) and overshooting under 30% load variations. |
| *Almughram et al. [128]* | Developed RL-HCPV using fuzzy QL and deep learning (ANN, LSTM) for predictive PV generation, EV SOC, and energy trading in smart homes with V2H. Reduced grid reliance by 38% (sunny days) and 24% (cloudy days), cutting electricity bills to USD 3.92/day. |
| *Zhou et al. [129]* | Integrated DDPG with a rule-based expert system for energy flexibility in a net-zero energy office building with PV and ESS. Used CART to analyze external influences. Achieved 7% cost reduction, 9.2% PV self-consumption increase, and 10.6% lower grid reliance |
| *Binyamin et al. [130]* | Combined multi-agent QL and ANN for smart home P2P energy trading with PV, ESS, and EVs. Optimized load scheduling, storage, and trading under real-time pricing. Achieved 9.3–16.07% household reward improvements and cost reductions under varying solar penetration |
| *Wang et al. [131]* | Proposed MAPPO and Imitation Learning for scalable hybrid energy management (PV, ESS, HVAC) in ZEHs. Centralized training, decentralized execution improved PV self-consumption (18.47%) and energy self-sufficiency (46.10%) while ensuring thermal comfort. |

**Table 10.** Summaries of Other RL applications.

| Author (Year) | Summary |
|---|---|
| *Anvari et al. [132]* | Developed an ontology-driven multi-agent system for energy management in residential microgrids integrating buildings, RES, and controllable loads. The system used Bayesian RL for real-time battery bank optimization, managing PV, WT, DHW, radiant floor systems, and micro-CHP. Achieved 5–13% cost reductions, improved occupant comfort, and enhanced system resilience under dynamic pricing schemes. |
| *Tomin et al. [133]* | Proposed a multi-criteria approach for designing and managing renewable energy communities using bi-level programming and RL, namely, the Monte Carlo Tree Search (MCTS) approach. Optimized the operation of PV, WT, biomass gasifiers, and storage in remote Japanese villages. Results showed 75% cost reductions in electricity tariffs, higher RES penetration, and balanced economic-environmental benefits |
| *Tomin et al. [134]* | Applied bi-level optimization with RL (MCTS) for designing and managing community microgrids in Siberian settlements. Integrated PV, WT, biomass gasifiers, and storage, optimizing energy market operations. Demonstrated a 20–40% LCOE reduction, improved electricity reliability, and operational flexibility. |

## 5. Evaluation

In order to provide a well-structured, comparative overview of existing RL solutions for RES-integrated BEMS, the evaluation centers on seven key attributes: (1) RL Type and Methodology, (2) Model Type, (3) Agent Type, (4) Target Output Type, (5) Baseline Control Type, (6) BEMS Combination Type, and (7) Building Type. Such facets were selected since they capture the most critical dimensions in designing, deploying, and benchmarking RL-based controllers. By dissecting the literature according to these attributes, the interested reader can gain a focused perspective on how RL methods align with different building environments, performance priorities, and design constraints—ultimately guiding more informed decisions about which RL approaches are best suited to specific BEMS needs.

### 5.1. RL Types and RL Methods

**Value-based** RL approaches have been the most widely studied in the literature on RES-integrated BEMS from 2015 to 2025 (see Figure 4). Among these, QL stands out as the most frequently adopted method due to its simplicity, flexibility, and suitability for discrete state and action spaces (see Figure 5). QL has proven particularly effective in applications requiring computational efficiency and adaptability to diverse RES components, such as PV systems and geothermal heat pumps (GHP) [84,98,99]. Its straightforward implementation, combined with its ability to optimize energy management within simulation environments, has made it the preferred choice for researchers [85,88,89,91–95].

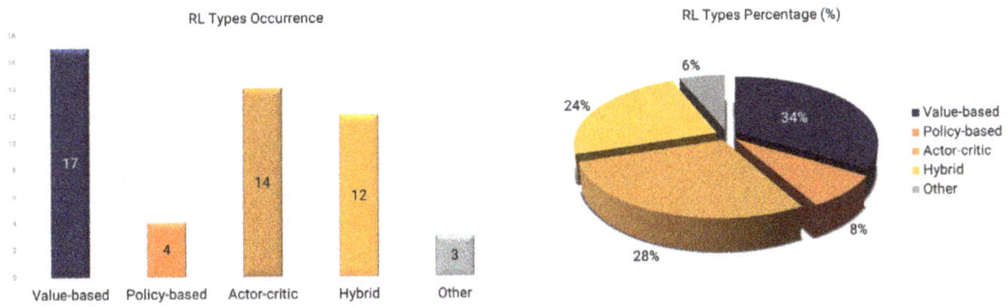


**Figure 4.** RL types occurrence (**Left**) and RL types percentage (%) (**Right**) in RES-integrated BEMS applications (2015–2025).

The dominance of QL in the literature highlights the field's inclination toward computationally inexpensive methods that may be deployed with minimal tuning. However, QL reliance on discrete action spaces limits the applicability of such a straightforward RL approach considering real-world scenarios where continuous control variables, such as temperature regulation or battery charge levels, play a crucial role. While extensions like function approximation and discretization strategies have been explored, they introduce additional trade-offs in terms of stability and convergence.

DQN, an extension of QL, has gained significant traction in recent years due to advancements in computational capabilities. The integration of deep ANNs as function approximators has enabled DQN to handle high-dimensional state-action spaces, making it more effective in complex energy optimization tasks. Studies [90,96–98,100] demonstrate DQN's ability to outperform traditional QL by improving convergence speed and decision-making accuracy in dynamic energy environments. However, challenges remain in terms of stability, sample efficiency, and generalization across diverse building types; DQN reliance on predefined reward structures poses challenges for real-world applications, contributing to a gradual shift toward alternative RL paradigms. Notably, literature findings indicate that only one value-based application incorporates real-life validation, utilizing the novel Fitted Q-Iteration (FQI) algorithm [86]. Such limited real-world validation of value-based methods underscores the difficulties in designing reward functions that generalize well across varying environmental conditions. The reliance on simulation environments often overlooks crucial aspects, such as sensor noise, hardware latency, and the stochastic behavior of occupants, which can significantly affect policy performance in deployment. This suggests that future research should explore methods to enhance transferability, such as domain adaptation or robust RL techniques.

**Policy-based** RL methods, such as Proximal Policy Optimization (PPO), exhibit a lower prevalence in RES-integrated BEMS research from 2015 to 2025 (see Figure 5). These approaches demand higher computational resources and require high-quality data for gradient-based optimization, making them less favorable in energy systems research, where modeling inaccuracies and data sparsity are common [103,104]. Their primary advantage lies in their ability to handle continuous action spaces and stochastic policies, making them well suited for complex control optimization. However, the dynamic and uncertain nature of building-integrated RES systems has hindered their widespread adoption, as their high computational cost and lack of robust, scalable frameworks present notable challenges. The lower adoption rate of policy-based methods likely stems from the difficulty in balancing exploration and exploitation in dynamic energy systems. Unlike value-based methods, which rely on well-defined value functions, policy-based approaches appear to be more sensitive to hyperparameter tuning and sample efficiency. Future research efforts should focus on improving sample efficiency and stability through techniques such as trust region optimization and adaptive learning rates, which could make these methods more practical for real-time energy management.

**Actor-critic** methods, particularly the Twin Delayed Deep Deterministic Policy Gradient (TD3) [107,112,118], DDPG [105,108,113,116], and Soft Actor-Critic (SAC) [110,111,114, 115,117], are gaining increasing attention in the field of building-integrated RES control (see Figure 5). These methods bridge the gap between value-based and policy-based approaches, providing a balance between stability and efficiency in optimizing continuous and high-dimensional action spaces. Their rising adoption is driven by their robustness in handling real-world complexities, including nonlinear system dynamics, multi-agent interactions, and partially observable environments—common challenges in energy systems. TD3 and SAC, in particular, have demonstrated superior stability during training and more effective exploration strategies, enabling them to outperform earlier actor-critic variants in complex scenarios. The growing prominence of actor-critic methods reflects an effort within the research community to address the limitations of purely value-based or policy-based approaches, particularly in real-time applications and multi-objective optimization for RES integration. Inevitably, such increasing adoption suggests a shift toward more data-driven and adaptive control strategies in RES-integrated BEMS. Their ability to incorporate both policy optimization and value estimation offers a crucial advantage in dynamically evolving environments. However, one major challenge remains—their sensitivity to reward shaping and hyperparameter selection. Research in this area should explore automated tuning mechanisms, such as meta-RL or RL with evolutionary strategies, to enhance adaptability and reduce the need for manual intervention.

**Hybrid** RL applications represent an emerging and rapidly growing niche in the field of RES-integrated BEMS. Such approaches combine RL algorithms with complementary techniques, such as fuzzy logic control (FLC) [121,125,128], ANN [124,130], model predictive control (MPC) [127], as well as RBC [125,129], to leverage the strengths of multiple methodologies for improved performance. The most commonly observed combinations include RL with FLC, RBC, and ANNs, each offering distinct advantages for specific energy management challenges. More specifically, integrating QL with FLC provides an intuitive approach to handling uncertainties and nonlinearities in RES systems, making it particularly effective in residential applications where energy demand and supply fluctuations are significant [121]. Similarly, coupling RL and DRL methods with ANNs enables the development of adaptive controllers capable of learning from large, complex datasets while accounting for dynamic system behaviors [124,130]. Additionally, hybrid frameworks such as TD3/MPC offer the benefits of predictive model-based control while utilizing the policy optimization strengths of actor-critic methods [123]. This makes them highly suitable for multi-objective optimization in both residential and commercial settings, further solidifying their potential as an innovative research direction in RL-based energy management.

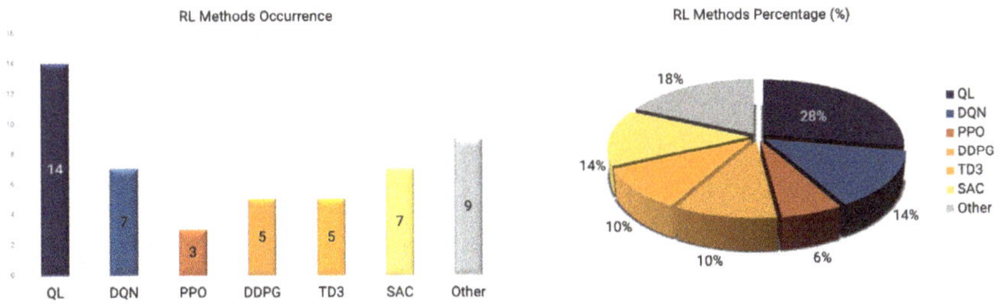


**Figure 5.** RL methodologies occurrence (**Left**) and RL methodologies percentage (%) (**Right**) in RES-integrated BEMS applications (2015–2025).

The rise of hybrid RL approaches signifies a recognition of the limitations inherent in standalone RL models when applied to real-world RES-integrated BEMS. The incorporation of auxiliary control methods mitigates stability and generalization issues, yet it also introduces additional complexity in model integration and interpretability. Future research should explore standardized frameworks for hybrid RL implementation, enabling seamless adaptation across various energy systems without an excessive tuning and computational burden.

### *5.2. Model Types*

Unlike model-free RL, which learns optimal policies through direct interaction and trial-and-error without relying on a predefined system model, model-based RL integrates a predictive model of the environment. This allows the agent to simulate future outcomes, enabling more efficient decision-making and planning. However, despite the growing adoption of RL, only a smaller subset of studies explore model-based RL approaches [86,103,120,122,133,134] due to their higher data efficiency and predictive capabilities (see Figure 6). To this end, the wider implementation of model-based RL for BEMS control is constrained by the complexity of accurately modeling building dynamics and computational overhead.

However, model-based RL may prove to be particularly advantageous in energy systems with predictable physical dynamics, such as HVAC thermal responses or battery charge/discharge cycles. It should be noted though that in real-world BEMS, where occupant behavior and external weather conditions introduce significant uncertainty, even minor modeling errors may lead to suboptimal control strategies. In order to enhance the applications of model-based RL, future work could further explore hybrid frameworks that fuse data-driven adaptation with physics-based models, offering a balance between computational tractability and robustness to real-world variability. Recent studies have begun addressing such challenges by integrating hybrid approaches, combining model-based planning with deep learning techniques to improve scalability and real-world applicability.

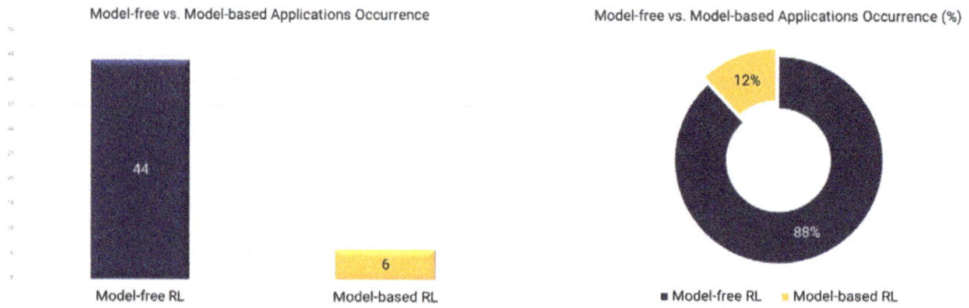


**Figure 6.** Model types occurrence (**Left**) and model types percentage (%) (**Right**) in RES-integrated BEMS applications (2015–2025).

### 5.3. Agent Type

The high prevalence of multi-agent RL in RES-integrated energy systems is evident, with approximately 40% of the evaluated studies employing MARL-based control architectures (see Figure 7). This trend reflects the increasing complexity of modern building energy frameworks, where multiple interconnected energy systems must be managed simultaneously. Similar to other engineering domains that require distributed optimization [135,136], the rise of smart buildings and IoT-driven automation has further accelerated the need for adaptive, self-learning control solutions [137].

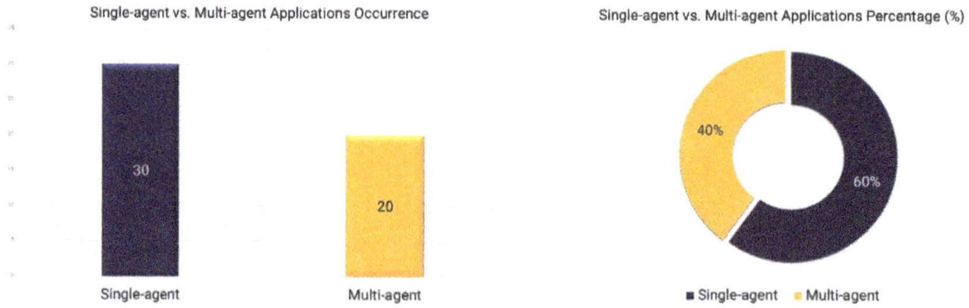


**Figure 7.** Agent-based applications occurrence (**Left**) and agent-based applications percentage (%) (**Right**) in RES-integrated BEMS applications (2015–2025).

As buildings increasingly integrate PV systems, battery storage, HVAC units, and smart loads, optimizing their operation necessitates decentralized or distributed decision-making to account for diverse constraints and dynamic interactions. Each of these subsystems presents distinct operational characteristics, making MARL a compelling solution for energy management. In a MARL framework, decentralized agents—each representing a specific energy component—independently learn and adapt to their local environment while collectively working toward system-wide efficiency objectives, such as cost minimization and energy optimization [97,98,102,116,133]. Moreover, MARL is particularly well suited for handling multi-objective trade-offs, such as balancing thermal comfort with demand-side management or maximizing self-consumption while minimizing grid reliance [101,110,114]. The increasing availability of computational resources and RL frameworks has further facilitated its adoption, solidifying MARL as a fundamental methodology for managing complex, integrated energy systems. Another key advantage of MARL is its robustness in handling uncertainties associated with RES integration. Given the inherent variability of PV and solar thermal output due to weather conditions, MARL enables adaptive control strategies that optimize real-time energy allocation [118,123]. Additionally, it enhances flexibility by dynamically adjusting energy use based on real-time occupancy data, electricity pricing, or predictive demand models [100,117]. These advantages position MARL as a crucial approach for the future of intelligent, decentralized energy management in buildings.

According to the evaluation, value-based approaches, and more specifically, MARL QL, represent the most commonly applied approaches found in [84,85,91,121,130] (Figure 8), along with DQN [90,97,123]. The higher occurrence of value-based methods, particularly QL, in research may be attributed to their straightforward implementation, off-policy nature, and ability to handle discrete action spaces effectively. QL-based MARL frameworks, including DQN, are widely utilized due to their sample efficiency and ease of adaptation to decentralized learning environments. Such methods facilitate independent learning, making them suitable for multi-agent systems where communication constraints and non-stationarity are challenges.

Actor-critic MARL algorithms are also highly prevalent having SAC [114,115,117] and MCTS [122,133,134] as the main representatives. SAC, in particular, is emerging as a dominant choice due to its entropy-regularized objective, which enhances exploration and stability in continuous action spaces. SAC's ability to balance exploration and exploitation effectively makes it particularly useful in MARL scenarios with complex, high-dimensional control tasks. Less-often applied MARL methodologies include DDPG [116], PPO [131], and A2C [106] likely due to their sensitivity to hyperparameters and exploration inefficiencies in multi-agent environments.

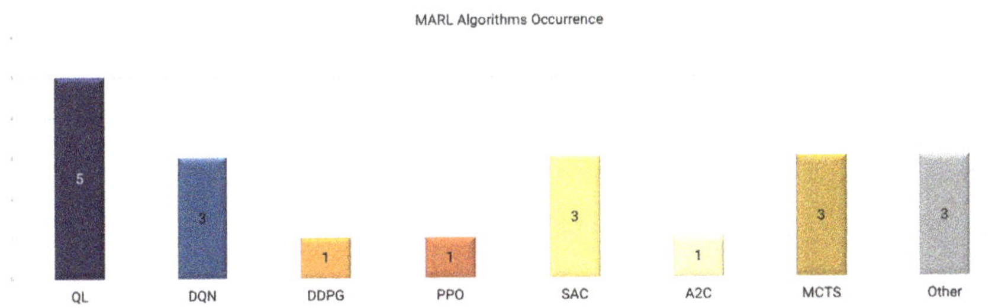


**Figure 8.** Multi-agent RL methodologies occurrence in RES-integrated BEMS applications (2015–2025).

In sum, while multi-agent RL demonstrates clear advantages for complex, decentralized BEMS optimization, it also introduces challenges, such as inter-agent coordination and potential non-stationarity in multi-agent environments. In practice, heterogeneous agents may learn conflicting objectives unless carefully designed reward structures or consensus mechanisms are in place. Future research could focus on hierarchical agent architectures or communication protocols that balance local autonomy with global coordination, ensuring scalable and robust energy management in increasingly interconnected building ecosystems.

### *5.4. Value and Action Network Types*

The fundamental distinction in MARL approaches arises from the structure of value networks (critics) and action networks (policies), where different methodologies have been employed to balance decentralized decision-making and global coordination. In this subsection, we explore some of the prevailing trends in MARL applications for BEMS by examining the structural and training paradigms of value and action networks, as well as the implications for system-wide energy optimization.

- **Value Networks in MARL Applications:** Value networks in MARL serve as the critic, helping agents estimate the expected return for state-action pairs. Two dominant approaches emerge in the existing research: fully decentralized critics, where each agent maintains its own independent Q-network, and centralized critics, which leverage global information to enhance coordination. Studies such as [84,90,91] demonstrate that fully separate critics are advantageous when agents operate independently with minimal reliance on global coordination. Such approaches enable robust, distributed learning but often suffer from non-stationarity, as agents adjust their strategies independently, causing instability in learning dynamics. In contrast, centralized critics, as employed in [85,102,109,116,130], integrate joint state-action evaluations to improve cooperation between agents. The ability to assess system-wide performance leads to more stable and coordinated energy management strategies. However, such practice may introduce scalability challenges, as sharing global state information may be computationally intensive and require significant communication overhead. The trade-off between decentralized and centralized critics is thus driven by the scale of deployment and the necessity for real-time agent collaboration.

- **Action Networks in MARL Applications:** Action networks—or policy networks—determine how agents select actions. The choice between independent and coordinated policy networks is crucial in defining MARL strategies for BEMS. In fully decentralized settings, as seen in [84,90,91,97,115], each agent learns and optimizes its policy independently, often through deep QL or actor-critic models. This method ensures resilience and local autonomy, allowing each agent to optimize its own objectives. However, the lack of explicit cooperation mechanisms may lead to conflicts in decision-making, where independent policies may not align with overall energy efficiency goals. On the other hand, coordinated policy networks found in [85,102,116,131], introduce mechanisms to balance local autonomy with system-wide optimization. Such approaches often adopt a centralized training, decentralized execution (CTDE) paradigm, allowing agents to learn with shared knowledge while maintaining independent execution capabilities. The integration of federated RL [109] or shared reward functions [131] further enhances cooperative learning by aligning incentives among distributed agents without imposing direct communication constraints.

The following Figure 9 portrays the occurrence of the different structure, training and coordination approaches in the multi-agent research works: In Figure 9-Left, the structure and its subtypes occurrence are portrayed (one centralized critic with separate policy networks/fully separate critics and policies for each agent/shared critic and policy across all agents/hybrid types); In Figure 9-Center, the training and its subtypes occurrence are portrayed (centralized training, decentralized execution (CTDE)/fully decentralized training/fully centralized training/hybrid training); In Figure 9-Right, the coordination and its subtypes occurrence are portrayed (implicit coordination/explicit coordination/emergent coordination/predefined roles or hierarchical coordination).

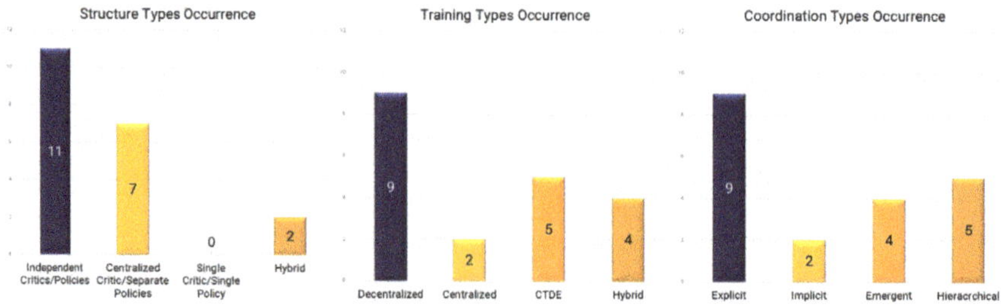


**Figure 9.** MARL structure (**Left**), training (**Center**), and coordination (**Right**) occurrences in RES-integrated BEMS applications (2015–2025).

According to the measurements, to create MARL structures, the majority of works either use a centralized critic with separate policies (7 occurrences) or fully independent critics/policies (11 occurrences). Hybrid approaches are rarer, and no study reported here fully shares a single critic and single policy across all agents (see Figure 9-Left). With regard to training, the fully decentralized training and CTDE represent the most common approaches, reflecting a desire to balance local autonomy with some centralized coordination. Mixed/hybrid training is slightly less frequent, while purely centralized training is relatively uncommon. Last but no least, with regard to coordination, explicit coordination and emergent coordination together cover most cases, but hierarchical/predefined-role approaches also often appear. Only a small subset relies purely on implicit coordination (e.g., a shared global reward or shared critic) without any direct information exchange.

### 5.5. Reward Function Type

In RL, the reward function represents the core mechanism that defines the agent's learning objective. By providing feedback on actions taken, it guides the agent toward optimal energy management strategies. To this end, a well-designed reward function is crucial for ensuring convergence, stability, and adaptability in RL-based control systems, particularly in building energy management, microgrid optimization, and demand-response applications.

Across the studies reviewed in this paper, a wide variety of reward function structures have been proposed, reflecting the complexity and multi-dimensional nature of integrated energy applications. The reviewed RL applications for RES-integrated BEMS exhibit a variety of reward functions tailored to specific operational objectives. These functions can be categorized broadly into several key architectures. The **weighted sum of multiple objectives** is the most prevalent, employed by numerous authors, such as [92,94,96–99,102,105–108,111–114,117–119,123,125,126,131,133,134]. These studies typically integrate diverse factors, such as energy costs, peak demand, occupant comfort, carbon emissions, renewable energy utilization, and storage management, into a scalar reward through tunable weights reflecting relative priorities. Such practice is highly flexible and allows for balancing various competing objectives; however, it requires careful tuning of weight parameters, which may result in a complicated optimization problem. Another common formulation found in the literature concerns the **penalty-based reward**, seen in the works of [84,88,93,95,103,106,120]. This type explicitly applies penalties for violating operational constraints, occupant discomfort, or suboptimal battery and appliance operations. Such an approach may provide clear enforcement of constraints, making it useful for safety and reliability, but harsh penalties may lead to unstable learning and overly conservative policies. Only one study employed **normalized or relative improvements** rewards, emphasizing comparison against baselines or rule-based strategies to quantify improvements explicitly [129]. This category is relatively rare, while it portrays an effective solution for benchmarking against predefined baselines. However, while this method ensures meaningful progress, tracking it may struggle when baselines are not well defined or adaptive. Furthermore, **sparse reward** functions calculating aggregated daily or episodic results were adopted by [86,90,101,115,122], typically to guide agents toward long-term optimal behaviors by focusing on overall outcomes rather than immediate feedback. This strategy commonly requires advanced exploration strategies, slowing the learning process. A notable sophisticated approach concerned the **hierarchical or multi-agent reward** structure identified in [102,116,117,121], aiming for coordination among multiple agents or subsystems for holistic optimization. A total of four studies were employed using this methodology, facilitating coordinated decision-making in complex systems. Typically, while such schemes are particularly useful in complicated BEMS, they certainly increase computational—and commonly, communicational—complexity and may require additional mechanisms to prevent conflicting objectives among agents.

Figure 10 illustrates the occurrence of the different types of reward functions for the integrated research applications. By analyzing these reward functions, it is evident that the majority of the integrated studies prefer to employ a weighted sum architecture due to its flexibility in adjusting the relative importance among multiple competing objectives, like cost efficiency, comfort, renewable energy maximization, and system stability. Moreover, occupant comfort consistently emerges as a critical factor, typically managed by explicit penalty terms that strongly discourage comfort deviations, thus reinforcing user satisfaction as a central theme [84,88,93,95,103,106,120]. Also, integration of renewable energy self-consumption and peak load reduction consistently appear, reflecting increasing grid responsiveness and sustainability objectives [98,111,117]. Lastly, there is a notable increase in employing advanced RL architectures, such as multi-agent and hierarchical methods [102,116,117,121], reflecting a trend towards more scalable, distributed, and cooperative control frameworks in future BEMS research.

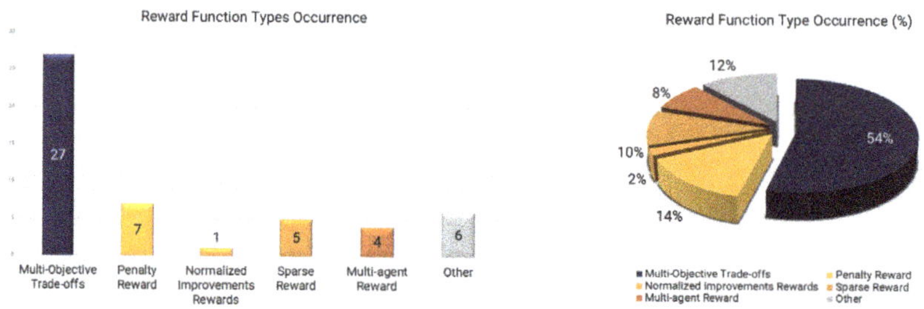


**Figure 10.** Reward function type occurrence in RES-integrated BEMS applications (2015–2025).

### 5.6. Target Output

RL applications for RES primarily focus on optimizing performance, efficiency, and adaptability. A key objective is ensuring occupant **comfort satisfaction** while balancing energy consumption, as demonstrated in studies that assess comfort satisfaction [91,93,98,99,119,126,132,133]. Additionally, **energy savings** and cost reduction serve as major evaluation metrics, assessing the ability of RL frameworks to minimize energy usage and operational expenses. As shown in Figure 11, these metrics appear in over 27 studies as primary justifications for RL-based optimization.

Another critical performance metric concerns **RES energy capture**, particularly PV self-sufficiency and self-consumption, which evaluates how effectively RES-generated energy is utilized on-site, minimizing reliance on external grid power. This metric is observed in nine studies [84,86,92,94,98,100,119,129,131]. **Grid reliance** represents another essential factor, encompassing load shifting, demand response, peak demand reduction, power balancing, and operational flexibility. Such metrics measure the ability of RL-based energy management strategies to optimize interactions with the electricity grid. Figure 11 indicates that this metric is prevalent in over 19 cases [89,90,94,95,97,100–102,109–111,113–115,117,118,121,123,127–129,134]. **Environmental footprint** represents another significant evaluation criterion, quantifying emissions reductions and overall sustainability improvements. However, relatively few studies explicitly incorporate this factor into their evaluation [116,133] (see Figure 11).

Additional key considerations include faster convergence, training time, computational efficiency, scalability, and data efficiency, which reflect the RL model's ability to learn optimal strategies quickly and effectively. Computational time is particularly relevant for real-world feasibility, assessing whether an RL-based control approach can operate efficiently in practical deployments. These factors are extensively evaluated in studies focused on model performance (**model-related metrics**) [89,95,106,111,115]. **Other** relevant considerations include system resilience, adaptability, and compliance with regulatory frameworks [112,121,122].

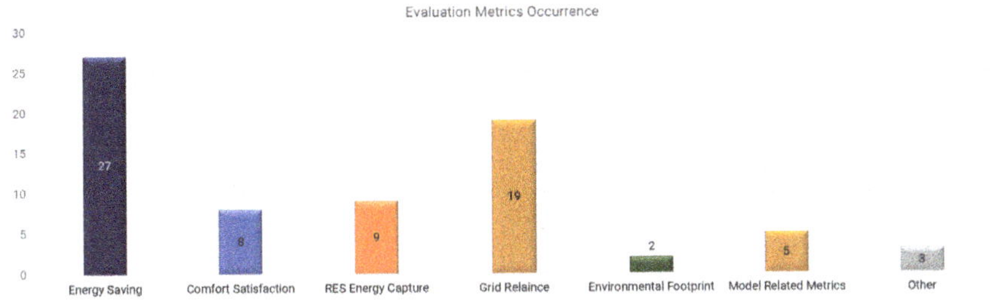


**Figure 11.** Building types occurrence (**Left**) and percentage (%) (**Right**) in RES-integrated BEMS applications (2015–2025).

Despite the clear emphasis on occupant comfort and energy savings, the practical implementation of RL-based BEMS faces challenges arising from the trade-offs among comfort, cost, and grid reliance. Small improvements in one metric can negatively affect others—for instance, aggressively minimizing costs might compromise thermal comfort or require higher grid dependency. Furthermore, the heterogeneous nature of metrics (e.g., environmental footprint vs. energy capture) often necessitates complex multi-objective reward structures, which can exacerbate difficulties in policy convergence. Future work could investigate adaptive weighting schemes or hierarchical RL approaches that dynamically balance these competing performance indicators in real time.

### 5.7. Baseline Control

Establishing baseline control approaches is crucial for benchmarking the effectiveness of RL-based energy management strategies. RBC portrays the most commonly used baseline approach for comparison, accounting for more than 50% of all evaluated cases (see Figure 12, Right). Although RBC is widely adopted, its simplistic nature makes it a limited benchmark for advanced RL techniques. Alternative RL approaches have also been used as baselines in nine studies [85,102,105–107,109,112,117,126], with DDPG emerging as the most frequently applied RL-based baseline [105,107,112,117,118,126] (see Figure 12, Left).

Optimization-based methods, including Mixed-Integer Linear Programming (MILP) [93,109,128] and MPC [95,119], have also been explored as baselines. Although less prevalent than RBC, these techniques provide structured decision-making models that contrast with RL's adaptive nature. Evolutionary algorithms, such as Genetic Algorithms (GAs) and Particle Swarm Optimization (PSO), have been employed in specific cases [104,128], while classical control strategies, including Proportional-Integral (PI) controllers, have been evaluated in hybrid or low-complexity implementations [103,127,131] (see Figure 12, Left).

Other alternative baseline approaches have a more limited presence in the literature. For instance, MCTS [134] and peer-to-peer trading models [116] remain relatively underexplored in RL-based energy management. The limited number of studies incorporating these alternative decision-making paradigms suggests a promising direction for future research, particularly in enhancing RL-based optimization frameworks.

While RBC is generally favored for its simplicity and ease of implementation, its limitations reveal only a fraction of RL's potential benefits. Robust optimization-based methods, such as MPC, may offer more stringent benchmarks against which RL algorithms can be compared towards the demonstration of adaptive decision-making control under dynamic conditions. However, the relatively sparse use of advanced optimization techniques or novel approaches like MCTS indicates a gap in examining how RL truly stacks up against best-in-class baselines. Exploring these underutilized baselines could provide deeper insights into RL's capacity for handling complex multi-objective and real-time control scenarios.

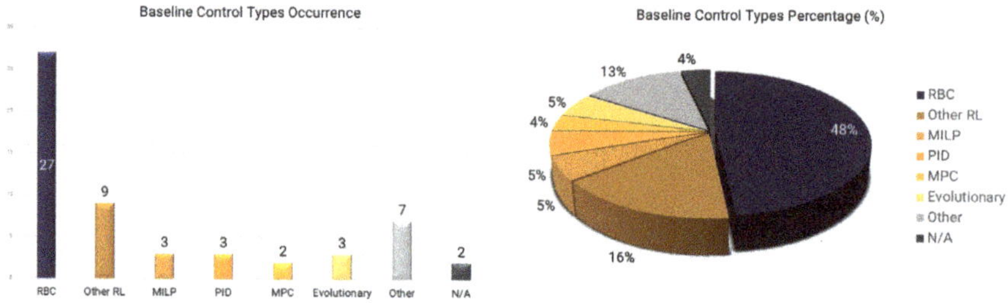


**Figure 12.** Baseline control methodologies occurrence (**Left**) and percentage (%) (**Right**) in RES-integrated BEMS applications (2015–2025).

### 5.8. RES Type

The dominance of PV integration over other RES types, such as SWH, WT, BIO, GHP, in the energy mix is driven by both technological and economic factors (See Figure 13-Left). PV systems offer a cost-effective, scalable, and easily integrated solution for buildings, whereas other RES technologies often require more extensive infrastructure, higher maintenance, and complex operational strategies. Additionally, advancements in battery storage and smart inverters have further enhanced the viability of PV for self-consumption and grid interaction, reinforcing its widespread adoption. This trend suggests that future RL research in building energy management will likely focus on advanced PV-driven optimization strategies, including real-time demand response, predictive control for ESS, and PV-based multi-agent coordination.

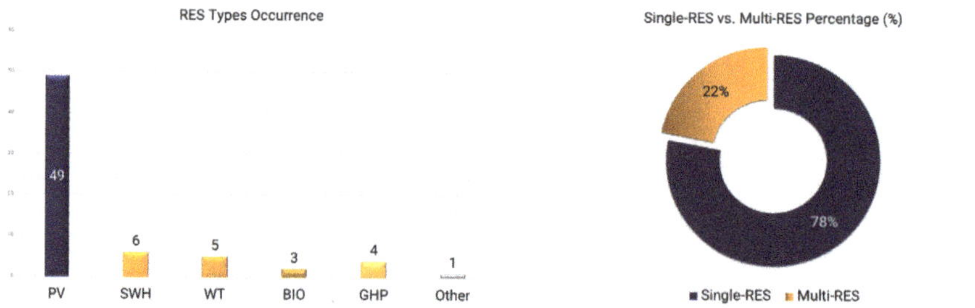


**Figure 13.** RES types occurrence (**Left**) and percentage of single-RES and multi-RES applications (%) (**Right**) in RES-integrated BEMS applications (2015–2025).

The focus on Hybrid-RES (Multi-RES) configurations (See Figure 13-Right) is closely linked to the adoption of MARL approaches, as managing multiple energy sources requires distributed control strategies capable of optimizing energy flow dynamically [84,114,123,133,134]. As smart buildings and microgrids grow in complexity, future RL research is expected to emphasize hybrid RES frameworks, developing adaptive, scalable, and cooperative RL models that efficiently integrate heterogeneous energy sources, storage solutions, and flexible loads. While PV systems understandably dominate RL-based BEMS research due to their cost-effectiveness and straightforward deployment, this narrow focus risks overlooking the synergies and resilience gains offered by hybrid-RES configurations. As building energy ecosystems become more intricate, future RL solutions must embrace multiple energy sources, exploiting their complementary generation profiles to enhance adaptability, reduce grid reliance, and address the nuanced trade-offs between cost, efficiency, and occupant comfort in real-world environments.

Another key trend emerging from the evaluation of existing studies is the prevalence of single-RES integration in building energy systems. This is primarily due to the lower complexity and reduced implementation costs associated with optimizing a single energy source, particularly PV. However, there is a growing interest in multi-RES (or hybrid-RES) systems, driven by the need for greater energy flexibility, resilience, and efficiency. Combining PV with other RES technologies, such as SWH, WT, BIO, or GHP, enhances self-sufficiency by leveraging complementary generation profiles, mitigating the variability of any single energy source.

### 5.9. BEMS Type

The analysis of existing studies highlights that PV and ESS are the most frequently combined energy components, emphasizing their role in enhancing self-consumption, grid independence, and energy flexibility [85,87,90,100,122,129,130] (see Figure 14). The integration of HVAC systems [102,105,106,108,109,116], as well as other smart appliances [93,95,121,124], further underlines the importance of demand-side optimization, given that heating and cooling systems contribute significantly to total energy consumption.

Although EVs have been studied less frequently in the literature, they present new opportunities for bidirectional energy flow through vehicle-to-grid (V2G) and vehicle-to-home (V2H) strategies. These strategies enhance grid flexibility but also introduce challenges related to stochastic availability and battery degradation [88,119,128] (see Figure 14). Lastly, the integration of WT and GHP, as observed in studies such as [133,134], suggests a growing interest in hybrid renewable energy systems. These complex multi-energy frameworks require RL-based optimization to effectively manage energy distribution and improve system efficiency within microgrid environments.

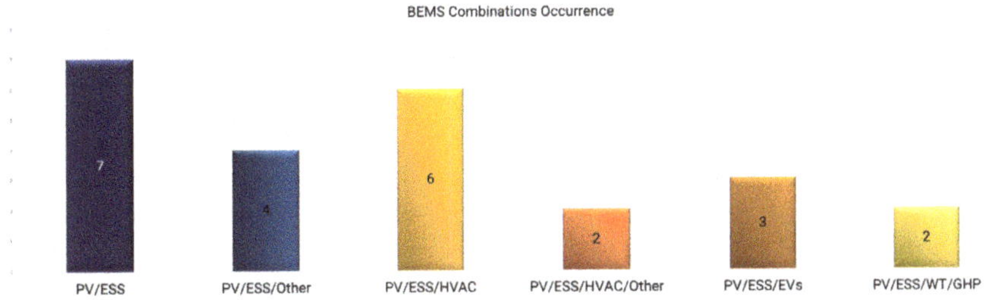


**Figure 14.** BEMS combination types occurrence in RES-integrated BEMS applications (2015–2025).

While PV and ESS dominate current BEMS research, the interactions among diverse energy assets—HVAC systems, EVs, and additional smart loads—call for more holistic RL-based control strategies that capture their interdependencies. The inclusion of EVs, in particular, necessitates tackling challenges related to battery health and uncertain charging patterns, moving beyond purely static optimization. As building ecosystems diversify, future RL implementations must evolve to coordinate these disparate elements, striking a delicate balance between occupant comfort, operational complexity, and system-wide resilience.

### 5.10. Building Type

The evaluation of existing studies reveals a clear dominance of residential buildings over commercial buildings in RL applications for RES-integrated BEMS (see Figure 15, Right). Unlike model-free control applications for HVAC systems, which are predominantly implemented in commercial buildings [31], the integration of RES within the energy mix is primarily observed in residential structures.

This trend is influenced by several key factors. Residential energy management typically involves simpler operational dynamics, making RL deployment more feasible. Homeowners and small-scale users increasingly adopt PV systems and battery storage, benefiting from demand-side flexibility and self-consumption optimization—both of which RL can effectively manage. Additionally, the rise of smart home technologies and IoT-based automation is expected to further accelerate RL adoption in residential settings.

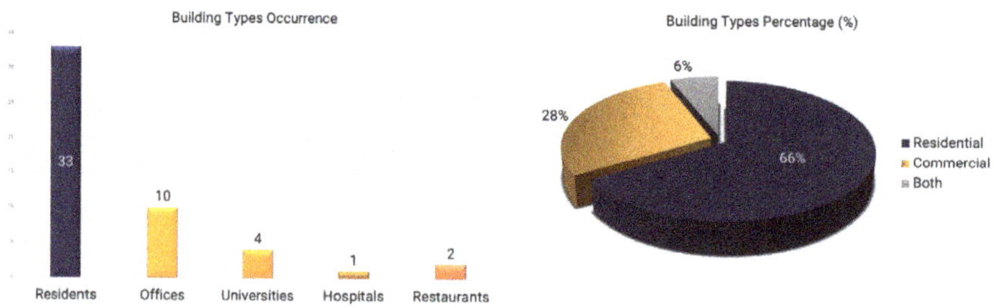


**Figure 15.** Building types occurrence (**Left**) and percentage (%) (**Right**) in RES-integrated BEMS applications (2015–2025).

In contrast, RL applications in commercial buildings, such as offices [98,109–112,117,118, 123,125,129], university campuses [89,99,103,108], hospitals [116], and restaurants [114,125] (see Figure 15, Left), remain relatively limited.

The higher complexity of energy demands in commercial buildings presents challenges for RL implementation. These environments typically involve HVAC systems, fluctuating occupancy patterns, and strict operational constraints, all of which necessitate more advanced control strategies. Additionally, commercial buildings often employ centralized energy management systems and rely on contract-based energy procurement, reducing the immediate need for RL-driven optimization. However, as RL research advances, applications in commercial settings are expected to expand, particularly in areas such as large-scale multi-zone control, grid-interactive buildings, and hybrid RES integration. In these cases, RL has the potential to optimize load shifting, demand response, and multi-agent coordination, significantly improving energy efficiency and operational resilience.

## 6. Discussion

The discussion section, divided into Current Trends and Future Directions, provides a structured reflection on key findings and their broader significance. The Current Trends subsection distills recurring themes and predominant approaches in RL applications for RES-integrated BEMS, offering insights into the present research focus. In Future Directions, open research challenges and unexplored areas are identified, equipping readers with strategic considerations for advancing both practical implementations and theoretical developments in the field.

### *6.1. Trends Identification*

✓ **Dominance of value-based RL and the rise of actor-critic methods:** The evaluation indicates that value-based RL, particularly QL, has been the most applied method due to its simplicity and effectiveness in handling discrete state-action spaces [84,88,89,91–94,99]. However, as RL applications become more complex, there is a noticeable shift toward actor-critic methods, such as TD3, DDPG, and SAC, which are better suited for high-dimensional, continuous control problems [105,107,113,115,117,118]. These methodologies integrate elements of building physics to enhance learning efficiency and stability while reducing convergence failures [86,103,111]. Despite their higher computational demands, actor-critic frameworks are emerging as a promising alternative for real-time energy optimization.

✓ **Hybrid RL for stability and reliability:** A growing number of studies integrate RL with complementary control techniques, such as FLC, MPC, or heuristic RBC [104,121,127,128,131,134] (see Figure 4). This hybridization ensures that RL agents operate within safe and reliable constraints, preventing failures in real-world scenarios. Hybrid RL approaches seem particularly effective in energy systems with slow thermal responses, where traditional control handles short-term fluctuations while RL optimizes long-term performance.

✓ **Widespread adoption of multi-agent reinforcement learning (MARL) for complex systems: Nearly 40 percent of studies now employ MARL to manage complex energy ecosystems where multiple subsystems require coordinated control [97,98,100,102,116,133]** (see Figure 7). This decentralized approach enhances adaptability to fluctuating renewable energy system generation and dynamic energy consumption patterns. However, MARL still faces challenges, such as increased communication overhead and coordination inefficiencies, necessitating advanced mechanisms for agent interaction and stability [110,117,123].

✓ **Multi-objective RL for balanced decision-making: RL applications increasingly incorporate multi-objective optimization to balance competing targets, such as energy cost, thermal comfort, carbon emissions, and peak demand reduction [93,101,107,110,117,118]** (see Figure 12). Rather than optimizing a single reward function, recent studies introduce weighted trade-offs and explicit constraints to ensure practical feasibility. Some works embed predefined comfort or $CO_2$ emission thresholds into the learning process, aligning RL policies with regulatory requirements.

✓ **Focus on federated, privacy-preserving, and real-time RL:** An increasing number of studies explore federated learning in RL-based BEMS, where multiple buildings train local models and share aggregated updates instead of raw data [106,109,116]. This approach enhances data privacy and security while supporting distributed learning. Additionally, there is a shift toward real-time RL applications capable of responding to changing conditions within minutes, moving beyond day-ahead scheduling toward sub-15-minute decision cycles [103,111,131].

### 6.2. Future Directions

➠ **Advancing hybrid RL and algorithmic innovations:** Future research should focus on more sophisticated hybrid RL approaches that blend model-based and model-free paradigms to enhance learning efficiency and decision robustness. Current studies highlight the trend of combining RL with predictive models, FLC, and MDP to mitigate convergence issues and accelerate policy adaptation [121,127,131,134]. Future advancements should ensure that RL agents leverage prior knowledge while maintaining adaptability to unforeseen conditions.

➠ **Scalable and hierarchical MARL:** As BEMS extend beyond individual buildings to microgrids and districts, flat MARL architectures may become inefficient due to increased communication overhead and coordination challenges. Hierarchical RL structures, where high-level policies coordinate multiple lower-level subsystems, can enhance scalability and stability [97,100].

➠ **Expanding storage utilization beyond batteries:** While battery storage is widely studied, other forms of ESS, such as thermal mass, phase-change materials, and distributed heating/cooling reserves, remain underexplored (see Figure 13, Left). Future research should focus on RL strategies that leverage thermal storage as a controllable resource, such as pre-cooling or intelligent heat storage management [98,133].

- **Accelerating RL adaptation across buildings and climates:** RL models trained for specific buildings often struggle when transferred to different environments due to variations in climate, occupancy, and infrastructure. Future research should develop meta-learning and domain-adaptation techniques to enable RL agents to rapidly adjust to new settings [90,101].
- **Ensuring safety and compliance with constraints:** For RL to be widely adopted in BEMS, safety assurances must be embedded directly into control policies. Techniques such as constrained RL, shielded RL, and Lyapunov-based critics will be essential to ensuring energy systems operate within predefined limits while preventing failures and regulatory non-compliance.
- **Transition from simulations to real-world deployment:** Despite significant progress in RL-driven energy management, most studies remain confined to simulation environments. Notably, only two works among the overall dataset implement real-life experiments considering RL in RES-integrated BEMS. Future research must prioritize real-world implementation with robust fallback mechanisms, ensuring safe operation under unexpected conditions [86].

## 7. Conclusions

This review involved a large-scale, systematic examination of RL control strategies for building energy management systems that integrate renewable energy system technologies from 2015 to 2025. Unlike previous surveys, this study categorized high-impact, peer-reviewed publications according to their RL method, building typology, and validation methodology, thereby offering an extensive perspective on the current state of research. Emphasis was placed on how decentralized agents coordinate in complex environments, how data-driven exploration can be blended with physically informed techniques, and how emerging hybrid approaches improve real-time adaptability and occupant comfort.

A key innovation of this review lies in its multi-level mapping of algorithmic structures, agent configurations, and operational objectives, highlighting where specific RL methods excel and where they face limitations. By linking these technical details to application outcomes, such as cost reduction, grid flexibility, and reduced carbon footprints, this study illuminates how RL can be practically deployed in increasingly complex building scenarios. Moreover, the compilation of MARL architectures and their coordination mechanisms provides actionable insights for practitioners aiming to design resilient, scalable energy systems, while also addressing occupant comfort and regulatory constraints.

Despite such advancements, a number of refinements could further accelerate RL-based BEMS toward large-scale practical use. Hybrid control schemes, for instance, can merge data-driven learning with model-based or heuristic strategies to shorten convergence times and reduce the likelihood of policy failures. Similarly, the development of enhanced exploration strategies for multi-agent systems would further mitigate non-stationarity and communication overhead, enabling robust coordination under dynamic conditions. Finally, domain adaptation and transfer learning approaches offer the promise of quickly repurposing trained policies across different building types and climates, minimizing the effort of retraining from scratch and fostering broader adoption.

Looking ahead, the next generation of RL-enabled BEMS would benefit from hierarchical MARL to manage control at multiple timescales, privacy-preserving architectures (e.g., federated RL) to facilitate data sharing without compromising user confidentiality, and constrained RL frameworks that embed essential safety and regulatory limits into the learning process. By addressing such research challenges, RL-driven energy management could transition from isolated demonstrations into wide-scale adoption, ultimately driving more sustainable, resilient, and occupant-centric building energy systems.

**Funding:** This research was partially funded by the SEED4AI project. The project is being implemented within the framework of the National Recovery and Resilience Plan "Greece 2.0", with funding from the European Union—NextGenerationEU (Implementing Body: Hellenic Foundation for Research and Innovation (HFRI))/ID: 16880. SEED4AI: https://seed4ai.ee.duth.gr/ accessed on 26 March 2025.

**Conflicts of Interest:** The authors declare no conflicts of interest.

## Abbreviations

| | |
|---|---|
| A2C | Advantage Actor-Critic |
| BEMS | Building Energy Management System |
| BIO | Biomass Energy Systems |
| DDPG | Deep Deterministic Policy Gradient |
| DDQN | Double Deep Q-Network |
| D3QN | Dueling Double Deep Q-Network |
| DHW | Domestic Hot Water |
| DQN | Deep Q-Network |
| DR | Demand Response |
| DRL | Deep Reinforcement Learning |
| ESS | Energy Storage System |
| EVs | Electric Vehicles |
| FLC | Fuzzy Logic Control |
| GA | Genetic Algorithm |
| GHP | Ground Heat Pump |
| HEMS | Home Energy Management System |
| HP | Heat Pump |
| HVAC | Heating, Ventilation, and Air Conditioning |
| LS | Lighting System |
| LSTM | Long Short-Term Memory |
| MARL | Multi-Agent Reinforcement Learning |
| MCTS | Monte Carlo Tree Search |
| MDP | Markov Decision Process |
| MPC | Model Predictive Control |
| ANN | Artificial Neural Network |
| NZEB | Net-Zero Energy Building |
| PAR | Peak-to-Average Ratio |
| PV | Photovoltaic |
| QL | Q-Learning |
| RBC | Rule-Based Control |
| RES | Renewable Energy System |
| RL | Reinforcement Learning |
| SAC | Soft Actor-Critic |
| SWH | Solar Water Heating |
| TD3 | Twin Delayed Deep Deterministic Policy Gradient |
| TSS | Thermal Storage System |
| V2G | Vehicle-to-Grid |
| V2H | Vehicle-to-Home |
| WT | Wind Turbine |

## References

1. Economidou, M.; Todeschi, V.; Bertoldi, P.; D'Agostino, D.; Zangheri, P.; Castellazzi, L. Review of 50 years of EU energy efficiency policies for buildings. *Energy Build.* **2020**, *225*, 110322. [CrossRef]
2. Pavel, T.; Polina, S.; Liubov, N. The research of the impact of energy efficiency on mitigating greenhouse gas emissions at the national level. *Energy Convers. Manag.* **2024**, *314*, 118671. [CrossRef]
3. Ye, J.; Fanyang, Y.; Wang, J.; Meng, S.; Tang, D. A Literature Review of Green Building Policies: Perspectives from Bibliometric Analysis. *Buildings* **2024**, *14*, 2607. [CrossRef]
4. Reddy, V.J.; Hariram, N.; Ghazali, M.F.; Kumarasamy, S. Pathway to sustainability: An overview of renewable energy integration in building systems. *Sustainability* **2024**, *16*, 638. [CrossRef]
5. Rehmani, M.H.; Reisslein, M.; Rachedi, A.; Erol-Kantarci, M.; Radenkovic, M. Integrating renewable energy resources into the smart grid: Recent developments in information and communication technologies. *IEEE Trans. Ind. Inform.* **2018**, *14*, 2814–2825. [CrossRef]
6. Harvey, L.D. Reducing energy use in the buildings sector: Measures, costs, and examples. *Energy Effic.* **2009**, *2*, 139–163. [CrossRef]
7. Chel, A.; Kaushik, G. Renewable energy technologies for sustainable development of energy efficient building. *Alex. Eng. J.* **2018**, *57*, 655–669. [CrossRef]
8. Farghali, M.; Osman, A.I.; Mohamed, I.M.; Chen, Z.; Chen, L.; Ihara, I.; Yap, P.S.; Rooney, D.W. Strategies to save energy in the context of the energy crisis: A review. *Environ. Chem. Lett.* **2023**, *21*, 2003–2039. [CrossRef]
9. Yudelson, J. *The Green Building Revolution*; Island Press: Washington, DC, USA, 2010.
10. Cao, X.; Dai, X.; Liu, J. Building energy-consumption status worldwide and the state-of-the-art technologies for zero-energy buildings during the past decade. *Energy Build.* **2016**, *128*, 198–213. [CrossRef]
11. Gielen, D.; Boshell, F.; Saygin, D.; Bazilian, M.D.; Wagner, N.; Gorini, R. The role of renewable energy in the global energy transformation. *Energy Strategy Rev.* **2019**, *24*, 38–50. [CrossRef]
12. Mogoș, R.I.; Petrescu, I.; Chiotan, R.A.; Crețu, R.C.; Troacă, V.A.; Mogoș, P.L. Greenhouse gas emissions and Green Deal in the European Union. *Front. Environ. Sci.* **2023**, *11*, 1141473.
13. Famà, R. REPowerEU. In *Research Handbook on Post-Pandemic EU Economic Governance and NGEU Law*; Edward Elgar Publishing: London, UK, 2024; pp. 128–143.
14. Jäger-Waldau, A.; Bodis, K.; Kougias, I.; Szabo, S. The New European Renewable Energy Directive-Opportunities and Challenges for Photovoltaics. In Proceedings of the 2019 IEEE 46th Photovoltaic Specialists Conference (PVSC), Chicago, IL, USA, 16–21 June 2019; IEEE: Piscataway, NJ, USA, 2019; pp. 0592–0594.
15. Fetsis, P. The LIFE Programme–Over 20 Years Improving Sustainability in the Built Environment in the EU. *Procedia Environ. Sci.* **2017**, *38*, 913–918.
16. Hassan, Q.; Algburi, S.; Sameen, A.Z.; Salman, H.M.; Jaszczur, M. A review of hybrid renewable energy systems: Solar and wind-powered solutions: Challenges, opportunities, and policy implications. *Results Eng.* **2023**, *20*, 101621. [CrossRef]
17. Kalogirou, S.A. Building integration of solar renewable energy systems towards zero or nearly zero energy buildings. *Int. J. Low-Carbon Technol.* **2015**, *10*, 379–385.
18. Orikpete, O.F.; Ikemba, S.; Ewim, D.R.E. Integration of renewable energy technologies in smart building design for enhanced energy efficiency and self-sufficiency. *J. Eng. Exact Sci.* **2023**, *9*, 16423-01e. [CrossRef]
19. Saloux, E.; Teyssedou, A.; Sorin, M. Analysis of photovoltaic (PV) and photovoltaic/thermal (PV/T) systems using the exergy method. *Energy Build.* **2013**, *67*, 275–285.
20. Michailidis, P.; Michailidis, I.; Kosmatopoulos, E. Review and Evaluation of Multi-Agent Control Applications for Energy Management in Buildings. *Energies* **2024**, *17*, 4835. [CrossRef]
21. Abdulraheem, A.; Lee, S.; Jung, I.Y. Dynamic Personalized Thermal Comfort Model: Integrating Temporal Dynamics and Environmental Variability with Individual Preferences. *J. Build. Eng.* **2025**, *102*, 111938.
22. Lu, X.; Fu, Y.; O'Neill, Z. Benchmarking high performance HVAC Rule-Based controls with advanced intelligent Controllers: A case study in a Multi-Zone system in Modelica. *Energy Build.* **2023**, *284*, 112854.
23. Drgoňa, J.; Picard, D.; Kvasnica, M.; Helsen, L. Approximate model predictive building control via machine learning. *Appl. Energy* **2018**, *218*, 199–216. [CrossRef]
24. Blinn, A.; Kue, U.R.S.; Kennel, F. A Comparison of Model Predictive Control and Heuristics in Building Energy Management. In Proceedings of the 2024 8th International Conference on Smart Grid and Smart Cities (ICSGSC), Shanghai, China, 25–27 October 2024; IEEE: Piscataway, NJ, USA, 2024; pp. 275–285.
25. Chen, Z.; Xiao, F.; Guo, F.; Yan, J. Interpretable machine learning for building energy management: A state-of-the-art review. *Adv. Appl. Energy* **2023**, *9*, 100123. [CrossRef]
26. Ukoba, K.; Olatunji, K.O.; Adeoye, E.; Jen, T.C.; Madyira, D.M. Optimizing renewable energy systems through artificial intelligence: Review and future prospects. *Energy Environ.* **2024**, *35*, 3833–3879. [CrossRef]

27. Shobanke, M.; Bhatt, M.; Shittu, E. Advancements and future outlook of Artificial Intelligence in energy and climate change modeling. *Adv. Appl. Energy* **2025**, *17*, 100211. [CrossRef]
28. Pergantis, E.N.; Al Theeb, N.; Dhillon, P.; Ore, J.P.; Ziviani, D.; Groll, E.A.; Kircher, K.J.; et al. Field demonstration of predictive heating control for an all-electric house in a cold climate. *Appl. Energy* **2024**, *360*, 122820. [CrossRef]
29. Manic, M.; Wijayasekara, D.; Amarasinghe, K.; Rodriguez-Andina, J.J. Building energy management systems: The age of intelligent and adaptive buildings. *IEEE Ind. Electron. Mag.* **2016**, *10*, 25–39. [CrossRef]
30. Zia, M.F.; Elbouchikhi, E.; Benbouzid, M. Microgrids energy management systems: A critical review on methods, solutions, and prospects. *Appl. Energy* **2018**, *222*, 1033–1055. [CrossRef]
31. Michailidis, P.; Michailidis, I.; Vamvakas, D.; Kosmatopoulos, E. Model-Free HVAC Control in Buildings: A Review. *Energies* **2023**, *16*, 7124. [CrossRef]
32. Mariano-Hernández, D.; Hernández-Callejo, L.; Zorita-Lamadrid, A.; Duque-Pérez, O.; García, F.S. A review of strategies for building energy management system: Model predictive control, demand side management, optimization, and fault detect & diagnosis. *J. Build. Eng.* **2021**, *33*, 101692.
33. Pergantis, E.N.; Dhillon, P.; Premer, L.D.R.; Lee, A.H.; Ziviani, D.; Kircher, K.J. Humidity-aware model predictive control for residential air conditioning: A field study. *Build. Environ.* **2024**, *266*, 112093. [CrossRef]
34. Pergantis, E.N.; Premer, L.D.R.; Lee, A.H.; Ziviani, D.; Groll, E.A.; Kircher, K.J.; et al. Latent and Sensible Model Predictive Controller Demonstration in a House During Cooling Operation. *ASHRAE Trans.* **2024**, *130*, 177–185.
35. Michailidis, I.T.; Schild, T.; Sangi, R.; Michailidis, P.; Korkas, C.; Fütterer, J.; Müller, D.; Kosmatopoulos, E.B. Energy-efficient HVAC management using cooperative, self-trained, control agents: A real-life German building case study. *Appl. Energy* **2018**, *211*, 113–125.
36. Michailidis, P.; Pelitaris, P.; Korkas, C.; Michailidis, I.; Baldi, S.; Kosmatopoulos, E. Enabling optimal energy management with minimal IoT requirements: A legacy A/C case study. *Energies* **2021**, *14*, 7910. [CrossRef]
37. Michailidis, I.T.; Sangi, R.; Michailidis, P.; Schild, T.; Fuetterer, J.; Mueller, D.; Kosmatopoulos, E.B. Balancing energy efficiency with indoor comfort using smart control agents: A simulative case study. *Energies* **2020**, *13*, 6228. [CrossRef]
38. Michailidis, P.; Michailidis, I.; Gkelios, S.; Kosmatopoulos, E. Artificial Neural Network Applications for Energy Management in Buildings: Current Trends and Future Directions. *Energies* **2024**, *17*, 570. [CrossRef]
39. Li, S.E. *Reinforcement Learning for Sequential Decision and Optimal Control*; Springer: Berlin/Heidelberg, Germany, 2023.
40. Vamvakas, D.; Michailidis, P.; Korkas, C.; Kosmatopoulos, E. Review and evaluation of reinforcement learning frameworks on smart grid applications. *Energies* **2023**, *16*, 5326. [CrossRef]
41. Recht, B. A tour of reinforcement learning: The view from continuous control. *Annu. Rev. Control Robot. Auton. Syst.* **2019**, *2*, 253–279.
42. Mohammadi, P.; Darshi, R.; Shamaghdari, S.; Siano, P. Comparative Analysis of Control Strategies for Microgrid Energy Management with a Focus on Reinforcement Learning. *IEEE Access* **2024**, *12*, 171368–171395. [CrossRef]
43. Ruelens, F.; Claessens, B.J.; Vandael, S.; De Schutter, B.; Babuška, R.; Belmans, R. Residential demand response of thermostatically controlled loads using batch reinforcement learning. *IEEE Trans. Smart Grid* **2016**, *8*, 2149–2159. [CrossRef]
44. Kadamala, K.; Chambers, D.; Barrett, E. Enhancing HVAC control systems through transfer learning with deep reinforcement learning agents. *Smart Energy* **2024**, *13*, 100131. [CrossRef]
45. Al Sayed, K.; Boodi, A.; Broujeny, R.S.; Beddiar, K. Reinforcement learning for HVAC control in intelligent buildings: A technical and conceptual review. *J. Build. Eng.* **2024**, *95*, 110085. [CrossRef]
46. Wang, Z.; Hong, T. Reinforcement learning for building controls: The opportunities and challenges. *Appl. Energy* **2020**, *269*, 115036. [CrossRef]
47. Gautam, M. Deep Reinforcement learning for resilient power and energy systems: Progress, prospects, and future avenues. *Electricity* **2023**, *4*, 336–380. [CrossRef]
48. Jiang, Z.; Risbeck, M.J.; Ramamurti, V.; Murugesan, S.; Amores, J.; Zhang, C.; Lee, Y.M.; Drees, K.H. Building HVAC control with reinforcement learning for reduction of energy cost and demand charge. *Energy Build.* **2021**, *239*, 110833. [CrossRef]
49. Yu, L.; Qin, S.; Zhang, M.; Shen, C.; Jiang, T.; Guan, X. A review of deep reinforcement learning for smart building energy management. *IEEE Internet Things J.* **2021**, *8*, 12046–12063. [CrossRef]
50. Liu, S.; Henze, G.P. Experimental analysis of simulated reinforcement learning control for active and passive building thermal storage inventory: Part 2: Results and analysis. *Energy Build.* **2006**, *38*, 148–161. [CrossRef]
51. Lazaridis, C.R.; Michailidis, I.; Karatzinis, G.; Michailidis, P.; Kosmatopoulos, E. Evaluating Reinforcement Learning Algorithms in Residential Energy Saving and Comfort Management. *Energies* **2024**, *17*, 581. [CrossRef]
52. Singh, Y.; Pal, N. Reinforcement learning with fuzzified reward approach for MPPT control of PV systems. *Sustain. Energy Technol. Assess.* **2021**, *48*, 101665. [CrossRef]

53. Wang, Z.; Xue, W.; Li, K.; Tang, Z.; Liu, Y.; Zhang, F.; Cao, S.; Peng, X.; Wu, E.Q.; Zhou, H. Dynamic combustion optimization of a pulverized coal boiler considering the wall temperature constraints: A deep reinforcement learning-based framework. *Appl. Therm. Eng.* **2025**, *259*, 124923. [CrossRef]
54. Perera, A.; Kamalaruban, P. Applications of reinforcement learning in energy systems. *Renew. Sustain. Energy Rev.* **2021**, *137*, 110618.
55. Gaviria, J.F.; Narváez, G.; Guillen, C.; Giraldo, L.F.; Bressan, M. Machine learning in photovoltaic systems: A review. *Renew. Energy* **2022**, *196*, 298–318. [CrossRef]
56. Fu, Q.; Han, Z.; Chen, J.; Lu, Y.; Wu, H.; Wang, Y. Applications of reinforcement learning for building energy efficiency control: A review. *J. Build. Eng.* **2022**, *50*, 104165.
57. Rezaie, B.; Esmailzadeh, E.; Dincer, I. Renewable energy options for buildings: Case studies. *Energy Build.* **2011**, *43*, 56–65.
58. Lin, Y.; Yang, W.; Hao, X.; Yu, C. Building integrated renewable energy. *Energy Explor. Exploit.* **2021**, *39*, 603–607.
59. Hayter, S.J. *Integrating Renewable Energy Systems in Buildings (Presentation)*; Technical report; National Renewable Energy Lab. (NREL): Golden, CO, USA, 2011.
60. Le, T.V.; ChuDuc, H.; Tran, Q.X. Optimized Integration of Renewable Energy in Smart Buildings: A Systematic Review from Scopus Data. In Proceedings of the 2024 9th International Conference on Applying New Technology in Green Buildings (ATiGB), Danang, Vietnam, 30–31 August 2024; IEEE: Piscataway, NJ, USA, 2024; pp. 397–402.
61. Chen, L.; Hu, Y.; Wang, R.; Li, X.; Chen, Z.; Hua, J.; Osman, A.I.; Farghali, M.; Huang, L.; Li, J.; et al. Green building practices to integrate renewable energy in the construction sector: A review. *Environ. Chem. Lett.* **2024**, *22*, 751–784.
62. Hayter, S.J.; Kandt, A. *Renewable Energy Applications for Existing Buildings*; Technical report; National Renewable Energy Lab. (NREL): Golden, CO, USA, 2011.
63. Vassiliades, C.; Agathokleous, R.; Barone, G.; Forzano, C.; Giuzio, G.; Palombo, A.; Buonomano, A.; Kalogirou, S. Building integration of active solar energy systems: A review of geometrical and architectural characteristics. *Renew. Sustain. Energy Rev.* **2022**, *164*, 112482.
64. Bougiatioti, F.; Michael, A. The architectural integration of active solar systems. Building applications in the Eastern Mediterranean region. *Renew. Sustain. Energy Rev.* **2015**, *47*, 966–982. [CrossRef]
65. Canale, L.; Di Fazio, A.R.; Russo, M.; Frattolillo, A.; Dell'Isola, M. An overview on functional integration of hybrid renewable energy systems in multi-energy buildings. *Energies* **2021**, *14*, 1078. [CrossRef]
66. Mnih, V.; Kavukcuoglu, K.; Silver, D.; Rusu, A.A.; Veness, J.; Bellemare, M.G.; Graves, A.; Riedmiller, M.; Fidjeland, A.K.; Ostrovski, G.; et al. Human-level control through deep reinforcement learning. *Nature* **2015**, *518*, 529–533. [CrossRef]
67. Sutton, R.S.; Barto, A.G. *Reinforcement Learning: An Introduction*; MIT Press: Cambridge, MA, USA, 2018.
68. Ahrarinouri, M.; Rastegar, M.; Seifi, A.R. Multiagent reinforcement learning for energy management in residential buildings. *IEEE Trans. Ind. Inform.* **2020**, *17*, 659–666.
69. Hernandez-Leal, P.; Kartal, B.; Taylor, M.E. A survey and critique of multiagent deep reinforcement learning. *Auton. Agents Multi-Agent Syst.* **2019**, *33*, 750–797.
70. Busoniu, L.; Babuska, R.; De Schutter, B. A comprehensive survey of multiagent reinforcement learning. *IEEE Trans. Syst. Man, Cybern. Part C (Appl. Rev.)* **2008**, *38*, 156–172.
71. Castellini, J.; Oliehoek, F.A.; Savani, R.; Whiteson, S. The representational capacity of action-value networks for multi-agent reinforcement learning. *arXiv* **2019**, arXiv:1902.07497.
72. Sheng, J.; Wang, X.; Jin, B.; Yan, J.; Li, W.; Chang, T.H.; Wang, J.; Zha, H. Learning structured communication for multi-agent reinforcement learning. *Auton. Agents Multi-Agent Syst.* **2022**, *36*, 50.
73. Canese, L.; Cardarilli, G.C.; Di Nunzio, L.; Fazzolari, R.; Giardino, D.; Re, M.; Spanò, S. Multi-agent reinforcement learning: A review of challenges and applications. *Appl. Sci.* **2021**, *11*, 4948. [CrossRef]
74. Xie, H.; Song, G.; Shi, Z.; Zhang, J.; Lin, Z.; Yu, Q.; Fu, H.; Song, X.; Zhang, H. Reinforcement learning for vehicle-to-grid: A review. *Advances in Applied Energy* **2025**, *17*, 100214.
75. Mohammadi, P.; Nasiri, A.; Darshi, R.; Shirzad, A.; Abdollahipour, R. Achieving Cost Efficiency in Cloud Data Centers Through Model-Free Q-Learning. In Proceedings of the International Conference on Electrical and Electronics Engineering, Marmaris, Turkey, 22–24 April 2024; Springer: Berlin/Heidelberg, Germany, 2024; pp. 457–468.
76. Watkins, C.J.; Dayan, P. Q-learning. *Mach. Learn.* **1992**, *8*, 279–292.
77. Zhang, H.; Seal, S.; Wu, D.; Bouffard, F.; Boulet, B. Building energy management with reinforcement learning and model predictive control: A survey. *IEEE Access* **2022**, *10*, 27853–27862. [CrossRef]
78. Wei, T.; Wang, Y.; Zhu, Q. Deep reinforcement learning for building HVAC control. In Proceedings of the 54th Annual Design Automation Conference 2017, Austin, TX, USA, 18–22 June 2017; pp. 1–6.
79. Wang, Y.; He, H.; Tan, X. Truly proximal policy optimization. In Proceedings of the Uncertainty in Artificial Intelligence, PMLR, Virtual, 3–6 August 2020; pp. 113–122.

80. Bolt, P.; Ziebart, V.; Jaeger, C.; Schmid, N.; Stadelmann, T.; Füchslin, R.M. A simulation study on energy optimization in building control with reinforcement learning. In Proceedings of the IAPR Workshop on Artificial Neural Networks in Pattern Recognition, Montréal, QC, Canada, 10–12 October 2024; Springer: Berlin/Heidelberg, Germany, 2024; pp. 320–331.
81. Sumiea, E.H.; Abdulkadir, S.J.; Alhussian, H.S.; Al-Selwi, S.M.; Alqushaibi, A.; Ragab, M.G.; Fati, S.M. Deep deterministic policy gradient algorithm: A systematic review. *Heliyon* **2024**, *10*, e30697.
82. Ding, F.; Ma, G.; Chen, Z.; Gao, J.; Li, P. Averaged Soft Actor-Critic for Deep Reinforcement Learning. *Complexity* **2021**, *2021*, 6658724. [CrossRef]
83. Dong, J.; Wang, H.; Yang, J.; Lu, X.; Gao, L.; Zhou, X. Optimal scheduling framework of electricity-gas-heat integrated energy system based on asynchronous advantage actor-critic algorithm. *IEEE Access* **2021**, *9*, 139685–139696.
84. Yang, L.; Nagy, Z.; Goffin, P.; Schlueter, A. Reinforcement learning for optimal control of low exergy buildings. *Appl. Energy* **2015**, *156*, 577–586. [CrossRef]
85. Raju, L.; Sankar, S.; Milton, R. Distributed optimization of solar micro-grid using multi agent reinforcement learning. *Procedia Comput. Sci.* **2015**, *46*, 231–239.
86. De Somer, O.; Soares, A.; Vanthournout, K.; Spiessens, F.; Kuijpers, T.; Vossen, K. Using reinforcement learning for demand response of domestic hot water buffers: A real-life demonstration. In Proceedings of the 2017 IEEE PES Innovative Smart Grid Technologies Conference Europe (ISGT-Europe), Torino, Italy, 26–29 September 2017; IEEE: Piscataway, NJ, USA, 2017; pp. 1–7.
87. Ebell, N.; Heinrich, F.; Schlund, J.; Pruckner, M. Reinforcement learning control algorithm for a pv-battery-system providing frequency containment reserve power. In Proceedings of the 2018 IEEE International Conference on Communications, Control, and Computing Technologies for Smart Grids (SmartGridComm), Aalborg, Denmark, 29–31 October 2018; IEEE: Piscataway, NJ, USA, 2018; pp. 1–6.
88. Remani, T.; Jasmin, E.; Ahamed, T.I. Residential load scheduling with renewable generation in the smart grid: A reinforcement learning approach. *IEEE Syst. J.* **2018**, *13*, 3283–3294.
89. Kim, S.; Lim, H. Reinforcement learning based energy management algorithm for smart energy buildings. *Energies* **2018**, *11*, 2010. [CrossRef]
90. Prasad, A.; Dusparic, I. Multi-agent deep reinforcement learning for zero energy communities. In Proceedings of the 2019 IEEE PES Innovative Smart Grid Technologies Europe (ISGT-Europe), Bucharest, Romania, 29 September–2 October 2019; IEEE: Piscataway, NJ, USA, 2019; pp. 1–5.
91. Xu, X.; Jia, Y.; Xu, Y.; Xu, Z.; Chai, S.; Lai, C.S. A multi-agent reinforcement learning-based data-driven method for home energy management. *IEEE Trans. Smart Grid* **2020**, *11*, 3201–3211.
92. Correa-Jullian, C.; Droguett, E.L.; Cardemil, J.M. Operation scheduling in a solar thermal system: A reinforcement learning-based framework. *Appl. Energy* **2020**, *268*, 114943. [CrossRef]
93. Chen, S.J.; Chiu, W.Y.; Liu, W.J. User preference-based demand response for smart home energy management using multiobjective reinforcement learning. *IEEE Access* **2021**, *9*, 161627–161637. [CrossRef]
94. Lissa, P.; Deane, C.; Schukat, M.; Seri, F.; Keane, M.; Barrett, E. Deep reinforcement learning for home energy management system control. *Energy AI* **2021**, *3*, 100043. [CrossRef]
95. Raman, N.S.; Gaikwad, N.; Barooah, P.; Meyn, S.P. Reinforcement learning-based home energy management system for resiliency. In Proceedings of the 2021 American Control Conference (ACC), Online, 25–28 May 2021; IEEE: Piscataway, NJ, USA, 2021; pp. 1358–1364.
96. Heidari, A.; Maréchal, F.; Khovalyg, D. Reinforcement Learning for proactive operation of residential energy systems by learning stochastic occupant behavior and fluctuating solar energy: Balancing comfort, hygiene and energy use. *Appl. Energy* **2022**, *318*, 119206. [CrossRef]
97. Lu, J.; Mannion, P.; Mason, K. A multi-objective multi-agent deep reinforcement learning approach to residential appliance scheduling. *IET Smart Grid* **2022**, *5*, 260–280. [CrossRef]
98. Shen, R.; Zhong, S.; Wen, X.; An, Q.; Zheng, R.; Li, Y.; Zhao, J. Multi-agent deep reinforcement learning optimization framework for building energy system with renewable energy. *Appl. Energy* **2022**, *312*, 118724.
99. Wang, L.; Zhang, G.; Yin, X.; Zhang, H.; Ghalandari, M. Optimal control of renewable energy in buildings using the machine learning method. *Sustain. Energy Technol. Assess.* **2022**, *53*, 102534. [CrossRef]
100. Cordeiro-Costas, M.; Villanueva, D.; Eguía-Oller, P.; Granada-Álvarez, E. Intelligent energy storage management trade-off system applied to Deep Learning predictions. *J. Energy Storage* **2023**, *61*, 106784. [CrossRef]
101. Mocanu, E.; Mocanu, D.C.; Nguyen, P.H.; Liotta, A.; Webber, M.E.; Gibescu, M.; Slootweg, J.G. On-line building energy optimization using deep reinforcement learning. *IEEE Trans. Smart Grid* **2018**, *10*, 3698–3708.
102. Vazquez-Canteli, J.R.; Henze, G.; Nagy, Z. MARLISA: Multi-agent reinforcement learning with iterative sequential action selection for load shaping of grid-interactive connected buildings. In Proceedings of the 7th ACM International Conference on Systems for Energy-Efficient Buildings, Cities, and Transportation, Virtual, 18–20 November 2020; pp. 170–179.

103. Chen, B.; Donti, P.L.; Baker, K.; Kolter, J.Z.; Bergés, M. Enforcing policy feasibility constraints through differentiable projection for energy optimization. In Proceedings of the Twelfth ACM International Conference on Future Energy Systems, Torino, Italy, 28 June–2 July 2021; pp. 199–210.
104. Jung, S.; Jeoung, J.; Kang, H.; Hong, T. Optimal planning of a rooftop PV system using GIS-based reinforcement learning. *Appl. Energy* **2021**, *298*, 117239.
105. Yu, L.; Xie, W.; Xie, D.; Zou, Y.; Zhang, D.; Sun, Z.; Zhang, L.; Zhang, Y.; Jiang, T. Deep reinforcement learning for smart home energy management. *IEEE Internet Things J.* **2019**, *7*, 2751–2762.
106. Lee, S.; Choi, D.H. Federated reinforcement learning for energy management of multiple smart homes with distributed energy resources. *IEEE Trans. Ind. Inform.* **2020**, *18*, 488–497.
107. Ye, Y.; Qiu, D.; Wang, H.; Tang, Y.; Strbac, G. Real-time autonomous residential demand response management based on twin delayed deep deterministic policy gradient learning. *Energies* **2021**, *14*, 531. [CrossRef]
108. Touzani, S.; Prakash, A.K.; Wang, Z.; Agarwal, S.; Pritoni, M.; Kiran, M.; Brown, R.; Granderson, J. Controlling distributed energy resources via deep reinforcement learning for load flexibility and energy efficiency. *Appl. Energy* **2021**, *304*, 117733.
109. Lee, S.; Xie, L.; Choi, D.H. Privacy-preserving energy management of a shared energy storage system for smart buildings: A federated deep reinforcement learning approach. *Sensors* **2021**, *21*, 4898. [CrossRef]
110. Pinto, G.; Piscitelli, M.S.; Vázquez-Canteli, J.R.; Nagy, Z.; Capozzoli, A. Coordinated energy management for a cluster of buildings through deep reinforcement learning. *Energy* **2021**, *229*, 120725. [CrossRef]
111. Pinto, G.; Deltetto, D.; Capozzoli, A. Data-driven district energy management with surrogate models and deep reinforcement learning. *Appl. Energy* **2021**, *304*, 117642. [CrossRef]
112. Gao, Y.; Matsunami, Y.; Miyata, S.; Akashi, Y. Operational optimization for off-grid renewable building energy system using deep reinforcement learning. *Appl. Energy* **2022**, *325*, 119783. [CrossRef]
113. Langer, L.; Volling, T. A reinforcement learning approach to home energy management for modulating heat pumps and photovoltaic systems. *Appl. Energy* **2022**, *327*, 120020. [CrossRef]
114. Pinto, G.; Kathirgamanathan, A.; Mangina, E.; Finn, D.P.; Capozzoli, A. Enhancing energy management in grid-interactive buildings: A comparison among cooperative and coordinated architectures. *Appl. Energy* **2022**, *310*, 118497. [CrossRef]
115. Nweye, K.; Sankaranarayanan, S.; Nagy, Z. MERLIN: Multi-agent offline and transfer learning for occupant-centric operation of grid-interactive communities. *Appl. Energy* **2023**, *346*, 121323. [CrossRef]
116. Qiu, D.; Xue, J.; Zhang, T.; Wang, J.; Sun, M. Federated reinforcement learning for smart building joint peer-to-peer energy and carbon allowance trading. *Appl. Energy* **2023**, *333*, 120526. [CrossRef]
117. Xie, J.; Ajagekar, A.; You, F. Multi-agent attention-based deep reinforcement learning for demand response in grid-responsive buildings. *Appl. Energy* **2023**, *342*, 121162. [CrossRef]
118. Qiu, Y.; Zhou, S.; Xia, D.; Gu, W.; Sun, K.; Han, G.; Zhang, K.; Lv, H. Local integrated energy system operational optimization considering multi-type uncertainties: A reinforcement learning approach based on improved TD3 algorithm. *IET Renew. Power Gener.* **2023**, *17*, 2236–2256. [CrossRef]
119. Deng, X.; Zhang, Y.; Jiang, Y.; Qi, H. A novel operation method for renewable building by combining distributed DC energy system and deep reinforcement learning. *Appl. Energy* **2024**, *353*, 122188. [CrossRef]
120. Kazmi, H.; D'Oca, S.; Delmastro, C.; Lodeweyckx, S.; Corgnati, S.P. Generalizable occupant-driven optimization model for domestic hot water production in NZEB. *Appl. Energy* **2016**, *175*, 1–15. [CrossRef]
121. Kofinas, P.; Dounis, A.I.; Vouros, G.A. Fuzzy Q-Learning for multi-agent decentralized energy management in microgrids. *Appl. Energy* **2018**, *219*, 53–67.
122. Chasparis, G.C.; Pichler, M.; Spreitzhofer, J.; Esterl, T. A cooperative demand-response framework for day-ahead optimization in battery pools. *Energy Inform.* **2019**, *2*, 29.
123. Gao, Y.; Matsunami, Y.; Miyata, S.; Akashi, Y. Multi-agent reinforcement learning dealing with hybrid action spaces: A case study for off-grid oriented renewable building energy system. *Appl. Energy* **2022**, *326*, 120021.
124. Ashenov, N.; Myrzaliyeva, M.; Mussakhanova, M.; Nunna, H.K. Dynamic cloud and ANN based home energy management system for end-users with smart-plugs and PV generation. In Proceedings of the 2021 IEEE Texas Power and Energy Conference (TPEC), College Station, TX, USA, 2–5 February 2021; IEEE: Piscataway, NJ, USA, 2021; pp. 1–6.
125. Deltetto, D.; Coraci, D.; Pinto, G.; Piscitelli, M.S.; Capozzoli, A. Exploring the potentialities of deep reinforcement learning for incentive-based demand response in a cluster of small commercial buildings. *Energies* **2021**, *14*, 2933. [CrossRef]
126. Huang, C.; Zhang, H.; Wang, L.; Luo, X.; Song, Y. Mixed deep reinforcement learning considering discrete-continuous hybrid action space for smart home energy management. *J. Mod. Power Syst. Clean Energy* **2022**, *10*, 743–754.
127. Nicola, M.; Nicola, C.I.; Selișteanu, D. Improvement of the control of a grid connected photovoltaic system based on synergetic and sliding mode controllers using a reinforcement learning deep deterministic policy gradient agent. *Energies* **2022**, *15*, 2392. [CrossRef]

128. Almughram, O.; Abdullah ben Slama, S.; Zafar, B.A. A reinforcement learning approach for integrating an intelligent home energy management system with a vehicle-to-home unit. *Appl. Sci.* **2023**, *13*, 5539. [CrossRef]
129. Zhou, X.; Du, H.; Sun, Y.; Ren, H.; Cui, P.; Ma, Z. A new framework integrating reinforcement learning, a rule-based expert system, and decision tree analysis to improve building energy flexibility. *J. Build. Eng.* **2023**, *71*, 106536. [CrossRef]
130. Binyamin, S.S.; Slama, S.A.B.; Zafar, B. Artificial intelligence-powered energy community management for developing renewable energy systems in smart homes. *Energy Strategy Rev.* **2024**, *51*, 101288.
131. Wang, Z.; Xiao, F.; Ran, Y.; Li, Y.; Xu, Y. Scalable energy management approach of residential hybrid energy system using multi-agent deep reinforcement learning. *Appl. Energy* **2024**, *367*, 123414. [CrossRef]
132. Anvari-Moghaddam, A.; Rahimi-Kian, A.; Mirian, M.S.; Guerrero, J.M. A multi-agent based energy management solution for integrated buildings and microgrid system. *Appl. Energy* **2017**, *203*, 41–56. [CrossRef]
133. Tomin, N.; Shakirov, V.; Kurbatsky, V.; Muzychuk, R.; Popova, E.; Sidorov, D.; Kozlov, A.; Yang, D. A multi-criteria approach to designing and managing a renewable energy community. *Renew. Energy* **2022**, *199*, 1153–1175. [CrossRef]
134. Tomin, N.; Shakirov, V.; Kozlov, A.; Sidorov, D.; Kurbatsky, V.; Rehtanz, C.; Lora, E.E. Design and optimal energy management of community microgrids with flexible renewable energy sources. *Renew. Energy* **2022**, *183*, 903–921. [CrossRef]
135. Michailidis, I.T.; Manolis, D.; Michailidis, P.; Diakaki, C.; Kosmatopoulos, E.B. Autonomous self-regulating intersections in large-scale urban traffic networks: A Chania City case study. In Proceedings of the 2018 5th international conference on control, decision and information technologies (CoDIT), Thessaloniki, Greece, 10–13 April 2018; IEEE: Piscataway, NJ, USA, 2018; pp. 853–858.
136. Michailidis, I.T.; Manolis, D.; Michailidis, P.; Diakaki, C.; Kosmatopoulos, E.B. A decentralized optimization approach employing cooperative cycle-regulation in an intersection-centric manner: A complex urban simulative case study. *Transp. Res. Interdiscip. Perspect.* **2020**, *8*, 100232. [CrossRef]
137. Michailidis, I.T.; Kapoutsis, A.C.; Korkas, C.D.; Michailidis, P.T.; Alexandridou, K.A.; Ravanis, C.; Kosmatopoulos, E.B. Embedding autonomy in large-scale IoT ecosystems using CAO and L4G-CAO. *Discov. Internet Things* **2021**, *1*, 8. [CrossRef]

MDPI AG
Grosspeteranlage 5
4052 Basel
Switzerland
Tel.: +41 61 683 77 34

*Energies* Editorial Office
E-mail: energies@mdpi.com
www.mdpi.com/journal/energies

www.ingramcontent.com/pod-product-compliance
Lightning Source LLC
Chambersburg PA
CBHW040741250926
51850CB00007BB/1374
* 9 7 8 3 7 2 5 8 7 3 4 4 9 *